Applied Agriculture

Co-ordinating Editor:
B. Yaron

Editors:
B.L. McNeal, F. Tardieu
H. Van Keulen, D. Van Vleck

Springer
Berlin
Heidelberg
New York
Barcelona
Budapest
Hong Kong
London
Milan
Paris
Santa Clara
Singapore
Tokyo

Isaac Ishaaya   Danny Degheele† (Eds.)

Insecticides with Novel Modes of Action

Mechanisms and Application

With 40 Figures and 48 Tables

Springer

Prof. Dr. I. Ishaaya
Dept. of Entomology
Agricultural Research Organization
The Volcani Center, Bet Dagan 50250
Israel

Prof. Dr. Danny Degheele†

ISSN 1433-7576
ISBN 3-540-63058-9 Springer-Verlag Berlin Heidelberg New York

Library of Congress Cataloging-in-Publication Data

Insecticides with novel modes of action: mechanisms and application/
Isaac Ishaaya, Danny Degheele, editors.
p. cm. — (Applied agriculture)
Includes bibliographical references and index.
ISBN 3-540-63058-9 (hardcover)
1. Insecticides. 2. Insecticides—Mechanism of action.
I. Ishaaya. I. II. Degheele, Danny, 1941-. III, Series.
SB951.5.166 1997 97-22924
632' .9517-dc21 CIP

Printed in India

Cover design: Design & Production, Heidelberg

Typesetting and Printing: Replika Press Pvt Ltd, Delhi-110 040 (India)

SPIN: 10542711 31/3137/5 4 3 2 1 0

Preface

The future of insect control looked very bright in the 1950s and 1960s with new insecticides constantly coming onto the market. Today, however, whole classes of pesticide chemistry have fallen by the wayside due to misuse which generated resistance problems reaching crisis proportions, severe adverse effects on the environment, and public outcry that has led to increasingly stricter regulation and legislation. It is with this background, demanding the need for safer, environmentally friendly pesticides and new strategies to reduce resistance problems, that this book was written.

The authors of the various chapters have a wealth of experience in pesticide chemistry, biochemical modes of action, mechanism of resistance and application, and have presented concise reviews. Each is actively involved in the development of new groups of pesticide chemistry which led to the development of novel insecticides with special impact in controlling agricultural pests. Emphasis has been given to insecticides with selective properties, such as insect growth regulators (chitin synthesis inhibitors, juvenile hormone mimics, ecdysone agonists), chloronicotinyl insecticides (imidacloprid, acetamiprid), botanical insecticides (neem, plant oils), pymetrozine, diafenthiuron, pyrrole insecticides, and others. The importance of these compounds, as components in integrated pest management programs and in insecticide resistance management strategies, is discussed. The data presented are essential in establishing new technologies and developing novel groups of compounds which will have impact on our future agricultural practices.

This book is intended to serve as a text for researchers, university professors, and graduate students involved in developing new groups of insecticides for crop protection and field crop management, and as a guide or reference book for farm advisors, pest control operators, and practical farmers.

In the preparation of the manuscript, the editors and the authors are indebted to the reviewers of the various chapters for valuable suggestions and criticism: K.R.S. Ascher (Israel), M. Cahill (UK), B. Darvas (Hungary), D. Degheele (Belgium), I. Denholm (UK), A.R. Horowitz (Israel), I. Ishaaya (Israel), A. McCaffery (UK), H. Oberlander (USA), D.L. Silhacek (USA), G. Smagghe (Belgium), P. Weintraub (Israel), and M.E. Whalon (USA). The editors are also thankful to Mrs. Svetlana Kontsedalov (Israel) for her patience in typing and organizing the various sections of this book and to Mrs. Eulalie Ishaaya-van Hoye for assisting in organizing the subject index.

Isaac Ishaaya

Contents

List of Contributors

M. Cahill
Department of Biological and Ecological Chemistry,
Integrated Approach to Crop Research (IACR)-Rothamstead, Harpenden,
Hertfordshire AL5 2JQ, United Kingdom

B. Darvas
Plant Protection Institute, Hungarian Academy of Sciences, P.O. Box 102,
1525 Budapest, Hungary

A. De Cock
Laboratory of Agrozoology, Faculty of Agricultural and Applied Biological
Sciences, University of Gent, Coupure Links 653, 9000 Gent, Belgium

I. Denholm
Department of Biological and Ecological Chemistry, Integrated Approach to
Crop Research (IACR)-Rothamstead, Harpenden, Hertfordshire AL5 2JQ, UK

R.A. Dybas
Merck Research Laboratories, Agricultural Research and Development Merck
& Co., Inc., P.O. Box 450, Hillsborough Road, Three Bridges,
New Jersey 08887, USA
Current address: 298 Reaville Road, Flemington, New Jersey 08822, USA

A. Elbert
Bayer AG, Institute for Insecticides, Crop Protection Development,
Centre Monheim 51368 Leverkusen, Germany

S.J. Fergusson
Novartis Crop Protection, Inc., 7145 58th Ave., Vero Beach, Florida 32967,
USA

C. Flückiger
Novartis Crop Protection, Inc., O.P. Box 18300, Greensboro, North
Carolina 27419-8300, USA

D. Fuog
Novartis Crop Protection, Inc., Basel, Switzerland

A.R. Horowitz
Department of Entomology, Agricultural Research Organization,
The Volcani Center, Bet Dagan 50250, Israel

D.A. Hunt
American Cyanamid Company, Agricultural Products Research Division,
P.O. Box 400, Princeton, New Jersey 08543-0400, USA

I. Ishaaya
Department of Entomology, Agricultural Research Organization,
The Volcani Center, Bet Dagan 50250, Israel

R.K. Jansson
Merck Research Laboratories, Agricultural Research and
Development, Merck & Co., Inc., P.O. Box 450, Hillsborough Road,
Three Bridges, New Jersey 08887, USA

M. Kostjukovsky
Department of Stored Products, Agricultural Research Organization,
The Volcani Center, Bet Dagan 50250, Israel

W. Leicht
Bayer AG, Institute for Insecticides, Crop Protection Development,
Centre Monheim 51368 Leverkusen, Germany

W.H. McGaughey
US Grain Marketing Research Laboratory, Agricultural Research Service,
United States Department of Agriculture Manhattan, Kansas, USA

R. Nauen
Bayer AG, Institute for Insecticides, Crop Protection Development,
Centre Monheim, 51368 Leverkusen, Germany

H. Oberlander
Center for Medical, Agricultural and Veterinary Entomology,
Agricultural Research Service, United States Department of Agriculture,
P.O. Box 14565, Gainesville, Florida 32604, USA

L.A. Polgár
Plant Protection Institute, Hungarian Academy of Sciences, P.O. Box 102,
1525 Budapest, Hungary

E. Shaaya
Department of Stored Products, Agricultural Research Organization,
The Volcani Center, Bet Dagan 50250, Israel

D.L. Silhacek
Center for Medical, Agricultural and Veterinary Entomology,
Agricultural Research Service, United States Department of Agriculture,
P.O. Box 14565, Gainesville, Florida 32604, USA

G. Smagghe
Laboratory of Agrozoology, Faculty of Agricultural and Applied Biological Sciences,
University of Gent, Coupure Links 653, 9000 Gent, Belgium

M.F. Treacy
American Cyanamid Company, Agricultural Products Research Division,
P.O. Box 400, Princeton, New Jarsey 08543-0400, USA

M.E. Whalon
Department of Entomology and Pesticide Research Center, Michigan State University, East Lansing, Michigan, USA

CHAPTER 1

Insecticides with Novel Modes of Action: An Overview

I. Ishaaya and A.R. Horowitz
Department of Entomology, Agricultural Research Organization, The Volcani Center, Bet Dagan 50250, Israel

1 Introduction

Conventional insecticides such as chlorinated hydrocarbons, organophosphates, carbamates and pyrethroids were successful in controlling insect pests during the past five decades, minimizing thereby losses in agricultural yields. Unfortunately, many of these chemicals are harmful to man and beneficial organisms and cause ecological disturbances. Although considerable efforts have been made to minimize the adverse environmental impact of pesticides and to maximize food production and health of the human population and domestic animals, there is today a great demand for safer and more selective insecticides affecting specifically harmful pests, while sparing beneficial insect species and other organisms. Furthermore, the rapidly developing resistance to conventional insecticides provides the impetus to study new alternatives and more ecologically acceptable methods of insect control as part of integrated pest management (IPM) programs.

One of these approaches which has captured worldwide attention is the development of novel compounds affecting developmental processes in insects, such as chitin synthesis inhibitors, juvenile hormone mimics, and ecdysone agonists. In addition, extensive efforts have been made to develop compounds acting selectively on some groups of insects by inhibiting or enhancing the activity of biochemical sites such as respiration (diafenthiuron), the nicotinyl acetyl chlorine receptor (imidacloprid and acetamiprid), and salivary glands of sucking pests (pymetrozine). Progress has been made to introduce improved biocontrol agents such as *Bacillus thuringiensis* for controlling lepidopteran, coleopteran, and dipteran pests, avermectins for controlling mites and other sucking pests. Neem extract, plant oils, detergents, and mineral oils are used today as insecticides for controlling selectively some groups of insect pests or as additives for enhancing the potency of some novel compounds. Most of the novel groups of insecticides will be dealt with in detail in the forthcoming chapters. This chapter is aimed at presenting an overview of novel groups and compounds, with special emphasis on their modes of action and their importance to serve as components in IPM programs for the benefit of agriculture and the environment.

2 Insect Growth Regulators

2.1 Chitin Synthesis Inhibitors

Chitin synthesis inhibitors consist of various compounds acting on insects of different orders by inhibiting chitin formation (Ishaaya and Casida 1974; Post et al. 1974), thereby causing abnormal endocuticular deposition and abortive molting (Mulder and Gijswijt 1973). They are composed of two substructural types: the benzoylphenyl ureas and buprofezin (Fig. 1). Studies with diflubenzuron [1-(4-chlorophenyl)-3-(2,6-diflubenzoyl) urea], the most thoroughly investigated compound of the benzoylphenyl ureas, revealed that the compound alters cuticle composition—especially that of chitin—thereby affecting the elasticity and firmness of the endocuticle (Grosscurt 1978; Grosscurt and Anderson 1980). The reduced level of chitin in the cuticle seems to result from inhibition of biochemical processes leading to chitin formation (Post et al. 1974; Hajjar and Casida 1979; Van Eck 1979). It is not clear whether inhibition of chitin synthetase is the primary biochemical site for the reduced level of chitin, since in some studies benzoylphenyl ureas do not inhibit chitin synthetase activity in cell-free systems (Cohen and Casida 1980; Mayer et al. 1981; Cohen 1985). Several reports published during the past decade indicate the possibility that benzoylphenyl ureas might affect the hormonal balance in insects, thereby resulting in physiological disturbances such as the inhibition of DNA synthesis (Mitlin et al. 1977; DeLoach et al. 1981; Soltani et al. 1984), alteration in carbohydrase and phenoxidase activities (Ishaaya and Casida 1974; Ishaaya and Ascher 1977) and suppression of microsomal oxidase activity (Van Eck 1979). Mauchamp and Perrineau (1987) indicated that protein and chitin microfibrils are not associated after treatment with benzoylphenyl ureas. The most recent hypothesis made by Zimowska et al. (1994) indicates that a protein fraction is most likely involved in the assembly of chitin microfibrils and diflubenzuron might affect this process (see also Oberlander and Silhacek, this Vol).

Benzoylphenyl ureas are compounds with selective properties, affecting the larval stage (Grosscurt 1978). They act mainly by ingestion, but in some species they suppress fecundity (Arambourg et al. 1977; Sarasua and Santiago-Alvarez 1983) and exhibit ovicidal and contact toxicity (Ascher and Nemny 1974; Holst 1975; Wright and Harris 1976; Horowitz et al. 1992). The main uses of diflubenzuron, the first commercial compound of this series, are against pests in forestry, horticulture and field crops, and in the home (Retnakaran and Wright 1987). The search for more potent acylureas has led to the development of new compounds such as chlorfluazuron (Haga et al. 1982), teflubenzuron (Becher et al. 1983), and hexaflumuron (Sbragia et al. 1983, Fig. 1). These novel compounds are very potent against important lepidopteran pests, such as the Egyptian cotton leafworm *Spodoptera littoralis* (Boisd.) in cotton and ornamentals (Ishaaya et al. 1986; Ishaaya and Klein 1990), whereas teflubenzuron is used against the grapevine moths, *Lobesia botrana* Schif. and *Cryptoblabes gnidiella* Mill. (Baum et al. 1990, 1992).

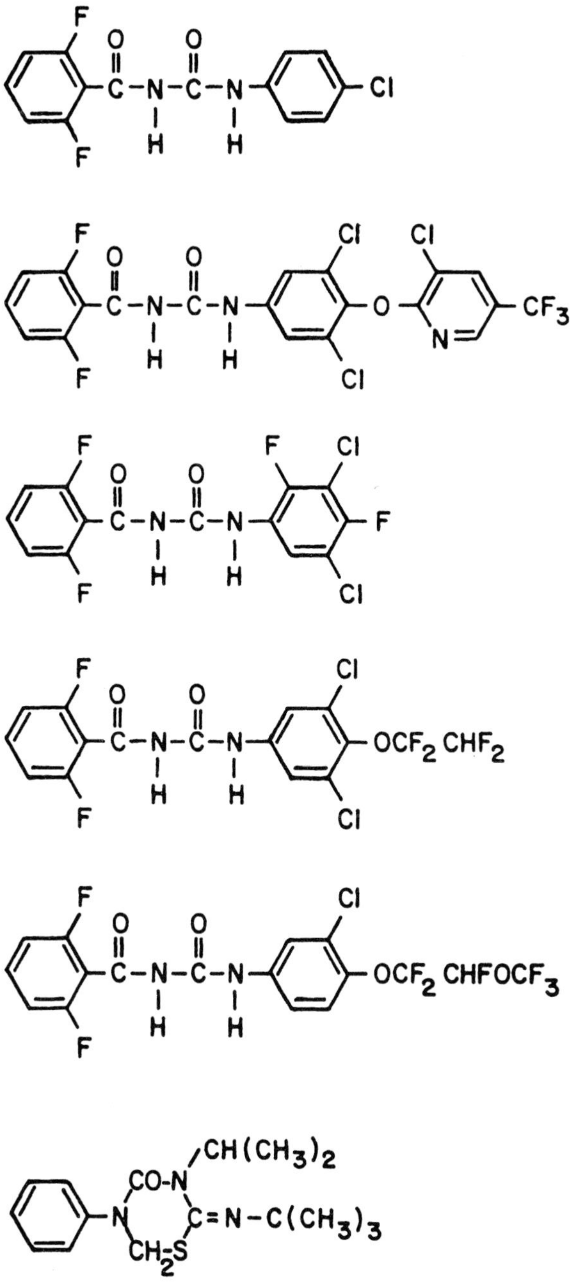

Fig. 1. Structure of some chitin synthesis inhibitors-five benzoylphenyl ureas and buprofezin. *From the top* Diflubenzuron, chlorfluazuron, teflubenzuron, hexaflumuron, novaluron, and buprofezin

Novaluron, 1-[3-chloro-4(1,1,2-trifluoro-2-trifluoro-methoxyethoxy) phenyl]-3-(2,6-difluorobenzoyl) urea (Fig. 1), is a novel benzoylphenyl urea that acts by both ingestion and contact. As such, it is a powerful suppressor of lepidopteran larvae such as *S. littoralis* and *Helicoverpa armigera* (Hübner) (by ingestion) and of cotton whitefly larvae *Bemisia tabaci* (Gennadius) (by contact). The compound is now under advanced development by Makhteshim Chemicals, Be'er Sheva, Israel. Preliminary studies (Ishaaya et al. 1996) indicated that the LC_{50} value of novaluron on third-instar *S. littoralis* fed on treated castor bean leaves is ~ 0.1 mg a.i. l^{-1}. This value resembles that of chlorfluazuron and is about ten-fold lower than that of teflubenzuron (Fig. 2). An application of 250 g a.i.ha^{-1} in cotton fields resulted in 100% mortality of both *S. littoralis and H. armigera*, upon exposure to treated leaves, up to day 8 after application, and about 60 and 30% mortality, respectively, at day 15. Novaluron affects, to a much greater extent, eggs and larvae of *B. tabaci*, than chlorfluazuron and teflubenzuron. Suppression of 100% egg hatch was obtained at a dipping concentration of 1 mg a.i. l^{-1}. Novaluron is far more active in suppressing developing stages of the leafminer *Liriomyza huidobrensis* (Blanchard) than teflubenzuron and chlorfluazuron. Suppression of over 80% adult formation was obtained at a concentration of 0.8 mg a.i. l^{-1} and a similar suppression of pupation and mine formation at a concentration of 20 mg a.i. l^{-1}. Addition of 0.1% surfactant to novaluron solution at a concentration of 0.8 mg a.i. l^{-1}, increased its potency considerably, resulting in 80 and 100% suppression of pupation and adult formation, respectively. These results indicate that novaluron is a potential compound for controlling lepidopteran pests, whiteflies, and agromyzid leafminers in field crops, vegetables, and ornamentals.

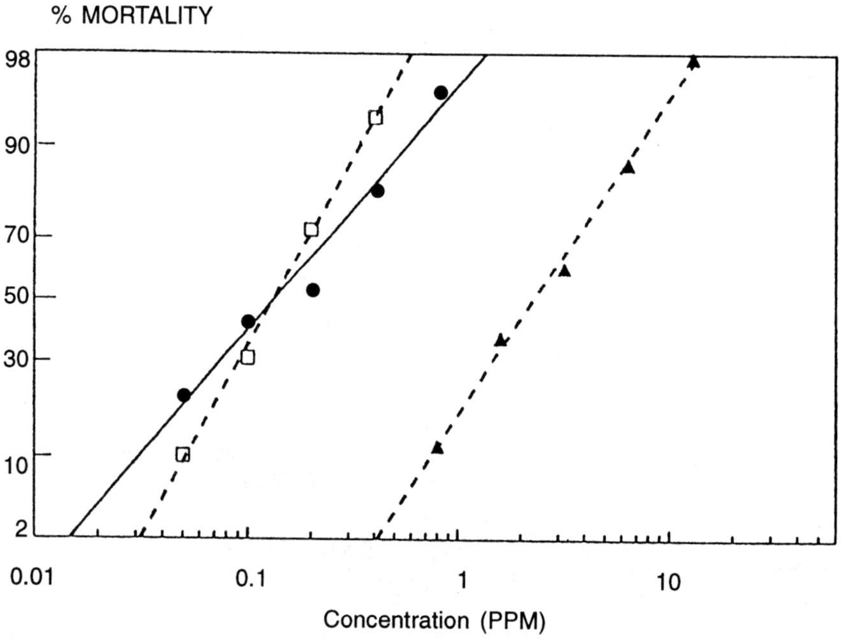

Fig. 2. Comparative toxicity of novaluron (□), chlorfluazuron (●), and teflubenzuron (▲) on third-instar *Spodoptera littoralis* larvae. Larvae were fed for 2 days on castor bean leaves treated with various concentrations of the test compound, and for an additional 4 days on untreated leaves. Mortality was then determined. Data are averages of 10–15 replicates of 10 larvae each

Benzoylphenyl ureas, in general, affect the larval stages which are actively synthesizing chitin. Hence, the adults of nontarget species, *e.g.*, parasites and predators, are seldom affected. Parasites of the house fly *Musca domestica L.* are not affected by diflubenzuron (Ables et al. 1975; Shepard and Kissam 1981). In some cases, parasite larvae inside treated hosts are sensitive to diflubenzuron but the adults are not affected (Granett and Weseloh 1975; Broadbent and Pree 1984). Predatory mites and adult predators are not appreciably affected when fed on treated host larvae (Anderson and Elliott 1982; Jones et al. 1983; Broadbent and Pree 1984). Hence benzoylphenyl ureas are considered important compounds in IPM programs (Ishaaya 1990). However, of some concern regarding this group of compounds in the environment is their effect on crustacean species: cladocerans are very sensitive, copepodes are susceptible to some extent, while ostracods are unaffected at levels of diflubenzuron used for controlling pests (Mulla et al. 1975; Apperson et al. 1978; Ishaaya 1990).

Parallel to the development of benzoylphenyl ureas, a chitin synthesis inhibitor, buprofezin (Applaud, 2-tert-butylimino-3-isopropyl-5-phenyl-1,3,5-thiadiazinan-4-one) has been developed (Kanno et al. 1981, Fig. 1). It acts specifically on some homopteran pests such as the greenhouse whitefly *Trialeurodes vaporariorum* (Westwood) (Yasui et al. 1985, 1987), the sweetpotato whitefly *B. tabaci* (Ishaaya et al. 1988; De Cock et al. 1990), the brown planthopper *Nilaparvata lugens* Stål. (Izawa et al. 1985; Nagata 1986), and the citrus scale insects *Aonidiella aurantii* (Maskell) and *Saissetia oleae* (Bernard) (Yarom et al. 1988; Ishaaya et al. 1989). Whiteflies are important pests in cotton and vegetables. In some cases they transmit viruses and are considered limiting factors for various agricultural crops. The brown planthopper is an important pest of rice in the Far East. Most of these pests have developed resistance to conventional insecticides and buprofezin is an important addition enabling the continued production of various agricultural commodities. Buprofezin suppresses embryogenesis and progeny formation of *B. tabaci* (Ishaaya et al. 1988). The estimated concentration for 50% inhibition of egg hatch applied to adults was 15 mg a.i. l^{-1} in the spray solution and 6 mg a.i l^{-1} for 50% cumulative larval mortality. Hence, the compound exerts its effect on the egg hatch and on the larval stage. It has no ovicidal activity, but suppresses embryogenesis through adults. The length of exposure of the whitefly female to buprofezin corresponds well with the suppression of egg hatch. Adult females exposed for up to 5 h to cotton seedlings treated with 62.5 mg a.i. l^{-1}, laid eggs with a viability similar to that of the control, but those exposed for a period of more than 24 h laid nonviable eggs (Ishaaya et al. 1988). A good correlation between the length of adult exposure to buprofezin and egg viability was observed also with *T. vaporariorum* (Yasui et al. 1987).

Part of buprofezin's efficiency resulted from its vapor phase (De Cock et al. 1990). Vapor eminating from sprayed cotton (500 g a.i.-ha^{-1}) caused mortalities of 90, 83, and 65% to larval whitefly when nontreated infested plants were placed in the field 1, 9, and 16 days after treatment, respectively (Ishaaya 1992). These results indicate that vapor phase toxicity of buprofezin enables control of whitefly larvae which are present on the lower surface of leaves, and are therefore difficult to control under standard spray conditions. Buprofezin is harmless to aphelinid parasitoids such as *Encarsia formosa* Gahan and *Cales noaki* Howard

(Garrido et al. 1984; Wilson and Anema 1988; Mendel et al. 1994) and to predacious mites (Anonymous 1987). In a recent research study, it was observed that only some developmental stages of aphelinid parasitoids were partially affected (Gerling and Sinai 1994); as such it may be considered a potential compound to be used in IPM programs. The compound was introduced in 1989, for controlling whiteflies in cotton fields in Israel (Horowitz and Ishaaya 1992) accompanied with an insecticide resistance management (IRM) program aimed at delaying resistance development to this important insecticide (Horowitz and Ishaaya 1994; Horowitz et al. 1994).

2.2
Juvenile Hormone Mimics

Extensive research has been done to evaluate physiological effects of the juvenile hormone (JH) and its analogs (Retnakaran et al. 1985; Schooley and Baker 1985). Williams (1967) suggested that compounds exhibiting JH activity may be used as selective control agents. Details of syntheses and structure-activity relationship have been extensively studied and reviewed (Slama et al. 1974; Henrick et al. 1976; Staal 1982; Retnakaran et al. 1985). However, most of the JH analogs synthesized at the early stages, such as farnesol derivatives, juvabione, methoprene, hydroprene, and kinoprene, were not sufficiently stable and active under field conditions.

Among the novel JH mimics with relatively high toxicity and potency on agricultural pests are fenoxycarb and pyriproxyfen (Fig. 3). Fenoxycarb, ethyl[2-(4-phenoxyphenoxy) ethyl] carbamate, is the first commercial compound being marketed for controlling agricultural pests under field conditions (Dorn et al. 1981; Masner et al. 1987). The compound exhibits ovicidal and ovolarvicidal activity against various insect species (Masner et al. 1987). Pyriproxyfen, 2-[1-methyl-2-(4-phenoxy-phenoxy) ethoxy] pyridine, is a fenoxycarb derivative in which a part of the aliphatic chain has been replaced by pyridyl oxyethylene. The compound is a potent JH mimic affecting the hormonal balance in insects and resulting, in some cases, in a strong suppression of embryogenesis, metamorphosis, and adult formation (Itaya 1987; Kawada 1988; Langley 1990; Koehler and Patterson 1991). The compound is considered a leading component for controlling whiteflies (Ishaaya and Horowitz 1992; Ishaaya et al. 1994) and scale insects (Peleg 1988). Dipping cotton or tomato seedlings infested with 0–1-day-old eggs in 0.1 mg a.i. l^{-1} pyriproxyfen resulted in over 90% suppression of egg hatch of both *B. tabaci* and *T. vaporariorum* (Ishaaya and Horowitz 1992;

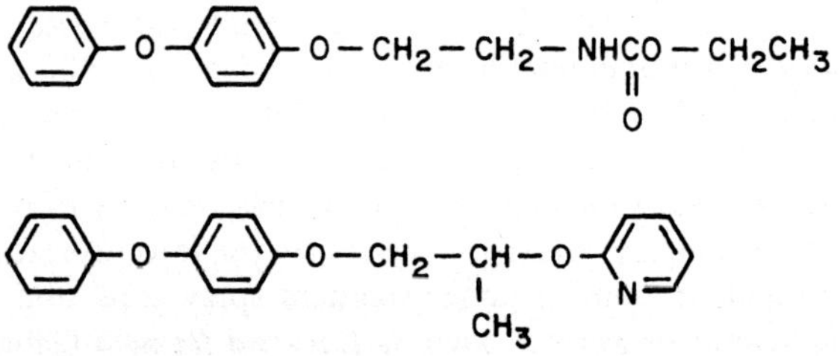

Fig. 3. Structure of the juvenile hormone mimics. *Above* Fenoxycarb; *below* pyriproxyfen

Ishaaya et al. 1994). Older eggs were affected to a lesser extent. Exposure of whitefly females to cotton or tomato seedlings treated with pyriproxyfen resulted in oviposition of nonviable eggs. The LC_{90} values for egg viability of *B. tabaci* and *T. vaporariorum* exposed to treated plants were 0.05 and 0.2 mg a.i. l^{-1}, respectively (Table 1). Treatment of Second-instar whitefly larvae with 0.04–5 mg a.i. l^{-1} resulted in normal development until the pupal stage; however, adult emergence was totally suppressed (Table 2). Inhibition of egg hatch on the lower surface of cotton leaves was observed when their upper surface was treated with 1-25 mg a.i. l^{-1}, indicating a pronounced translaminar activity. These findings indicate that pyriproxyfen is an efficient control agent of both *B. tabaci* and *T. vaporariorum* (Ishaaya and Horowitz 1995). The compound has been used successfully for controlling whiteflies in Israeli cotton fields since 1991 (Horowitz and Ishaaya 1994, 1996).

Pyriproxyfen has no appreciable effect on various parasitoids but is harmful to lady beetles (Mendel et al. 1994). Parasitoids are important natural enemies for controlling whiteflies in cotton and ornamentals, hence pyriproxyfen has been introduced successfully for controlling whiteflies in these crops. Careful

Table 1. Transovarial activity of pyriproxyfen on *Bemisia tabaci* and *Trialeurodes vaporariorum*. (Ishaaya and Horowitz 1995; reproduced with the permission of *Pesticide Science*)

Whitefly species	Slope ± SEM	LC_{50} (mg l^{-1}) (95% F.L.)	LC_{90} (mg l^{-1}) (95% F.L.)
B. tabaci	4.6	0.03	0.05
	(0.2)	(0.02–0.03)	(0.05–0.05)
T. vaporariorum	4.8	0.12	0.22
	(0.4)	(0.08–0.16)	(0.16–0.53)

(F.L., fiducial units.
Whitefly females were exposed for 48 *h* (*B. tabaci*) or for 24 *h* (*T. vaporariorum*) oviposition to plants treated with various concentrations of pyriproxyfen. Assays with *B. tabaci* were done with cotton seedlings and those with *T. vaporariorum* with tomato seedlings.

Table 2. Effect of pyriproxyfen on pupation and emergence of *B. tabaci* treated at the Second-instar. (Ishaaya and Horowitz 1992; reproduced with permission from the Entomological Society of America)

Concentration (mg l^{-1})	Number	Pupation ± SEM (%)	Emergence ± SEM (%)
Untreated	1014	96 ± 1a	95 ± 2a
0.04	455	95 ± 2a	0b
0.20	1191	92 ± 2a	0b
1.00	1153	95 ± 2a	0b
5.00	489	94 ± 2a	0b

Data are averages of 5–10 replicates of 80–140 larvae each. Values followed by the same letter do not differ significantly at $P = 0.05$.

evaluation should be made prior to the introduction of pyriproxyfen for controlling scale insects in citrus, since lady beetles are considered important natural enemies for various insect pests in citrus groves.

Extensive research is being done by several chemical companies aiming at synthesizing more potent and selective JH mimics. Hence, this group of insecticides is considered a potential component to be used in future IPM programs for controlling a diversity of insect pests in various agricultural crops.

2.3 Ecdysone Agonists

Physiological and biochemical processes occurring during metamorphosis in insects depend largely on the presence of juvenile and ecdysteroid hormones, the titer and activity of which are due to biosynthesis, enzymatic degradation and compartmentalization of these hormonally active compounds (Gilbert et al. 1980; Koolman and Karlson 1985). During the last decade, many investigations have been directed to elucidate the possible use of ecdysteroid receptors as target site for developing insecticides with novel modes of action (Robbins et al. 1970; Bergamasco and Horn 1980; Horn et al. 1981). Several substituted dibenzoyl hydrazine compounds acting as ecdysone agonists have been developed by Rohm and Haas Co. (Spring House, Pennsylvania, USA); the most well-known compounds of this series are RH-5849 and RH-5992 (tebufenozide) (Fig. 4). The first compound of this series, RH-5849, binds to the ecdysteroid receptors, thereby initiating the molting process (Wing 1988; Wing et al. 1988). The compound acts on lepidopteran pests such as *Manduca sexta* (L.) (Wing et al. 1988), *Plodia interpunctella* (Hübner) (Silhacek et al. 1990), *Spodoptera frugiperda* (Smith) (Monthéan and Potter 1992), *S. littoralis* (Smagghe and

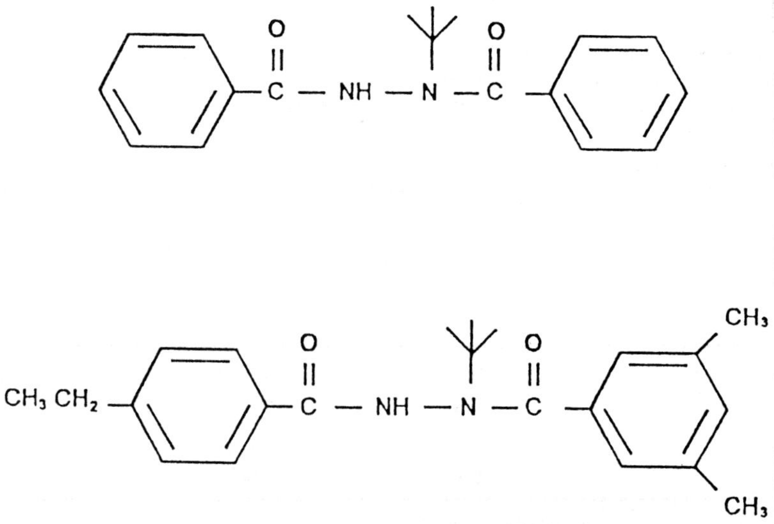

Fig. 4. Structure of some ecdysone agonists. *Above* RH-5849; *below* tebufenozide

Degheele 1992a, b; Ishaaya et al. 1995), and *S. exempta* Walker (Smagghe and Degheele 1993). The potency of tebufenozide against agricultural pests, has been demonstrated against several lepidopteran pests, such as the codling moth *Carpocapsa pomonella* (L.) in apple orchards (Heller et al. 1992; Brown 1994), the fall armyworm *S. frugiperda* in maize, the bollworm *Heliothis zea* (Boddie) (Chandler et al. 1992, Monthéan and Potter 1992), the armyworms *S. exigua* (Hübner), *S. littoralis* and *S. exempta* in vegetables, cotton, cereals, and rice (Chandler et al. 1992; Smagghe and Degheele 1994a, b), and the Indian meal moth *P. interpunctella* in wheat (Silhacek et al. 1990). (Detailed mode of action, selectivity and potency on various insect species can be found in Smagghe and Degheele, this Vol.).

Comparative toxicity of the most recent ecdysone agonists, tebufenozide, and RH-2485 (structure not yet disclosed), against Sixth-instar *S. littoralis* larvae indicates that RH-2485 was three- to Seven-fold more potent than tebufenozide against a susceptible laboratory strain and 7 to 14-fold against a pyrethroid-resistant strain (Table 3). The LC_{50} values for RH-2485 and tebufenozide in the Sixth-instar of the susceptible strain were 0.57 and 1.86 mg a.i. l^{-1}, respectively, and of the pyrethroid-resistant strain were 0.31 and 4.22 mg a.i. l^{-1}, respectively. Both compounds are powerful toxicants for controlling both susceptible and field-resistant strains of *S. littoralis*. These compounds have no adverse activity on the predators, *Podisus maculiventris* (Say) *Podisus nigrispinus* (Dallas), and *Orius insidiosus* (Say) (Smagghe and Degheele 1994a, 1995). Thus, this series of compounds may serve as potential components in IPM programs for controlling lepidopteran pests in vegetables and ornamentals.

Table 3. Comparative toxicity of RH-2485 and tebufenozide on a susceptible laboratory strain (S) and a pyrethroid-resistant field strain (R) of Sixth-instar *Spodoptera littoralis*. (Ishaaya et al. 1995; reproduced with permission from the secretariat of Phytoparasitica)

Biological parameters	S-strain	R-strain	RR[a]
RH-2485			
Slope ± SEM	3.1 ± 0.5	1.9 ± 0.3	
LC_{50}[b]	0.57	0.31	0.5
(95% F.L.)	(0.46–0.67)	(0.06–0.53)	
LC_{90}	1.49	1.51	
(95% F.L.)	(1.21–2.15)	(0.87–7.84)	1.0
Tebufenozide			
Slope ± SE	2.0 ± 0.3	3.2 ± 0.5	
LC_{50}	1.86	4.22	2.3
(95% F.L.)	(0.97–2.66)	(3.39–5.03)	
LC_{90}	7.93	10.5	1.3
(95% F.L.)	(5.22–20.0)	(8.35–15.4)	

F.L., Fiducial limits.
[a]RR, Resistance ratio, derived by dividing the LCs of the R-strain by those of the S-strain.
[b]LC values in mg a.i. l^{-1}.

3 Nicotinyl Insecticides

3.1 Imidacloprid and Acetamiprid

Nicotinyl insecticides interact with nicotinic acetylcholine receptors (nAChR) at the central and peripheral nervous systems, resulting in excitation and paralysis, followed by death. These compounds simulate the activity of acetylcholine which results from the inhibition of acetylcholine esterase by organophosphorous compounds. Novel nicotinoids of potential use in agriculture are imidacloprid and acetamiprid. Both of them interact with nAChR in a structure-activity relationship (Tomizawa et al. 1995a, b) and show a strong affinity to the insect receptor with a relatively low affinity to the vertebrate receptor, indicating a high selectivity toward insects as compared with the first compound in this series, nicotine (Tomizawa et al. 1995b). Hence, it has been suggested that imidacloprid and related compounds be called "*neonicotinoids*" (Yamamoto et al. 1995).

Imidacloprid and acetamiprid (Fig. 5) are the first commercial neonicotinoids developed for controlling agricultural pests. They are also called chloronicotinyl insecticides, indicating the biological importance of the *chlor* moiety in their chemical structure (Leicht 1993). Imidacloprid is a very important agent for controlling aphids, leafhoppers and whiteflies in vegetables and ornamentals (Elbert et al. 1991). It is considered a relatively polar material with good xylem mobility and hence it is suitable for seed treatment and soil application. Sublethal concentrations have strong effects on the feeding behavior of aphids, resulting in suppression of honeydew excretion, wandering, and, subsequently, death due to starvation (Neuen 1995; Devine et al. 1996).

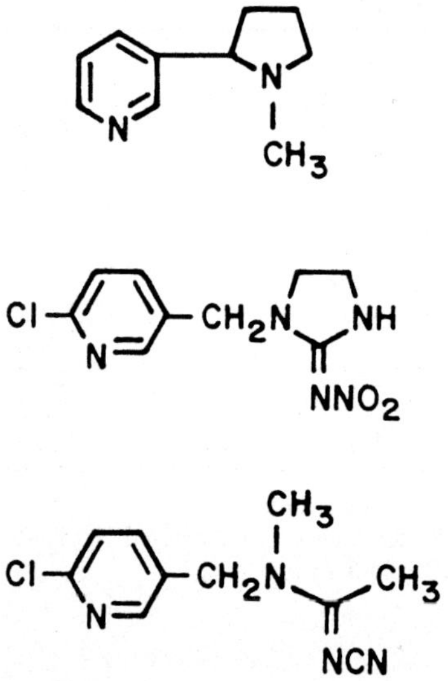

Fig. 5. Structure of nicotine-like compounds. *From top* Nicotine, imidacloprid, and acetamiprid

Comparative assays carried out in our standardized growth chamber conditions indicate that imidacloprid in soil application gave superior performance for controlling the whitefly *B. tabaci* as compared with acetamiprid, while acetamiprid was much more potent than imidacloprid in foliar application. The most impressive difference in potency was achieved when both compounds were compared under foliar application with cotton seedlings infested with 0–1-day–old *B. tabaci* eggs. According to LC_{50} and LC_{90} values, acetamiprid was 11- and 19-fold more potent than imidacloprid, respectively (Table 4). On the other hand, under similar foliar application both of them showed similar potency on adults and larvae. Under soil application, the residual toxicity of imidacloprid for controlling whitefly adults was somewhat higher than acetamiprid, reaching their highest toxicity 7–14 days after application (Table 5).

Table 4. Slopes and LC values of imidacloprid and acetamiprid on *Bemisia tabaci* eggs

Biological parameters	Imidacloprid	Acetamiprid	Ratio[a]
Slope ± SEM	1.6 ± 0.1	2.2 ± 0.2	
LC_{10} mg a.i. l^{-1}	0.29	0.05	6
(90% F.L.)	(0.01–0.87)	(0.02–0.12)	
LC_{50} mg a.i. l^{-1}	1.93	0.18	11
(90% F.L.)	(0.51–3.71)	(0.04–0.30)	
LC_{90} mg a.i. l^{-1}	12.8	0.68	19
(90% F.L.)	(6.8–44.5)	(0.42–1.82)	

F.L., fiducial linits.
[a]Derived by dividing the LCs of imidacloprid by those of acetamiprid.

Table 5. Residual toxicity of acetamiprid and imidacloprid on *Bemisia tabaci* adults after soil application

Compounds and concentration (mg a.i. l^{-1})	Mortality ± SEM at various days after application(%)			
	1	7	14	21
Acetamiprid				
0	9 ± 3a	11 ± 3a	17 ± 3a	13 ± 2a
5	25 ± 2b	65 ± 7b	49 ± 8b	30 ± 5b
25	39 ± 5c	84 ± 3c	76 ± 6c	62 ± 7c
Imidacloprid				
0	11 ± 4a	17 ± 3a	11 ± 3a	17 ± 3a
5	40 ± 3c	69 ± 1b	84 ± 10cd	67 ± 6c
25	47 ± 7c	93 ± 3c	96 ± 2d	79 ± 3c

Solutions of 25 ml containing 0.5 and 25 mg a.i. l^{-1} acetamiprid or imidacloprid were poured around the base of cotton seedlings kept in 0.25 l flower pots. Female adults held in leaf cages were exposed to the treated leaves at various days after soil application. Mortality was recorded after 48 h. Data are averages ± SEM of 5-6 replicates of 15 adults each.

Data followed by the same letter do not differ significantly at $P = 0.05$.

Table 6. Effect of foliar application of acetamiprid and imidacloprid on aphid population in a cotton field

Compounds and concentration ($g\ a.i.\ ha^{-1}$)	Aphids per leaf ± SEM at various days after application			
	3	10	17	24
Untreated	3.8 ± 0.9a	4.9 ± 1.6a	7.8 ± 1.4a	13.3 ± 2.2a
Imidacloprid (210)	2.3 ± 1.5ab	1.1 ± 0.4b	1.3 ± 0.4b	3.3 ± 1.1b
Acetamiprid (60)	0b	0c	0b	0c

Treatments were applied in an experimental plot of 0.4 ha cotton field in a randomized block design with 4 replicates of 5 × 8 m each. Spray volume was 500 $l\ ha^{-1}$ with the recommended application of each product. Ten randomly selected cotton leaves at maximal infestation (third–fifth node from the growing top point) were collected from each replicate and examined for aphid infestation. Data are averages ± SEM of 4 replicates of 10 leaves each.

Data followed by the same letter do not differ significantly at $P = 0.05$.

Foliar application of acetamiprid in a cotton field at a rate of 60 g a.i. ha^{-1} suppressed totally aphid development up to 24 days after application, exceeding the efficacy obtained with an application of 210 g a.i. ha^{-1} imidacloprid (Table 6). The results obtained thus far indicate that acetamiprid is more efficient than imidacloprid in controlling whiteflies and aphids under foliar application and the contrary may be true when the application is carried out through the soil. These variations, in the biological activity may result from a difference in the xylem mobility when applied through soil and in penetration through the leaves when applied on the plants.

Soil application of imidacloprid has, probably, no appreciable effect on beneficial insects. Imidacloprid does not affect Diplopoda, spiders, or predatory mites after foliar application (see Elbert et al.) this Vol. Further studies are required to evaluate the toxicity of acetamiprid and imidacloprid on parasitoids and lady beetles. Their relatively high selectivity towards aphids, leafhoppers, and whiteflies along with the apparent nonharmful effect on some natural enemies, enable their introduction as components in IPM programs for controlling sucking pests in field crops, vegetables and ornamentals.

4 Miscellaneous

4.1 Diafenthiuron

Diafenthiuron, 3-(2,6-diisopropyl-4-phenoxyphenyl)-1-tert-butyl-thiourea, is a new type of thiourea derivative (Fig. 6) which acts specifically on sucking pests such as mites, whiteflies, and aphids (Streibert et al. 1988; Anonymous 1989; Kadir and Knowles 1991a; Ishaaya et al. 1993). It has a favorable acute mammalian toxicity, the LD_{50} for acute oral and dermal toxicity in the rat being more than

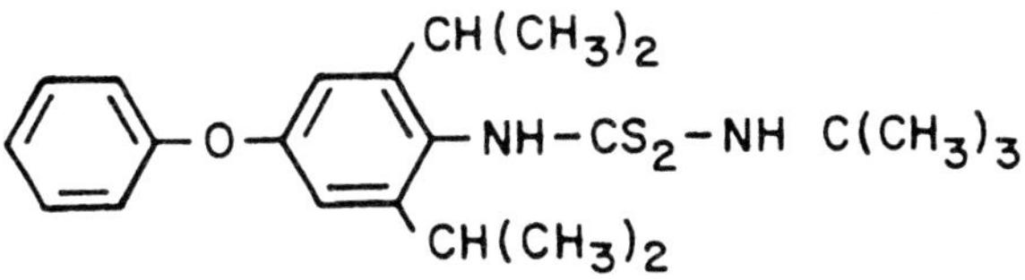

Fig. 6. Diafenthiuron

2000 mg a.i. kg^{-1}; coupled with relatively low toxicity to beneficial insects and predatory mites (Streibert et al. 1988; Anonymous 1989). It is considered harmless to the whitefly parasitoid *E. formosa* and the predatory bug *Orius niger* Wolff (Van de Veire and Degheele 1993). Diafenthiuron is photochemically converted within a few hours in sunlight to its carbodiimide derivative, which is a much more powerful acaricide/insecticide than diafenthiuron; it is then degraded by photolysis and volatilized over a period of a few days (Steinemann et al. 1990). The carbodiimide is a potent inhibitor of mitochondrial ATP synthesis and reacts covalently with the proteolipid subunit of the ATPase and porin which forms the outer membrane of the mitochondria (Ruder et al. 1992). The compound inhibits ATPase activities in preparations from the bulb mite *Rhizoglyphus echinopus* (Fumouze and Robin), the two-spotted spider mite *Tetranychus urticae Koch*, and the bluegill *Lepomis macrochirus Rafinesque* (Kadir and Knowles 1991a, b).

Studies carried out in our laboratory revealed that diafenthiuron is a powerful toxicant for controlling the sweetpotato whitefly *B. tabaci* under both laboratory (Ishaaya et al. 1993) and field (Horowitz et al. 1994) conditions. According to mortality curves, whitefly larvae were the most susceptible (LC_{50} for Second-instar = 6.5 mg a.i. l^{-1}) followed by adults (LC_{50} = 23 mg a.i. l^{-1}) and pupae (LC_{50} = 45 mg a.i. l^{-1}; Table 7). Whitefly eggs were least susceptible; inhibition of ~ 3.5% egg hatch was observed at a concentration of 125 mg a.i. l^{-1} (Ishaaya et al. 1993).

No appreciable cross resistance with pyriproxyfen or buprofezin could be detected with either *B. tabaci* (Ishaaya and Horowitz 1995) or with the greenhouse whitefly *T. vaporariorum* (De Cock et al. 1995). Thus, the compound can be

Table 7. Comparative toxicity of diafenthiuron on different stages of the sweetpotato whitefly *Bemisia tabaci*. (Ishaaya et al. 1993; reproduced with permission from the secretariat of Phytoparasitica)

Stage	Slope SEM	LC_{50} (mg a.i. l^{-1}) (95% F.L.)	LC_{90} (mg a.i. l^{-1}) (95% F.L.)
Second-instar	1.46 ± 0.11	6.5 (1.5–20.9)	49.2 (16.4–[a])
Pupa	3.73 ± 0.28	45.0 (36.6–54.8)	99.1 (77.7–146.0)
Adult	1.98 ± 0.17	23.0 (11.9–38.7)	102.4 (55.0–600.5)

[a]Not enough data for 95% F.L. analysis.
Mortality curves were done with 4–5 concentrations. Each concentration was carried out with at least 5 replicates of 40–200 individuals.

alternated with compounds acting on different biochemical sites in IRM programs aiming at delaying or preventing development of resistance in field crop pests to novel insecticides.

The relatively high potency of diafenthiuron for controlling whiteflies, aphids, and mites along with its low toxicity to mammals and the environment renders this compound an important component in IPM programs for controlling the above pests in various field crops.

4.2 Pymetrozine

Pymetrozine is a new insecticide with a novel mode of action. It affects the nerves controlling the salivary pump of some sucking pests and causes irreversible cessation of feeding within a few hours after application, followed by starvation and death (Schwinger et al. 1994). The compound is a powerful toxicant against aphids [*Aphis gossypii* Glover and *Myzus persicae* (Sulzer)], whiteflies (*B. tabaci* and *T. vaporariorum*) and planthoppers (*N. lugens*). The compound has a systemic and translaminar activity and can be used in soil or foliar application (Flückiger et al. 1992a, b). In some cases, host plants have major effects on the potency of pymetrozine. Assays carried out in our laboratory indicate that pymetrozine is a much more powerful toxicant to whiteflies when applied on Bulgarian beans as compared with cotton plants. The LC_{90} value of pymetrozine on *B. tabaci* adults when applied on Bulgarian bean seedlings (3 mg a.i. l^{-1}) is about 20-fold lower than that obtained when the compound was applied on cotton seedlings (58.6 mg a.i. l^{-1}; Table 8). Thus, the compound may have different potency on *B. tabaci* when applied on various host plants. Plant constituents seem to affect its translaminar activity, hence the compound should be tested for its potency on aphids or whiteflies on each host plant separately.

Table 8. Comparative toxicity of pymetrozine on *Bemisia tabaci* adults exposed to Bulgarian bean and cotton seedlings treated with various concentrations of the test compound. LC values were compared and the relative toxicity was determined

Biological parameters	Adults exposed to bean seedlings	Adults exposed to cotton seedlings	Relative toxicity cotton/bean
Slope ± SEM	2.38 ± 0.36	2.66 ± 0.37	
LC_{10}[a]	0.25	6.4	26
(95% F.L.)	(0.11–0.42)	(4.1–8.3)	
LC_{50}	0.88	19.3	22
(95% F.L.)	(0.58–1.17)	(16.2–23.3)	
LC_{90}	3.04	58.6	19
(95% F.L.)	(2.32–4.40)	(43.3–96.7)	

Mortality curves were done with 4–5 concentrations. Each concentration was carried out with at least 5 replicates of 15 adult females each.
F.L., fiducial limits.
[a]LC values in mg a.i. l^{-1}.

Pymetrozine has no appreciable effect on natural enemies and the environment and as such it is considered a potential component in IPM programs for controlling aphids and whiteflies and for suppressing virus transmission in various vegetable crops (for more details, see Fuog et al., this Vol.).

4.3 Biological Insecticides

Over the last decade, much attention has been given to biological insecticides used either directly or produced through formulation processes. Among the most important groups in this series are the avermectins and the *Bacillus thuringiensis* insecticides.

4.3.1 *Avermectins*

The avermectins are a group of macrocyclic lactones isolated from fermentation of the soil microorganism *Streptomyces avermitilis* Burg. These compounds act as agonists for gamma-amino butyric acid (GABA)-gated chloride channels (Mellin et al. 1983; Albrecht and Sherman 1987). They bind with high affinity to sites in the head and muscle neuronal membranes of various insect species (Deng and Casida 1992; Rohrer et al. 1995). Studies with nematodes showed that avermectins interact with specific glutamate-gated chloride channels distinct from the GABA-sensitive channels (Schaeffer and Haines 1989; Arena et al. 1992). Thus, it is suggested that avermectins may bind to a common site in all gated chloride channels (Arena 1994).

Abamectin, developed for agricultural use by Merck Sharp & Dohme (New Jersey, USA), consists of about 80% avermectin B1a and 20% avermectin B1b (Fisher and Mrozik 1989). The compound is specifically toxic to phytophagous mites and to a select panel of insect species, but it is markedly less potent against some lepidopteran and homopteran species (Dybas 1989; Lasota and Dybas 1991). In general, abamectin is less toxic to beneficial arthropods (e.g., honey bees, parasitoids, and predators), especially when exposure to treated plants occurs beyond 1 day after application (Hoy and Cave 1985; Dybas 1989; Zhang and Sanderson 1990). The compound degrades rapidly when exposed to sunlight (Wislocki et al. 1989). Despite its rapid photodecomposition following application, abamectin provides residual activity in the field due to its translaminar activity (Wright et al. 1985; Dybas 1989). In addition, various authors reported that the use of mineral oils or surfactants in combination with abamectin extended its residual toxicity to phytophagous mites (Wright et al. 1985; Mizell et al. 1986; Dybas 1989), probably resulting from a higher translaminar activity. Assays carried out in our laboratory (Horowitz et al. 1997) indicated that addition of 0.5% Ultra Fine (UF) oil considerably enhanced the potency of abamectin (Vertimec) on the sweetpotato whitefly *B. tabaci*. The residual toxicity of 1 mg a.i. l^{-1} abamectin applied to cotton seedlings under growth chamber conditions, without exposure to sunlight, resulted in adult mortality that declined from 86 to 39% over 28 days. Addition of 0.5% UF oil resulted in a considerably higher

Table 9. Residual toxicity of emamectin benzoate (MK-244) against first-instar *Helicoverpa armigera* in a cotton field

Concentration (g^{-1} ha^{-1})	Larval mortality at various days after treatment ± SEM (%)				
	1	3	7	10	16
0 (control)	0	0	0	0	0
200	100	99 ± 1	95 ± 3	89 ± 6	22 ± 7
800	100	100	100	100	65 ± 9

Cotton rows (10 m each) were sprayed on 12 September 1995, until run-off at a rate of 200 and 800 g ha^{-1} (1.9% EC). Leaves were collected periodically after application. First-instar larvae (0–24–h–old) were exposed to treated leaves. Mortality was determined after 6 days. Data are averages ± SEM of 10 replicates of 10 larvae each.

adult mortality of 100 and 88% at day 0 and 28 after application, respectively. A similar pattern of enhancement was obtained in outdoor assays (Horowitz et al. 1997). Thus, abamectin, in combination with a suitable mineral oil, is a potential agent for controlling *B. tabaci*.

A new avermectin derivative, emamectin benzoate (MK-244 1.9% EC), developed by Merck Sharp & Dohme, New Jersey, USA, acts specifically on lepidopteran species (LC_{50} less than 1 ppm) (Cox et al. 1995). Assays carried out in our cotton field indicated that emamectin benzoate is highly potent against the armyworm *H. armigera*. A concentration of 800 g ha^{-1}, i.e. (~ 15 g a.i. ha^{-1}) resulted in 100% mortality up to day 10 after application and 65% mortality at day 16 after application (Table 9). Thus, novel derivatives of the avermectin group acting specifically against some groups of insects are considered potential components in IPM programs of various agricultural crops for more details see Jansson and Dybas this Vol.).

4.3.2
Bacillus thuringiensis

Bacillus thuringiensis (Bt) kills insects primarily through the action of *d*-endotoxin, a proteinous constituent produced during sporulation; it affects the insect midgut epithelium upon ingestion (Höfte and Whiteley 1989; Gill et al. 1992; Navon 1993). In general, sporulated cells containing the δ-endotoxin, in formulated forms, are used as insecticides for controlling lepidopteran pests in various agricultural crops. Most conventional *Bt* products are based on the subspecies *kurstaki* HD-1 which was introduced in 1971 by Abbott laboratory, followed by Sandoz, Novo, Ecogen, and Monsanto, for controlling lepidopteran pests (Navon et al. 1990; Navon 1993). Sandoz introduced the HRD12 strain to improve the control of *Spodoptera* species. A product obtained from *Bt aizawai* is used specifically for controlling the wax moth *Galleria mellonella* L. (Thompson 1989; Navon 1993). Various formulations of *Bt kurstaki* are available for controlling lepidopteran pests, such as Dipel (HD-1, Abbott), Thuricide (HD-1, Sandoz), Biobit (HD-1, Novo), Javelin (NRD12, Sandoz), and MVP (Monsanto).

In addition, transgenic plants producing Bt-endotoxin as a systemic insecticide are now available for introduction in various agricultural systems. Several transgenic plants, e.g., cotton, potatoes, and tomatoes, showing resistance or tolerance to major lepidopteran pests, such as *Helicoverpa, Trichoplusia, Pectinophora, Spodoptera,* and *Ostrinia* species, are available (Fischhoff et al. 1987; Perlak et al. 1990; Ely 1993). The advantage of recombinant DNA technology makes it likely that the use of *Bt* protein as a resistance factor in transgenic plants will expand in the future. Extensive research is still required to learn how to cope with resistance problems which may rise after the use of transgenic plants.

Bt insecticides are considered safe to the environment with a little or no harm to natural enemies. Hence, they are considered important components in IPM programs. The use of recombinant DNA technology bears great promise for pest control, but only if tempered with thorough research relating to environmental risks (Meadows 1993; for more details, see Whalon and McGaughey, this Vol.).

4.4 Neem Extract

Neem extract contains several active ingredients. Azadirachtin is one of the major constituents, exhibiting about 90% of the neem extract activity. Azadirachtin, structurally similar to the insect molting hormone ecdysone, interacts with the corpus cardiacum, thereby blocking the activity of the molting hormone. As such, the compound acts as an insect growth regulator, suppressing fecundity, molting, pupation, and adult formation. Other constituents showing high potency on insects and structurally related to azadirachtin are the salannin, meliantrol, and nimbin (Schmutterer 1995).

Neem extract is a potential insecticide, acting mainly by ingestion and affecting a diversity of insects (Jacobson 1988; National Research Council 1992; Ascher 1993; Schmutterer 1995). The extract seems to have no harmful effect on beneficials such as spiders, lady beetles, and parasitoids, and is considered a potential component in IPM programs.

5 Conclusions

Novel insecticides with selective properties have been introduced during the last decade for controlling agricultural pests. Among the insect growth regulators, the chitin synthesis inhibitors, buprofezin, and benzoylphenyl ureas are used for controlling planthoppers and whiteflies, and lepidopterans, respectively; the juvenile hormone mimic, pyriproxyfen for controlling whiteflies and scale insects; and the ecdysone agonists are now considered potential components for controlling some lepidopteran pests. Among the non-insect growth regulators, diafenthiuron, imidacloprid, acetamiprid, and pymetrozine are used, among others, for controlling aphids and whiteflies. In addition, neem oil and *Bt* toxin are considered important components in various IPM programs.

Compounds with different modes of action for controlling whiteflies, such as buprofezin, pyriproxyfen, diafenthiuron (De Cock et al. 1995; Ishaaya and Horowitz 1995), and imidacloprid (unpubl. results), have no cross resistance and hence can be used in alternation in IRM programs, aiming at delaying resistance development to these important compounds. A variety of avermectin derivatives, such as abamectin, milbemectin, and emamectin, are in use in some countries and under progress development in others, aiming at controlling selectively agricultural pests, such as mites, lepidopteran, and whiteflies.

Novel insect control agents with different modes of action acting selectively against groups of insect pests are potential components in present and future IPM programs, enabling the continuous production of field crops for the benefit of the farmers and the environment.

Acknowledgments

The authors wish to thank Sara Yablonski, Zmira Mendelson, Svetlana Kontsedalov, and Yusef Mansour for many years of technical assistance and collaborative research. Special thanks are due to Phylis Weintraub for reading the text and to Svetlana Kontsedalov for organizing and typing the manuscript.

References

Ables JR, West RP, Shepard M (1975) Response of the house fly and its parasitoids to dimilin (TH-6040). J Econ Entomol 68: 622–624

Albrecht CP, Sherman M (1987) Lethal and sublethal effects of avermectin B1 on three fruit fly species (Diptera: Tephritidae). J Econ Entomol 80: 344–347

Anderson DW, Elliott RH (1982) Efficacy of diflubenzuron against the codling moth. *Laspeyresia pomonella* (Lepidoptera: Olethrentidae) and impact on orchard mites. Can Entonol 114: 733–737

Anonymous (1987) Applaud, a new pesticide (insect growth regulator): technical information. Nihon Nohyaku, Tokyo

Anonymous (1989) Polo (diaphenthiuron, CGA 106630), Technical Data Sheet. Ciba-Geigy, Basle, pp 1–18

Apperson CS, Schaefer CH, Colwell AE, Werner GH, Anderson NL, Dupras EF Jr, Longanecker DR (1978) Effect of diflubenzuron on *Chaoborus astictopus* and nontarget organisms and persistence of diflubenzuron in lentil habitat. J Econ Entomol 71: 521–527

Arambourg Y, Pralavario R, Dolbeau C (1977) Premières observations sur li'action du diflubenzuron (PH 6040) sur la fecondité, la longevité et la viabilitè des oeufs de *Ceratitis capitata* Wield. (Dipt Trypetidae). Rev Zool Agric Pathol Veg 76: 118–126

Arena JP (1994) Expression of *Caenorhabditis elegans* mRNA in *Xenophus oocytes*: a model system to study the mechanism of action of avermectins. Parasitol Today 10: 35–37

Arena JP, Liu KK, Paress PS, Schaeffer JM, Cully DF (1992) Expression of a glutamate-activated chloride current in *Xenophus oocytes* injected with *Caenorhabditis elegans* RNA: evidence for modulation by avermectin. Mol Brain Res 15: 339–348

Ascher KRS (1993) Non-conventional insecticidal effects of pesticides available from the neem tree, *Azadirachta indica*. Arch Insect Biochem Physiol 22: 433–449

Ascher KRS, Nemny NE (1974) The ovicidal effect of PH 60-40 [1-(4-chlorophenyl)-3-(2,6-difluorobenzoyl)-urea] in *Spodoptera littoralis* Boisd. Phytoparasitica 2: 131–133

Baum D, Yablonski S, Ishaaya I (1990) Biological mode of action of benzoylphenyl ureas on the grapevine moth *Lobesia botrana* Den & Schiff (Lepidoptera: Tortricidae). Abstr 7th Int Congr Pestic Chem (IUPAC), Hamburg, vol 1, p 375

Baum D, Yablonski S, Ishaaya I (1992) The ovicidal effect of some benzoylphenyl ureas on the grape berry moth *Lobesia botrana*. Phytoparasitica 20: 83

Becher HM, Becker P, Prokic-Immel R, Wirtz W (1983) CME, a new chitin synthesis inhibiting insecticide. Brighton Crop Prot Conf, vol 1; pp. 408–415

Bergamasco R, Horn DHS (1980) The biological activities of ecdysteroids and ecdysteroid analogues. In: Hoffman JA (ed) Progress in ecdysone research. Elsevier, Amsterdam, pp 299–324

Broadbent AB, Pree DJ (1984) Effects of diflubenzuron and BAY SIR 8514 on beneficial insects associated with peach. Environ Entomol 13: 133–136

Brown JJ (1994) Effects of a nonsteroidal ecdysone agonist, tebufenozide, on host/parasitoid interaction. Arch Insect Biochem Physiol 26: 235–248

Chandler LD, Pair SD, Harrison WE (1992) RH-5992, a new insect growth regulator active against corn earworm and fall armyworm (Lepidoptera: Noctuidae). J Econ Entomol 85: 1099–1103

Cohen E (1985) Chitin synthetase activity and inhibition in different insect microsomal preparations. Experientia 41: 470–472

Cohen E, Casida JE (1980) Inhibition of *Tribolium* gut synthetase. Pestic Biochem Physiol 13: 129–136

Cox DL, Knight AL, Biddinger DG, Lasota JA, Pikounis B, Hull LA, Dybas RA (1995) Toxicity and field efficacy of avermectins against codling moth (Lepidoptera: Tortricidae) on apples. J Econ Entomol 88: 708–715

De Cock A, Ishaaya I, Degheele D, Veierov D (1990) Vapor toxicity and concentration-dependent persistence of buprofezin applied to cotton foliage for controlling the sweetpotato whitefly (Homoptera: Aleyrodidae). J Econ Entomol 83: 1254–1260

De Cock A, Ishaaya I, Van De Veire M, Degheele D (1995) Response of buprofezin-susceptible and -resistant strains of *Trialeurodes vaporariorum* (Homoptera: Aleyrodidae) to pyriproxyfen and diafenthiuron. J Econ Entomol 88: 763–767

DeLoach JR, Meola SM, Mayer RT, Thompson JM (1981) Inhibition of DNA synthesis by diflubenzuron in pupae of the stable fly *Stomoxys calcitrans* (L.). Pestic Biochem Physiol 15: 172–180

Deng Y, Casida JE (1992) House fly head GABA-gated chloride channel: toxicologically relevant binding site for avermectins coupled to site for ethynyl-bicycloorthobenzoate. Pestic Biochem Physiol 43: 116–122

Devine G, Harling Z, Scarr AWS, Devonshire A (1996) Lethal and sublethal effects of imidacloprid on nicotine-tolecant *Myzus nicotianae* and *Myzus Persicae*. Pestic Sci 48: 57–62

Dorn S, Frischknecht ML, Martinez V, Zurflüh R, Fischer U (1981) A novel non-neurotoxic insecticide with a broad activity. Z Pflanzenkr Pflanzenschutz 88: 269–275

Dybas RA (1989) Abamectin use in crop protection. In: Campbell WC (ed) Ivermectin and abamectin. Springer, Berlin Heidelberg, New York, pp 287–310

Elbert A, Becker B, Hartwig J, Erdalen C (1991) Imidacloprid—a new systemic insecticide. Pflanzenschutz-Nachr 44: 113–136

Ely J (1993) The engineering of plants to express *Bacillus thuringiensis* δ-endotoxins. In: Entwistle PF, Cory JS, Bailey MJ, Higgs S (eds) *Bacillus thuringiensis*, an environmental biopesticide: theory and practice. John Wiley, Chichester, pp 105–124

Fischer MH, Mrozik H (1989) Chemistry. In: Campbell WC (ed) Ivermectin and abamectin. Springer, Berlin Heidelberg, New York, p 1–23

Fischhoff DA, Bowdisch KS, Perlak FJ, Marrone PG, McCormick SH, Niedermeyer JG, Dean DA, Kusano-Kretzmer K, Mayer EJ, Rochester DE, Rogers SG, Fraley RT (1987) Insect tolerant transgenic tomato plants. Biol Technology 5: 807–813

Flückiger CR, Kristinsson H, Senn R, Rindlisbacher A, Buholzer H, Voss G (1992a) CGA 215'944—a novel agent to control aphids and whiteflies. Brighton Crop Prot Conf—Pests and diseases, vol 1; pp 43–50

Fückiger CR, Senn R, Buholzer H (1992b) CGA 215'944—opportunities for use in vegetables. Brighton Crop Prot Conf—Pests and diseases, vol 3, pp 1187–1192

Garrido A, Beitia F, Gruenholz P (1984) Effects of PP 618 on immature stages of *Encarsia formosa* and *Cales noaki* (Hymenoptera: Aphelinidae). Brighton Crop Prot Conf—Pests and diseases, pp 305–310

Gerling D, Sinai P (1994) Buprofezin effects on two parasitoid species of whitefly (Homoptera: Aleyrodidae). J Econ Entomol 87: 842–846

Gilbert LI, Bollenbacher WE, Goodman W, Smith SL, Agui N, Granger N, Sedlak BJ (1980) Hormones controlling insect metamorphosis. Recent Prog Horm Res 36: 401–449

Gill SS, Cowles EA, Pietrantonio PV (1992) The mode of action of *Bacillus thuringiensis* endotoxins. Annu Rev Entomol 37: 615–636

Granett J, Weseloh RM (1975) Dimilin toxicity to the gypsy moth larval parasitoid, *Apanteles melanoscelus*. J Econ Entomol 68: 577–580

Grosscurt AC (1978) Effect of diflubenzuron on mechanical penetrability, chitin formation, and structure of the elytra of *Leptinotarsa decemlineata*. J Insect Physiol 24: 827–831

Grosscurt AC, Anderson SO (1980) Effect of diflubenzuron on some chemical and mechanical properties of the elytra of *Leptinotarsa decemlineata*. Proc K Ned Akad Wet 83C: 143–150

Haga T, Tobi T, Koyanagi T, Nishiyama R (1982) Structure activity relationships of a series of benzoyl-pyridyloxyphenyl-urea derivatives. Abstr, 5th Int Congr Pestic Chem (IUPAC), August 1982, Kyoto, p IId–7

Hajjar NP, Casida JE (1979) Structure activity relationships of benzoylphenyl ureas as toxicants and chitin synthesis inhibitors in *Oncopeltus fasciatus*. Pestic Biochem Physiol 11: 33–45

Heller JJ, Mattioda H, Klein E, Sagenmüller A (1992) Field evaluation of RH-5992 on lepidopterous pests in Europe. Brighton Crop Prot Conf—Pests and diseases Nov 1992, vol 1, pp 59–66

Henrick CA, Willy WE, Staal GB (1976) Insect juvenile hormone activity of alkyl (2E, 4E)-3, 7, 11-trimethyl-2, 4-dodecadienoates. Variations in the ester function and the carbon chain. J Agric Food Chem 24: 207–218

Höfte H, Whiteley HR (1989) Insecticidal crystal proteins of *Bacillus thuringiensis*. Microbiol Rev 53: 242–255

Holst H (1975) Die fertilit‰tsbeeinflussende Wirkung des neuen Insektizids DDD 60–40 bei *Epilachna varivestis* Muls. (Col.: Coccinellidae), und *Leptinotarsa decemlineata* Say (Col.: Chrysomelidae). Z Pflanzenkr Pflanzenschutz 82: 1–7

Horn DHS, Galbraith MN, Kelly BA, Kinnear JF, Martin MD, Middleton EJ, Virgonia CTF (1981) Moulting hormones L III. The synthesis and biological activity of some ecdysone analogues. Aust J Chem 34: 2607–2618

Horowitz AR, Ishaaya I (1992) Susceptibility of the sweetpotato whitefly (Homoptera: Aleyrodidae) to buprofezin during the cotton season. J Econ Entomol 85: 318–324

Horowitz AR, Ishaaya I (1994) Managing resistance to IGRs in the sweetpotato whitefly (Homoptera: Aleyrodidae). J Econ Entomol 87: 866–871

Horowitz AR, Ishaaya I (1996) Chemical control of *Bemisia tabaci*—management and application. In: Gerling G, Mayer RT (eds) Bemisia 1995: taxonomy, biology, damage, control and management. Intercept, Andover, pp 537–556

Horowitz AR, Klein M, Yablonski S, Ishaaya I (1992) Evaluation of banzoylphenyl ureas for controlling the spiny bollworm, *Earias insulana* (Boisd.) in cotton. Crop Prot. 11: 465–469

Horowitz AR, Forer G, Ishaaya I (1994) Managing resistance in *Bemisia tabaci* in Israel with emphasis on cotton. Pestic Sci 42: 113–122

Horowitz AR, Mendelson Z, Ishaaya I (1997) Effect of abamectin mixed with mineral oil on the sweetpotato whitefly (Homoptera: Aleyrodidae). J Econ Entomol 90: 349–353

Hoy MA, Cave FE (1985) Laboratory evaluation of avermectin as a selective acaricide for use with *Metaseiulus occidentalis* (Nesbitt) (Acarina: Phytoseiidae). Exp Appl Acarol 1: 139–152

Ishaaya I (1990) Benzoylphenyl ureas and other selective control agents—mechanism and application. In: Casida JE (ed) Pesticides and alternatives. Elsevier, Amsterdam, pp 365–376

Ishaaya I (1992) Selective insect control agents—mechanism and application. In: Otto D, Weber B (eds) Insecticides: mechanism of action and resistance. Intercept, Andover, pp 127–133

Ishaaya I, Ascher KRS (1977) Effect of diflubenzuron on growth and carbohydrate hydrolases of *Tribolium castaneum*. Phytoparasitica 5: 149–158

Ishaaya I, Casida JE (1974) Dietary TH 6040 alters cuticle composition and enzyme activity of house fly larval cuticle. Pestic Biochem Physiol 4: 484–490

Ishaaya I, Horowitz AR (1992) Novel phenoxy juvenile hormone analog (pyriproxyfen) suppresses embryogenesis and adult emergence of sweetpotato whitefly (Homoptera: Aleyrodidae). J Econ Entomol 85: 2113–2117

Ishaaya I, Horowitz AR (1995) Pyriproxyfen, a novel insect growth regulator for controlling whiteflies: mechanism and resistance management. Pestic Sci 43: 227–232

Ishaaya I, Klein M (1990) Response of susceptible laboratory and resistant field strains of *Spodoptera littoralis* (Lepidoptera: Noctuidae) to teflubenzuron. J Econ Entomol 83: 59–62

Ishaaya I, Navon A, Gurevitz E (1986) Comparative toxicity of chlorfluazuron (IKI-7899) and cypermethrin to *Spodoptera littoralis*, *Lobesia botrana* and *Drosophila melanogaster*. Crop Prot 5: 385–388

Ishaaya I, Mendelson Z, Melamed-Majar V (1988) Effect of buprofezin on embryogenesis and progeny formation of sweetpotato whitefly (Homoptera: Aleyrodidae). J Econ Entomol 81: 781–784

Ishaaya I, Blumberg D, Yarom I (1989) Buprofezin—a novel IGR for controlling whiteflies and scale insects. Meded Fac Landbouwwet Rijksuniv Gent 54: 1003–1008

Ishaaya I, Mendelson Z, Horowitz AR (1993) Toxicity and growth suppression exerted by diafenthiuron in the sweetpotato whitefly *Bemisia tabaci*. Phytoparasitica 21: 199–204

Ishaaya I, De Cock A, Degheele D (1994) Pyriproxyfen, a potent suppresser of egg hatch and adult formation of the greenhouse whitefly (Homoptera: Aleyrodidae). J Econ Entomol 87: 1185–1189

Ishaaya I, Yablonski S, Horowitz AR (1995) Comparative toxicity of two ecdysteroid agonists, RH-2485 and RH-5992, on susceptible and pyrethroid-resistant strains of the Egyptian cotton leafworm, *Spodoptera littoralis*. Phytoparasitica 23: 139–145

Ishaaya I, Yablonski S, Mendelson Z, Mansour Y, Horowitz AR (1996) Novaluron (MCW-275), a novel benzoylphenyl urea, suppressing developing stages of lepidopteran, whitefly and leafminer pests. Brighton Crop Prot Conf—Pests and diseases Nov 1996, 1013–1020

Itaya N (1987) Insect juvenile hormone analogue as an insect growth regulator. Sumitomo Pyrethroid World 8: 2–4

Izawa Y, Uchida M, Sugimoto T, Asai T (1985) Inhibition of chitin biosynthesis by buprofezin analogs in relation to their activity controlling *Nilaparvata lugens* Stål. Pestic Biochem Physiol 24: 343–347

Jacobson M (Ed) (1988) Focus on phytochemical pesticides: the neem tree, vol 1. CRC Press, Boca Raton

Jones D, Snyder M, Granett J (1983) Can insecticides be integrated with biological control agents of *Trichoplusia ni* in celery? Entomol Exp Appl 33: 290–296

Kadir HA, Knowles CO (1991a) Toxicological studies of the thiourea diafenthiuron in diamondback moths (Lepidoptera: Yponomeutidae), two-spotted spider mites (Acari: Tetranychidae), and bulb mite (Acari: Acaridae). J Econ Entomol 84: 780–784

Kadir HA, Knowles CO (1991b) Inhibition of ATP dephosphorylation by acaricides with emphasis on the anti-ATPase activity of the carbodiimide metabolite of diafenthiuron. J Econ Entomol 84: 801–805

Kanno H, Ikeda K, Asai T, Maekawa S (1981) 2-tert-butylimino-3-isopropyl-5-phenyl-perhydro-1, 3, 5-thiodiazin-4-one (NNI 750), a new insecticide. Brighton Crop Prot Conf, vol 1. pp 56–69

Kawada H (1988) An insect growth regulator against cockroaches. Sumitomo Pyrethroid World 11: 2–4

Koehler PG, Patterson RJ (1991) Incorporation of pyriproxyfen in a German cockroach (Dictyoptera: Blattellidae) management program. J Econ Entomol 84: 917–921

Koolman J, Karlson P (1985) Regulation of ecdysteroid titer: degradation. In: Kerkut GA, Gilbert LI (eds) Comprehensive insect physiology, Biochemistry and pharmacology, vol 7. Pergamon Press, Oxford, pp 343–361

Langley P (1990) Control of the tsetse fly using a juvenile hormone mimic, pyriproxyfen. Sumitomo Pyrethroid World 15: 2–5

Lasota JA, Dybas RA (1991) Avermectin, a novel class of compounds: implications for use in arthropod pest control. Annu Rev Entomol 36: 91–117

Leicht W (1993) Imidacloprid—a chloronicotinyl insecticide. Pestic Outlook 4: 17–21
Masner P, Angst M, Dorn S (1987) Fenoxycarb, an insect growth regulator with juvenile hormone activity: a candidate for *Heliothis virescens* (F.) control on cotton. Pestic Sci 18: 89–94
Mauchamp B, Perrineau O (1987) Chitin biosynthesis after treatment with benzoylphenyl ureas. In: Wright JE, Retnakaran A (eds) Chitin and benzoylphenyl ureas. Dr W Junk, Dordrecht, pp 101–109
Mayer RT, Chen AC, DeLoach JR (1981) Chitin synthesis inhibiting insect growth regulators do not inhibit chitin synthase. Experientia 37: 337–338
Meadows MP (1993) *Bacillus thuringiensis* in the environment: ecology and risk assessment. In: Entwistle PF, Cory JS, Bailey MJ, Higgs S (eds) *Bacillus thuringiensis*, an environmental biopesticide: theory and practice. John Wiley, Chichester, pp 193–220
Mellin TN, Busch RD, Wang CC (1983) Postsynaptic inhibition of invertebrate neuromuscular transmission by avermectin Bla. Neuropharmacology 22: 89–96
Mendel Z, Blumberg D, Ishaaya I (1994) Effect of some insect growth regulators on natural enemies of scale insects (Homoptera: Coccoidea). Entomophaga 39: 199–209
Mitlin N, Wiygul G, Haynes JW (1977) Inhibition of DNA synthesis in boll weevils (*Anthonomus grandis* Boheman) sterilized by dimilin. Pestic Biochem Physiol 7: 559–563
Mizell RF, Schiffhaner DE, Taylor JL (1986) Mortality of *Tetranychus urticae* Koch (Acari: Tetranychidae) from abamectin residues: effects of host plant, light and surfactants. J Econ Entomol 21: 329–337
Monthèan C, Potter DE (1992) Effects of RH-5849, a novel insect growth regulator, on Japanese beetle (Coleoptera: Scarabaeidae) and fall armyworm (Lepidoptera: Noctuidae) in turfgrass. J Econ Entomol 85: 507–513
Mulder R, Gijswijk MT (1973) The laboratory evaluation of two promising new insecticides which interfere with cuticle deposition. Pestic Sci 4: 737–745
Mulla MS, Majori G, Darwazeh HA (1975) Effects of the insect growth regulator Dimilin or TH 6040 on mosquitoes and some non-target organisms. Mosq News 35: 211–216
Nagata T (1986) Timing of buprofezin application for control of the brown planthopper, *Nilaparvata lugens* Stål. (Homoptera: Delphacidae). Appl Entomol Zool 21: 357–362
National Research Council (1992) Neem, a tree for solving global problems. National Academy Press, Washington, DC, 139 pp
Navon A (1993) Control of lepidopteran pests with *Bacillus thuringiensis.* In: Entwistle PF, Cory JS, Bailey MJ, Hidds S (eds) *Bacillus thuringiensis*, an environmental biopesticide: theory and practice. 6. John Wiley, Chichester, pp 125–146
Navon A, Klein M, Braun S (1990) *Bacillus thuringiensis* potency bioassays against *Heliothis armigera, Earias insulana* and *Spodoptera littoralis* larvae based on standardized diets. J Invertebr Pathol 55: 387–393
Neuen R (1995) Behaviour modifying effects of low systemic concentrations of imidacloprid on *Myzus persicae* with special reference to an antifeeding response. Pestic Sci 44: 145–153
Peleg BA (1988) Effect of a new phenoxy juvenile hormone analog on Çalifornia red scale (Homoptera: Diaspididae), Florida wax scale (Homoptera: Coccidae) and the ectoparasite *Aphytis holoxanthus* DeBache (Hymenoptera: Aphelinidae). J Econ Entomol 81: 88–92
Perlak FJ, Deaton RW, Armstrong TA, Fuchs RL, Sims SR, Greenplate JT, Fischhoff DA (1990) Insect resistant cotton plants. Bio/Technology 8: 939–943
Post LC, de Jong BJ, Vincent WR (1974) 1-(2,6-disubstituted benzoyl)-3-phenylurea insecticides: inhibitors of chitin synthesis. Pestic Biochem Physiol 4: 473–483
Retnakaran A, Wright JE (1987) Control of insect pests with benzoylphenyl ureas. In: Wright JE, Retnakaran A (eds) Chitin and benzoylphenyl ureas. Dr W Junk, Dordrecht, pp 205–282
Retnakaran A, Granett J, Ennis T (1985) Insect growth regulators. In: Kerkut GA, Gilbert LI (eds) Comprehensive insect physiology, biochemistry and pharmacology. vol 12. Pergamon Press, Oxford, pp 529–601
Robbins WE, Kaplanis JN, Thompson MJ, Shortino TJ, Joyner SC (1970) Ecdysone and synthetic analogs: molting hormone activity and inhibitive effects on insect growth, metamorphosis and reproduction. Steroids 16: 105–125

Rohrer SP, Birzin ET, Costa SD, Arena JP, Hayes EC, Schaeffer JM (1995) Identification of neuron-specific ivermectin binding sites in *Drosophila melanogaster* and *Schistocerca americana*. Insect Biochem Mol Biol 25: 11–17

Ruder FJ, Benson JA, Kayser H (1992) The mode of action of the insecticide/acaricide diafenthiuron. In: Otto D, Weber B (eds) Insecticides: mechanism of action and resistance. Intercept, Andover, pp 263–276

Sarasua MJ, Santiago-Alvarez C (1983) Effect of diflubenzuron on the fecundity of *Ceratitis capitata*. Entomol Exp Appl 33: 223–225

Sbragia R, Bisarbi-Ershadi B, Rigterink RH (1983) XRD-473, a new acylurea insecticide effective against *Heliothis*. Brighton Crop Prot Conf, vol 1. pp 417–424

Schaeffer JM, Haines HW (1989) Avermectin binding in *Caenorhabditis elegans*: a two-state model for the avermectin binding site. Biochem Pharmacol 38: 2329–2338

Schmutterer H (ed) (1995) Neem tree—source of unique natural products for integrated pest management, medicine industry and other purposes. VCH, Weinheim, 696 pp

Schooley DA, Baker FC (1985) Juvenile hormone biosynthesis. In: Kerkut GA, Gilbert LI (eds) Comprehensive insect physiology, biochemistry and pharmacology. vol 17. Pergamon Press, Oxford, pp 363–389

Schwinger M, Harrewijn P, Kayser H (1994) Effect of pymetrozine (CGA 215'944), a novel aphicide on feeding behavior of aphids. Proc 8th IUPAC Int Congr Pestic Chem, Washington, DC, vol 1. 230

Shepard M, Kissam JB (1981) Integrated control of house flies on poultry farms: treatment of house fly resting surfaces with diflubenzuron plus releases of the parasitoids, *Muscidifurax raptor*. J GA Entomol Soc 16: 222–227

Silhacek DL, Oberlander H, Procheron P (1990) Action of RH-5849, a non-steroidal ecdysteroid mimic, on *Plodia interpunctella* (Hübner) *in vivo* and *in vitro*. Arch Insect Biochem Physiol 15: 201–212

Slama K, Romanuk M, Sorm F (1974) Insect hormones and bioanalogues. Springer, Berlin Heidelberg, New York

Smagghe G, Degheele D (1992a) Effects of RH-5849, the first nonsteroidal ecdysteroid agonist, on larvae of *Spodoptera littoralis* (Boisd.) (Lepidoptera: Noctuidae). Arch Insect Biochem Physiol 21: 119–128

Smagghe G, Degheele D (1992b) Effect of the nonsteroidal ecdysteroid agonist RH-5849 on reproduction of *Spodoptera littoralis* (Boisd.) (Lepidoptera: Noctuidae). Parasitica 48: 23–29

Smagghe G, Degheele D (1993) Metabolism, pharmacokinetics, and toxicity of the first nonsteroidal ecdysteroid agonist RH-5849 to *Spodoptera exempta* (Walker), *Spodoptera exigua* (Hübner) and *Leptinotarsa decemlineata* (Say). Pestic Biochem Physiol 46: 149–160

Smagghe G, Degheele D (1994a) Action of a novel nonsteroidal ecdysteroid mimic, tebufenozide (RH-5992), on insects of different orders. Pestic Sci 42: 85–92

Smagghe G, Degheele D (1994b) Action of the nonsteroidal ecdysteroid mimic RH-5849 on larval development and adult reproduction of insects of different orders. Invertebr Reprod Dev 25: 227–236

Smagghe G, Degheele D (1995) Selectivity of nonsteroidal ecdysteroid agonists RH-5849 and RH-5992 to nymphs and adults of the predatory soldier bugs, *Podisus nigrispinus* and *Podisus maculiventris* (Hemiptera: Pentatomidae). J Econ Entomol 88: 40–45

Soltani N, Besson MT, Delachambre J (1984) Effect of diflubenzuron on the pupal–adult development of *Tenebrio molitor* L. (Coleoptera: Tenebrionidae): growth and development, cuticle secretion, epidermal cell density and DNA synthesis. Pestic Biochem Physiol 21: 256–264

Staal GB (1982) Insect control with growth regulators interfering with the endocrine system. Entomol Exp Appl 31: 15–23

Steinemann A, Stamm E, Frei B (1990) Chemodynamics in research and development of new plant protection agents. Pestic Outlook 1(3): 3–7

Streibert HP, Drabek J, Rindlisbacher A (1988) CGA 106630—a new type of acaricide/insecticide for the control of the sucking pest complex in cotton and other crops. Brighton Crop Prot Conf—Pests and diseases, vol 1. pp 25–33

Thompson WT (1989) Agricultural chemicals, book 1. Thompson, Fresno, pp 64–70

Tomizawa M, Otsuka H, Miyamoto T, Eldefrawi ME, Yamamoto I (1995a) Pharmacological characteristics of insect nicotinic acetylcholine receptor with its ion channel and the comparison of the effect of nicotinoids and neonicotinoids. J Pestic Sci 20: 57–64

Tomizawa M, Otsuka H, Miyamoto T, Yamamoto I (1995b) Pharmacological effects of imidacloprid and its related compounds on the nicotinic acetylcholine receptor with its ion channel from the *Torpedo* electric organ. J Pestic Sci 20: 49–56

Van de Veire M, Degheele D (1993) Side effects of diafenthiuron on the greenhouse whitefly parasitoid *Encarsia formosa* and the predatory bug *Orius niger* and its possible use in IPM in greenhouse vegetables. Meded Fac Landbouwwet Rijksuniv Gent 53: 509–514

Van Eck WH (1979) Mode of action of two benzoylphenyl ureas as inhibitors of chitin synthesis in insects. Insect Biochem 9: 295–300

Williams CM (1967) Third-generation pesticides. Sci Am 217: 13–17

Wilson D, Anema BP (1988) Development of buprofezin for control of whitefly *Trialeurodes vaporariorum* and *Bemisia tabaci* on glasshouse crops in the Netherlands and the UK. Brighton Crop Prot Conf—Pests and diseases, pp 175–180

Wing KD (1988) RH-5849, a nonsteroidal ecdysone agonist: effects on a *Drosophila* cell line. Science, Wash D C 241: 467–469

Wing KD, Slawecki RA, Carlson GR (1988) RH-5849, a nonsteroidal ecdysone agonist: effects on larval lepidoptera. Science, Wash D C 241: 470–472

Wislocki PG, Grosso LS, Dybas RA (1989) Environmental aspects of abamectin use in crop protection. In: Campbell WC (ed) Ivermectin and abamectin. Springer, Berlin Heidelberg, New York, pp 182–200

Wright JE, Harris RL (1976) Ovicidal activity of Thompson-Hayward TH 6040 in the stable fly and horn fly after surface contact by adults. J Econ Entomol 69: 728–730

Wright DJ, Loy A, Green ASJ, Dybas RA (1985) The translaminar activity of abamectin (MK-936) against mites and aphids. Meded Fac Landbouwwet Rijksuniv Gent 50: 633–637

Yamamoto I, Yabuta G, Tomizawa M, Saito T, Miyamoto T, Kagabu S (1995) Molecular mechanism of selective toxicity of nicotinoids and neonicotinoids. J Pestic Sci 20: 33–40

Yarom I, Blumberg D, Ishaaya I (1988) Effect of buprofezin on California red scale (Homoptera: Diaspididae) and mediterranean black scale (Homoptera: Coccidae). J Econ Entomol 81: 1581–1585

Yasui M, Fukada M, Maekawa S (1985) Effect of buprofezin on different developmental stages of the greenhouse whitefly, *Trialeurodes vaporariorum* (Westwood) (Homoptera: Aleyrodidae). Appl Entomol Zool 20: 340–347

Yasui M, Fukada M, Maekawa S (1987) Effect of buprofezin on reproduction of the greenhouse whitefly, *Trialeurodes vaporariorum* (Westwood) (Homoptera: Aleyrodidae). Appl Entomol Zool 22: 266–271

Zhang Z, Sanderson JP (1990) Relative toxicity of abamectin to the predatory mite *Phytoseiulus persimilis* (Acari: Phytoseiidae) and the two spotted spider mite (Acari: Tetranychidae). J Econ Entomol 83: 1783–1790

Zimowska G, Mikolajczyk P, Silhacek DL, Oberlander H (1994) Chitin synthesis in *Spodoptera frugiperda* wing imaginal discs. II. Selective action of chlorfluazuron on wheat germ agglutinin binding and cuticle ultrastructure. Arch Insect Biochem Physiol 27: 89–108

CHAPTER 2

Ecdysone Agonists: Mechanism and Biological Activity

G. Smagghe and D. Degheele
Laboratory of Agrozoology, Faculty of Agricultural and Applied Biological Sciences, University of Gent, Coupure Links 653, 9000 Gent, Belgium

1 Introduction

Substituted dibenzoyl hydrazines were described as the first non-steroidal ecdysone agonists in 1988 at Rohm and Haas Research Laboratories (Spring House, Pennsylvania, USA) and are characterized by the prototype compound RH-5849 (1,2-dibenzoyl,1-tert-butyl hydrazine) (Fig. 1; Wing 1988; Wing et al. 1988). The discovery that these synthetic compounds mimic the natural insect moulting hormones or ecdysteroids as 20E (Fig. 1) in both fruitfly *Drosophila melanogaster* (Meigem) K_c cells and tobacco hornworm *Manduca sexta* (L.) larvae generated great interest from several perspectives. Such compounds open up new avenues for endocrinological research and crop protection, since they mimic the ecdysteroid mode of action by true competitive binding on the ecdysteroid receptors (EcRs). As such, they introduce a new class of insect growth regulators (IGRs) exhibiting a novel mode of action, based on more insect-specific biochemical sites of action, which makes them interesting due to an enhanced safety to non-target organisms, the likelihood for more favourable public perception and the potential for combating resistance. So, they prove to have a potential use in crop pest control, distinct from juvenile hormone analogues (JHAs) and benzoylphenylurea IGRs. The first compound in this series to be commercialized is tebufenozide [RH-5992, 3,5-dimethylbenzoic acid 1-1(1,1-dimethylethyl)-2(4-ethylbenzoyl) hydrazine] (Fig. 1).

In this chapter, RH-5849 and tebufenozide. are evaluated especially in terms of their ecdysteroid-mimicking specificity and their possible use as novel selective insect pest control agents.

2 Ecdysteroid-Specific Mode of Action

2.1 Moulting and Feeding

Wing et al. (1988) firstly reported that RH-5849 induces premature moulting in the absence of a known source of endogenous ecdysteroids when tested on ligated larvae of *M. sexta*, leading to head capsule apolysis and cessation of

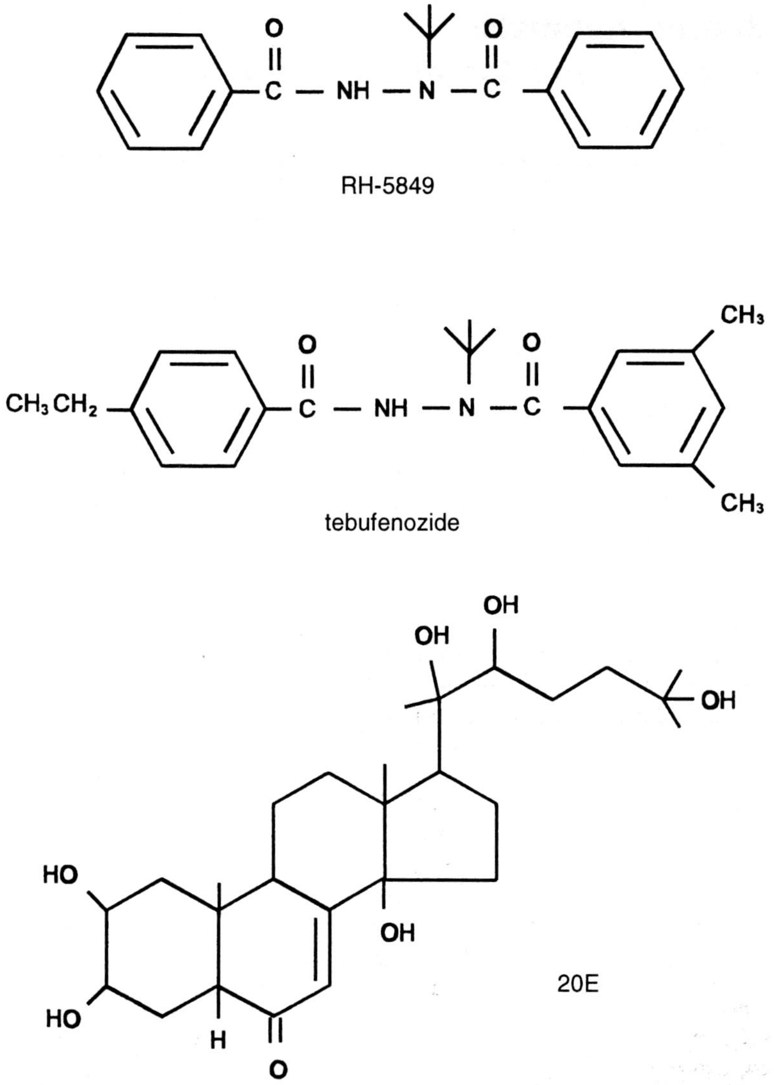

Fig. 1. Chemical structure of both non-steroidal ecdysone agonists, RH-5849 and tebufenozide (RH-5992), and that of the natural insect moulting hormone 20-hydroxyecdysone (20E)

feeding activity and weight gain in intact larvae. At present, these early findings have been confirmed in various other Lepidoptera including *Plodia interpunctella* (Hübner) (Silhacek et al. 1990), *Ostrinia nubilalis* Hübner (Gadenne et al. 1990), *Spodoptera eridania* (Cramer) (Hsu 1991), *Spodoptera littoralis* (Boisduval) (Smagghe and Degheele, 1992b), *Pieris brassicae* L. and *Mamestra brassicae* L. (Darvas et al. 1992; Smagghe and Degheele 1994a,b), *Spodoptera frugiperda* (J. E. Smith) (Chandler et al. 1992; Monthéan and Potter 1992), *Cydia pomonella* L. (Heller et al. 1992; Brown 1994; Charmillot et al. 1994b), *Spodoptera litura*

F. (Park et al. 1992; Tateishi et al. 1993), *Choristoneura fumiferana* (Clemens) (Retnakaran and Oberlander 1993), *Spodoptera exempta* (Walker) and *Spodoptera exigua* (Hübner) (Smagghe and Degheele 1994a, b), *Chilo suppressalis* (Walker) (Oikawa et al. 1994a, b) and *Galleria mellonella* (L.) (Smagghe and Degheele 1994a, b; Slama 1995). Furthermore, Wing et al. (1988) proved that RH-5849 is a physiological inducer of moulting, acting directly on the target tissues and not as a result of an elevation of endogenous ecdysteroids, as was also demonstrated in *Spodoptera* and *Leptinotarsa* (Smagghe et al. 1995). However, stimulation of precocious moulting by either non-steroidal agonists, RH-5849 and tebufenozide, is usually incomplete and therefore fatal. Already 6h after treatment of *S. exigua*, ecdysial space formation and secretion of a new epicuticle were started, which revealed a sequential change similar to normal moulting. However, as exemplified in Fig. 2, normal procuticle secretion was interrupted since the high number of endocuticular lamellae that normally appears in controls was conspicuously absent or only contained a very low number of lamellae (Smagghe et al. 1994, 1996b).

These electron microscopic observations indicate a hyperecdysteroid activity and confirm the moulting accelerating mode of action of tebufenozide resulting in a forced, untimely synthesis of cuticle via epidermal cell activation, followed by inhibition of the post-apolysis processes and normal ecdysis. Hence, a complete moult leading to extra-supernumerary larval stages can be obtained in cases when applying RH-5849 or tebufenozide and simultaneously elevating the juvenile hormone (JH) titre by a combined application with a JHA, such as methoprene or pyriproxyfen (Silhacek et al. 1990; Smagghe and Degheele 1994c). The JHA could counteract some deleterious effects of the ecdysone agonist such as premature incomplete moulting and cessation of feeding. This phenomenon provided increasing opportunities for investigation of physiological effects of nonsteroidal ecdysone agonists; however, many questions remain to be answered about their mode of action.

2.2 Specificity for Insecta

The major moulting hormone in arthropods is known to be 20E; however, other ecdysteroids may serve as the primary moulting hormone in some species (Horn and Bergamasco 1985; Feldlaufer 1989; Rees 1989). In this respect, it is an intriguing question whether RH-5849 and tebufenozide may act as universal ecdysone agonists. Such knowledge is particularly important for the evaluation of potential effects on non-target organisms, e.g. beneficials in integrated pest management (IPM) programs.

Both ecdysone agonists display an ecdysteroid-like mode of action, especially in larval Lepidoptera. In addition, RH-5849 and tebufenozide also have an ecdysteroid-mimicking activity in some larval Coleoptera, such as *Leptinotarsa decemlineata* (Say) (Darvas et al. 1992; Smagghe and Degheele 1994a, b) and some aquatic crustacean larvae (Clare et al. 1992; Kreutzweiser et al. 1994), but not in all orders of insects. Darvas et al. (1992) reported malformed structures and suppression of weight by RH-5849 in the dipteran species *Neobellieri bullata* Parker and the heteropteran insect *Oncopeltus fasciatus* Dallas. Further-

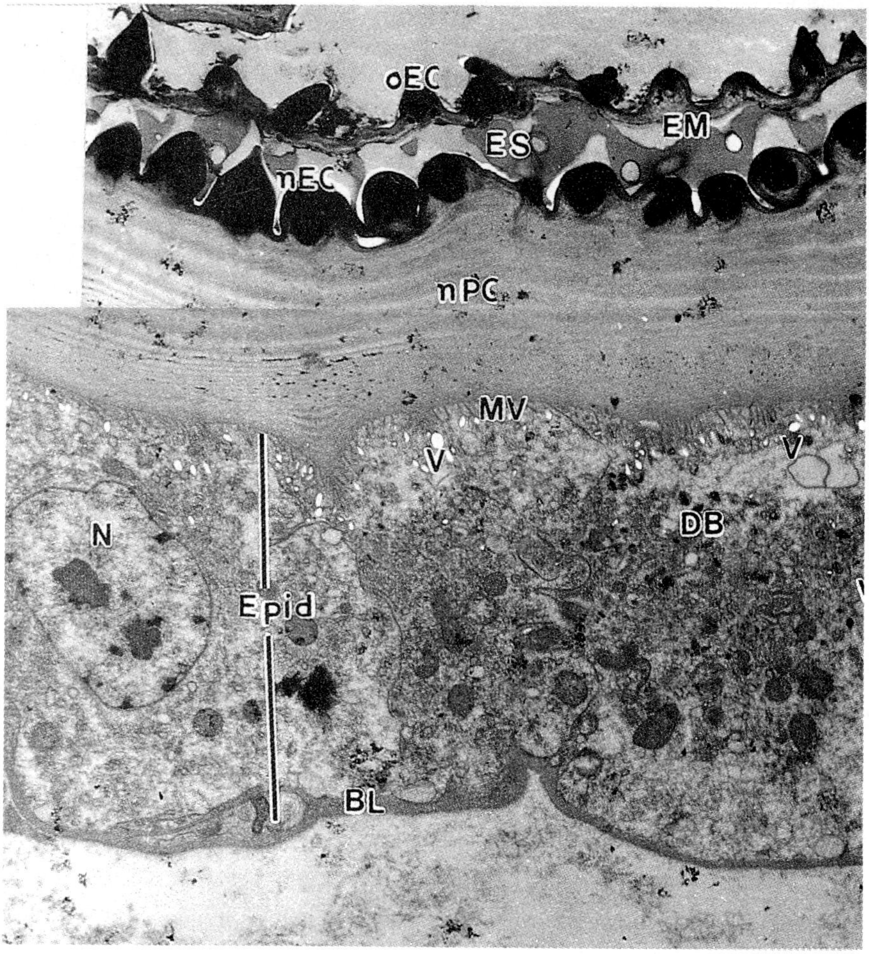

Fig. 2. Integument of a last-instar *S. podoptera exigua* larva, 48 h after treatment with tebufenozide at 10 mg/l, showing the presence of a double cuticle, the conspicuous absence of a high number of procuticular lamellae underneath the newly secreted epicuticle and signs of epidermal cell degeneration. *BL* Basal lamina; *DB* dense body; *Epid* epidermal cell; *EM* ecdysial membrane, *ES* ecdysial space; *MV* epidermal microvilli; *N* epidermal nucleus; *nEC* epicuticle of the newly secreted cuticle; *nPC* procuticle of the new cuticle; *oEC* old epicuticle of the particularly digested cuticle; *V* vacuoles and vesicles; times ×3,000.

more, they found only a very low activity of RH-5849 against the homopteran species *Acyrthosiphon pisum* (Harris), and no activity against the cockroaches *Blattella germanica* (L.) and *Periplaneta americana* (L.). Smagghe and Degheele (1994a,b) reported no adverse activity of RH-5849 and tebufenozide on *Locusta migratoria migratorioides* (R&F) larvae and on different developmental stages of the predatory bugs *Podisus maculiventris* (Say), *Podisus nigrispinus* (Dallas) and *Orius insidiosus* (Say) (Smagghe and Degheele 1994a, b, 1995b). Likewise, the target specificity of tebufenozide, especially, for Lepidoptera was strengthened when treating specimens of more than 150 different insect species of various orders (Carlson et al. 1994). Whether such differences in susceptibility are due to variations in transport and metabolism, to possible differences at EcR level or to differing abilities to mimic specific ecdysteroids in the different species is unclear at present.

2.3 Organ Culture and Cellular Effects

In addition to the earlier cited *in vivo* work, various groups investigated the direct ecdysteroid-like tissue and cellular action of RH-5849 and tebufenozide, as compared with natural hormones, on established cell lines and organ cultures *in vitro*. Wing (1988) reported that the response of K_c cells of *Drosophila* treated with RH-5849 consisted of a cessation of proliferation, clumping and the formation of extended processes, effects that were indistinguishable from the results of exposure to 20E; however, RH-5849 was about 100 times less potent than 20E. Independently, similar results were obtained using the established cell line IAL-PID_2 originating from *P. interpunctella* imaginal discs (Silhacek et al. 1990). RH-5849 inhibited proliferation in 50% of treated cells at a concentration of about 10 µM. More recent experiments showed that tebufenozide caused 50% inhibition of cell proliferation at 0.2 µM, which corresponded well with the I_{50} value of 20E (Oberlander et al. 1995). Next to alterations of cell morphogenesis, acetylcholinesterase (AChE) activity was elevated in cells of *D. melanogaster* and *Chironomus tentans* F. cultured with 20E; similarly, an increased enzyme activity was noted following exposure to RH-5849 (Wing 1988; Spindler-Barth et al. 1991). Furthermore, Spindler-Barth et al. (1991) and Quack et al. (1995) found that RH-5849 and tebufenozide inhibited the spontaneous synthesis of chitin in *C. tentans* cells; that inhibitory activity was similar to 20E. More typically, IAL-PID_2 cells responded to both 20E and RH-5849, by a similar increased uptake of GlcNAc (N-acetyl-ghicosamine), a precursor of chitin, although they did so at different concentrations (Silhacek et al. 1990).

In organ cultures of imaginal wing discs of *P. interpunctella*, RH-5849 stimulated evagination and chitin synthesis via an enhanced incorporation of GlcNAc; in these experiments, RH-5849 was 10 to 100 times less potent than 20E (Silhacek et al. 1990). Similarly, development was promoted in cultured imaginal discs of *Spodoptera*, *Leptinotarsa* and *Galleria* by ecdysteroids and non-steroidal agonists, although at differing concentrations (Smagghe and Degheele 1995a; Smagghe et al. 1996a). In addition, while short exposure to RH-5849, tebufenozide and 20E was necessary to stimulate chitin synthesis and morphogenesis, continuous application was inhibitory. Such inhibitory effects may therefore explain the conspicuous absence of lamellate endocuticle in treated larvae showing premature moulting (Retnakaran and Oberlander 1993; Smagghe et al. 1994, 1996b). Likewise, Oikawa et al. (1993) demonstrated that 20E, RH-5849 and some other related compounds enhanced the incorporation of GlcNAc into cultured fragments of *C. suppressalis* depending on dosage from 0.1 to 1 µM. At higher concentrations or upon exposure for longer than 48 h, the degree of enhancement of incorporation was significantly lower. This inhibition after prolonged exposure may be analogous to the negative effect on chitin synthesis cited earlier. Similarly, Ashok and Dutta-Gupta (1991) reported that acid phosphatase activity in a *Corcyra* fat body culture was stimulated in response to 20E and RH-5849; however, a higher concentration caused a lower degree of increase, while a very high dose failed to produce any change in enzyme activity. Interestingly, it is known that exposure of target cells/organs to very high concentrations of hormone or hormone agonist in vertebrates modifies their response apparently due to a state of desensitization. In most cases, this phenomenon was claimed to result from alternations in the hormone receptor interactions and/or reduction in the number of receptors; however, this mechanism is still poorly understood in insects. In general, the differential potency of action *in vitro* and *in vivo* of 20E and RH-5849/tebufenozide should essentially be attributed to a higher compound stability of either nonsteroid.

2.4 Molecular Biology Studies

The activation of EcRs by RH-5849 and tebufenozide involves their ability to induce molecular actions as ecdysteroids do. Retnakaran et al. (1995) reported that treatment of *Manduca* epidermis with either 20E or tebufenozide resulted in stimulation of mRNA of MHR_3; the latter compound being 10 times more active than the former. MHR_3 is a DNA binding protein in *Manduca* belonging to the steroid hormone receptor superfamily and a putative epidermal transcription factor expressed by ecdysteroids at moulting. In addition, the expression of a specific intermoult larval endocuticular protein gene, LCP_{14}, was inhibited by 20E and in a more persistent manner by tebufenozide (Retnakaran et al. 1995). Hence, tebufenozide prevented the expression of DOPA-decarboxylase mRNA. In these tests, tebufenozide's effect was stronger and more persistent as compared with 20E, suggesting greater compound stability and/or an enhanced binding affinity to target receptor sites (Retnakaran et al. 1995).

Furthermore, Palli et al. (1995) indicated that the midgut, fat body and epidermal cells of *C. fumiferana* were responsive to tebufenozide from the beginning to the end of the last larval stadium based on mRNA expression of *Choristoneura* hormone receptor 3. Otherwise, tebufenozide could only induce precocious moulting in larvae that had ingested the compound during the early part of the stadium, before the appearance of the ecdysteroid peak. One possible reason for this age-dependent effect may be that ecdysteroids regulate moulting and metamorphosis by controlling the expression of genes by different mechanisms. Herewith, different isoforms of the EcRs present before and after the commitment peak of ecdysteroids were observed in *Drosophila* (Talbot et al. 1993) and may be involved in differential EcR binding affinities of tebufenozide. Additionally, it is hypothesized that tebufenozide is unable to reinitiate a cascade of events once the natural occurring ecdysteroid peak has initiated the moulting sequence. So, further molecular analysis is required to elucidate the action of ecdysteroids and nonsteroidal analogues on the expression of specific genes involved in moulting, and to evaluate whether their activity is completely identical.

2.5 Receptor Binding

As RH-5849 and tebufenozide possess an ecdysteroid-mimicking activity, they should operate by the same mode of action as ecdysteroids; they should bind to specific EcR binding proteins that are members of the steroid hormone receptor superfamily (Bidmon and Sliter 1990; Jindra 1994). First evidence for the interaction of RH-5849 with EcRs was reported by Wing (1988) indicating that RH-5849 competitively displaced radiolabelled PoA from EcR extracts of *Drosophila* cells. In his experiments, it was also found that K_c cells that were incubated for 4 weeks with either 20E or RH-5849 became resistant to either ecdysteroid hormone and agonist. This cross-resistance suggested that RH-5849 and 20E competed for the same sites of action, i.e. the EcRs. In addition, both resistant strains showed a sharp reduction in ^{3}H-labelled PoA binding capacity as compared with untreated cells. Furthermore, Wing et al. (1988) described a positive correlation between the activity of 24 analogues of RH-5849 in inducing precocious moulting in *M. sexta* larvae and their ability to bind EcRs of a *Drosophila* K_c cell line, which should support the relevance that RH-5849 and related compounds act through receptor binding.

During the last few years, investigations in some laboratories have evaluated the binding affinity of EcRs for RH-5849 and tebufenozide in comparison with 20E by measuring the displacement of ^{3}H-PoA. Wing (1988) found that RH-5849 was about 100 times less effective than 20E in displacing labelled PoA in crude cytosolic EcR preparations from *Drosophila* K_c cells. RH-5849 competed with radiolabelled PoA for EcR sites extracted from nuclear extracts of *G. mellonella*; RH-5849 was approximately as effective as 20E, but 25 times less potent than ecdysone (Sobek et al. 1993). Similar results were obtained by Spindler-Barth et al. (1991) using the *C. tentans* cell line, but RH-5849 was only 4 times less effective than 20E. Recent binding assays demonstrated that tebufenozide was approximately 10 to 100 times more potent than 20E in displacing ^{3}H-PoA: equation binding constants of tebufenozide were in the order of that of PoA, 1 nM in cytosolic cell extracts of *D. melanogaster* (Carlson et al. 1994) and of *C. tentans* (Quack et al. 1995). In organ cultures of imaginal wing discs of *S. exigua*, Smagghe and Degheele (1995a) found that 50% competition of radiolabelled PoA resulted at 1397 nM of RH-5849 and at 32 nM of tebufenozide. In *Galleria* discs, binding affinities of RH-5849 and tebufenozide reached similar levels of 911 and 22 nM, respectively (Smagghe et al. 1996a). Interestingly, they also reported that in coleopteran discs of *L. decemlineata*, 50% receptor competition for PoA occurred at 740 nM of RH-5849, while a much higher concentration of tebufenozide was needed (1316 nM). The relevance of the latter receptor studies with discs of *S. exigua*, *G. mellonella* and *L. decemlineata* was supported as a good relationship was found with disc evagination data. Previous toxicity assays also demonstrated a more than 50-fold tolerance in *Leptinotarsa* larvae for tebufenozide as compared with RH-5849 and data on distribution and metabolism into the insect body indicated that such change in toxicity cannot essentially be explained by metabolic and pharmacokinetic variations (Smagghe and Degheele 1993, 1994d).

Therefore, the insect selective toxicity of such dibenzoyl hydrazine-based compounds, especially among insects originating from different orders, seems to be related with selective binding on the EcRs. However, up until now, no clear conclusions can be drawn of whether 20E and substituted dibenzoyl hydrazines are perceived in a similar manner for EcR binding by insects although they possess a similar ecdysteroid activity. On this matter, Chan et al. (1990) and Mohammed-Ali et al. (1995) expected that ligand and receptor interactions depend on the molecular crystal ligand structure. Recently, Oikawa et al. (1994a,b), Nakagawa et al. (1995) and Smagghe et al. (unpubl. data) found that the larvicidal potency of differently substituted dibenzoyl hydrazines depended on the conformational influence of the substituents and their positions on the benzoyl-benzene rings, thus affecting the molecular shape of the ligand binding molecule. Nevertheless, further research should provide knowledge needed for a better understanding of the nature of EcR activation as a basis for biological activity and the importance of the ligand crystal structure on binding.

2.6 Neurotoxic Mechanism

As first reported by Aller and Ramsay (1988) and Hsu (1991) and later in more detail by Salgado (1992a,b), RH-5849 caused a rapid onset of hyperactivity leading to paralysis, at least in some insect species, e.g. cockroaches, Coleoptera and Diptera. Otherwise, neurotoxicity could only be obtained in special cases in Lepidoptera by the injection of very high doses, but it has never been observed in feeding and topical application assay, and premature moulting due to an ecdysonergic activity occurs at much lower doses.

Immediately after injection in *Periplaneta* adults, moderate doses caused mild trembling of the legs after which the insects recovered completely. At higher doses, insects became hyperactive and they were prostrate with all their appendages moving incessantly. Furthermore, movement was significantly diminished and all insects were completely paralyzed after 6 h (Salgado 1992a,b). Salgado's extensive electrophysiological experiments proved that such apparent neurotoxic activity is mediated by a blockage of K^+ channels in nerve and muscle and would not involve EcR activation. In this connection, RH-5849 and analogous compounds may provide the opportunity to assess the impact of ecdysteroids on ion channel regulation. Such experiments may also aim to evaluate the relative importance of these two distinct sites of action of RH-5849 and analogous compounds and to clarify their dual action, as compared with ecdysteroids. Hereby, understanding the physiological mechanisms of neurotoxicity of RH-5849 and analogues is important from the point of view of insect control, especially in species where neurotoxicity is often the first evident sign.

2.7 Action in Adult Insects

Aller and Ramsay (1988) and Wing and Ramsay (1989) were the first to report that RH-5849 affects reproduction in various insect orders via cessation of oviposition. Further-more, Lawrence (1992) and Smagghe and Degheele (1992a, 1994a, b) demonstrated that RH-5849 and tebufenozide have a drastic effect on egg-laying in adults of Lepidoptera, Coleoptera and Diptera within 1-2 days of treatment. In addition, treated adults showed agitation and hyperactivity followed by cessation of feeding, movement and copulation. In all cases, dissection of females which had stopped oviposition suggested that new oocyte formation was inhibited. The ovaries showed signs of degeneration suggesting that (re)sorption of the oocytes was induced. Electrophoretic analysis carried out by Lawrence (1992) and Smagghe et al. (unpubl. data) indicated that the level of vitellogenins in the haemolymph of treated *Anastrepha* and *Leptinotarsa* females was higher than in untreated controls, suggesting resorption of vitellogenins. Likewise, Dhadialla and Raikhel (1994) reported the effectiveness of RH-5849 and tebufenozide as compared with 20E in sustaining vitellogenesis synthesis in vitellogenic fat bodies of female *Aedes aegypti* (L.), and indicated the complexity of the regulation of vitellogenesis. In addition, laboratory assays of Charmillot et al. (1994a) showed a detrimental effect on reproduction when either male or female moths of *Lobesia* and *Eupoecilia* were in contact with residues of tebufenozide. On the activity of tebufenozide in male adults, codling moth abdomen experiments of Friedländer and Brown (1995) revealed that the compound interferes with spermatogenesis. Taken together, such non-steroidal ecdysone agonists provoke a significant reduction in reproduction that is useful in insect control. Otherwise, details on their action site(s) with the reproduction system generally remain unclear at present; however, it is believed that such compounds may provide a metabolically stable tool in unravelling the role of ecdysteroids in insect reproduction.

3
IGRs in Insect Control

Efforts to identify mimics of the insect moulting hormones have led to the evaluation of ecdysteroids or closely related steroidal analogues for the last two decades. Although active ecdysteroids can be obtained from plant and animal sources, the main reasons for the failure to develop them as insecticides has been that their structures are too complex for economic production, their corpulent size and hydrophilic nature prohibits their penetration into the insect cuticle, and insects have powerful mechanisms to eliminate and catabolize steroidal structures (Horn and Bergamasco 1985; Koolman and Karlson 1985). During the last 10 years, the need for new insecticides with a selective activity spectrum and preferentially a new mode of action, has been recognized because of the increasing problems of insect resistance to broad-spectrum conventional insecticides, and environmental hazards. In this way, synthetic agonists of 20E, i.e. RH-5849 and tebufenozide, represent a new concept in insect growth control and induce the development of a new class of IGRs, following the chitin synthesis inhibitors and JHAs. Up until now, there has been strong evidence from laboratory and field results that tebufenozide may be of immense practical importance in the control of agriculturally important insect pests, especially Lepidoptera. Significant data for practical use of dibenzoyl hydrazines were firstly obtained with RH-

5849. However, it does not seem likely that RH-5849 will be developed for commercial use, although it showed promise in the control of the Colorado potato beetle *Leptinotarsa*. Furthermore, the potency of tebufenozide for agriculture has been demonstrated against some lepidopterous pests, including the codling moth *C. pomonella* in apple orchards (Heller et al. 1992; Mattioda et al. 1993; Brown 1994; Charmillot et al. 1994b), the vine moth *L. botrana* and the summerfruit tortrix *Adoxophyes orana* in vineyards (Mattioda et al. 1993; Charmillot et al. 1994a, b), the fall armyworm *S. frugiperda* in maize and the corn earworm, *Heliothis (Helicoverpa) zea* (Boddie) (Chandler et al. 1992; Monthéan and Potter 1992), the cabbage moth *M. brassicae*, the rice stemborer *C. suppressalis*, armyworms *S. exigua*, *S. littoralis* and *S. exempta*, in vegetables, cotton, cereals and rice (Chandler et al. 1992; Oikawa et al. 1994a, b; Smagghe and Degheele 1994b), and the Indian meal moth *P. interpunctella* in grain storage commodities (Oberlander et al. 1995).

In addition, tebufenozide's technical bulletin demonstrates its excellent field performance against various important lepidopteran pests in apple, cotton, forestry, rice pads, vegetables and grapes (Rohm and Haas 1994). It appears that this compound has already and will soon have more substantial practical application in pest Lepidoptera control. Meanwhile, tebufenozide has received registration as Mimic-Confirm in North-America for various food crops and forestry, in South-East Asia, i.e. the Philippines, Malaysia, Indonesia, Thailand and South-Korea, for rice, vines, vegetables and soy beans and in Japan for rice, apples, sugar-beet and tea. In France and Switzerland, tebufenozide's registration applications include particularly apples and vine grapes, and additionally cabbages and other leaf vegetables. In Belgium, the compound is registered for codling moth control in fruit orchards starting 1996. It was noteworthy that RH-5849 and tebufenozide rapidly caused precocious lethal moulting either by coating leaves, mixing into (semi-)artificial diet and administering topically. In addition, RH-5849 was effective systemically when applied as a soil treatment to tomato plants on which *Leptinotarsa* was fed (Wing and Aller 1990) or when applied to maize seedlings through irrigation water that was offered to *Spodoptera* larvae (Smagghe 1991). Clearly, tebufenozide is more potent on and more selective to Lepidoptera as compared with RH-5849. Thus, the unique mode of action of tebufenozide is its high specificity to Lepidoptera coupled with a high stability, suggesting that this compound may be a useful alternative to broad-spectrum chemicals in forest, fruit-tree, vegetables and rice pest management.

In addition, RH-5849 and tebufenozide, especially tebufenozide, were found in laboratory and field trials to have a high level of selectivity towards various beneficial insects and mites that are generally used as biological control agents of various (non)lepidopterous pests in IPM programmes. For example, tebufenozide had no adverse effects on different predatory mites, braconid parasites and pentatomid/heteropteran predators (Rohm and Haas 1989; 1994; Heller et al. 1992; Mattioda et al. 1993; Brown 1994; Shimizu 1994; Biddinger and Hull 1995; Jacas et al. 1995; Smagghe and Degheele 1995b). The target selectivity of tebufenozide for Lepidoptera has been confirmed after determining its pest-control spectrum using more than 150 different insect species of various orders (Carlson et al. 1994). Hereby, the safety of tebufenozide to other predatory beneficials will be tested in the 1996 International Organization for Biological and Integrated control of No xious Aninals and Plants (IOBC) working group 'Pesticides and Beneficial Organisms' programme (Rohm and Haas 1994). In addition, laboratory and field assays confirmed its selectivity towards *Apis mellifera* that are responsible for pollination in glasshouses.

RH-5849 and tebufenozide possess ideal characteristics to be used as environmentally safe insecticides as they are essentially non-toxic to mammals, birds and fishes and they score negative in the Ames mutagenicity test according to the technical bulletin (Rohm and Haas 1989, 1994) and Wing and Aller (1990). Otherwise, it is expected that such non-steroidal agonists would affect the development of some Crustacea, which is expected to be ecdysteroid-dependent. Only at very high concentrations of RH-5849 and tebufenozide was toxicity recorded in *Daphnia* spp. (Rohm and Haas 1989, 1994; Wing and Aller 1990) and crab zoeae (Clare et al. 1992; Kreutzweiser et al. 1994). However, clear answers on the processes leading to selective toxicity of such hydrazine–based ecdysteroid agonists cannot be given at present. Perhaps a different incorporation, distribution and metabolism in the insect body may have caused the observed differences in insect susceptibility. In addition, EcR binding specificity may account for the variations in insecticide effectiveness, as discussed in Section 2.5.

As the mode of action of RH-5849 and tebufenozide is completely unrelated to that of other classes of insecticides, the induction of cross-resistance is unlikely to occur. As such, tebufenozide and related compounds are useful in insecticide resistance management (IRM) programmes. Hereby, Smagghe and Degheele (1997) reported that tebufenozide was highly toxic against laboratory and Egyptian multi-resistant field strain larvae of *S. littoralis,* as compared with a few organophosphates (OPs), carbamates, synthetic pyrethroids and benzoylphenylureas (BPUs). The latter data revealed a low (four-fold) tolerance for tebufenozide in field strains as compared with susceptible laboratory strain larvae and no significant signs of cross-resistance were noted; field larvae were especially resistant to carbamate insecticides (50- to 280-fold), about 30-fold more resistant to chlorpyrifos and five-fold less susceptible to BPUs. Similarly, Ishaaya et al. (1995) reported that an Israeli field *S. littoralis* strain, being over 100-fold resistant to OPs and pyrethroids, showed a low tolerance of five-fold to tebufenozide. Both groups expect that this low level of tolerance for tebufenozide might be attributed to an enhanced level of detoxifying enzymes in the insect body, especially in the gut, so that less active parent insecticide molecules reach the biochemical target sites of action (EcRs) after passage through the insect body after treatment. In addition, recent experiments of Smagghe and Degheele (1997) illustrated that tolerance was not induced when tebufenozide was continuously applied over various generations of *S. littoralis.* However, in order to prevent development of resistance, it is recommended (Rohm and Haas 1994) to alternate tebufenozide with other groups of insecticides, especially when regular spraying is required to control the target pest. Hence, the results obtained so far strengthen the notion that tebufenozide and related compounds may be useful in the control of lepidopterous pests resistant to conventional insecticides and current IGRs. Nevertheless, further research on resistance and cross-resistance is required before final conclusions can be drawn and is in progress in our laboratory.

Acknowledgements

Dr. G. Smagghe was supported by a post-doctoral project 950162 from the IWT (Flemish Institute for encouragement of scientific-technological research in industry).

References

Aller HE, Ramsay JR (1988) RH-5849—a novel insect growth regulator with a new mode of action. Proc Brighton Crop Prot Conf—Pests and diseases November 1988, Brighton, vol 5, pp 511–518

Ashok A, Dutta-Gupta A (1991) *In vitro* effects of nonsteroidal ecdysone agonist RH-5849 on fat body acid phosphatase activity in rice moth, *Corcyra cephalonica* (Insecta). Biochem Int 24: 69–75

Biddinger DJ, Hull LA (1995) Effects of several insecticides on the mite predator, *Stethorus punctum* (Coleoptera: Coccinellidae), including insect growth regulators and abamectin. J Econ Entomol 88: 358–366

Bidmon H-J, Sliter TJ (1990) The ecdysteroid receptor. Invertebr Reprod Dev 18: 13–27

Brown JJ (1994) Effects of a nonsteroidal ecdysone agonist, tebufenozide, on host/parasitoid interactions. Arch Insect Biochem Physiol 26: 235–248

Carlson GR, Dhadialla TS, Thompson C, Ramsay R, Thirugnanam M, James W, Slawecki R (1994) Insect toxicity, metabolism and receptor binding characteristics of the non-steroidal ecdysone agonist, RH-5992. Proc 11th Ecdysone Worksh 43, June-July 1994, Ceske Budejovice, Czech Republic

Chan TH, Ali A, Britten JF, Thomas AW, Strunz G, Salonius A (1990) The crystal structure of 1,2-dibenzoyl-1-*tert*-butylhydrazine, a nonsteroidal ecdysone agonist, and its effects on spruce budworm (*Choristoneura fumiferana*). Can J Chem 68: 1176–1181

Chandler LD, Pair SD, Harrison WE (1992) RH-5992, a new insect growth regulator active against corn earworm and fall armyworm (Lepidoptera: Noctuidae). J Econ Entomol 85: 1099–1103

Charmillot P-J, Favre R, Pasquier D, Rhyn M, Scalco A (1994a) Effet du régulateur de croissance d'insectes (RCI) tébufénozide sur les oeufs, les larves et les papillons de vers de grappe *Lobesia botrana* Den.et Schiff. et *Eupoecilia ambiguella* Hb. Mitt Schweiz Entomol Ges 67: 393–402

Charmillot P-J, Pasquier D, Alipaz NJ (1994b) Le tébufénozide, un nouveau produit sélectif de lutte contre le carpocapse *Cydia pomomella* L. et la tordeuse de la pélure *Adoxophyes orana* F. v. R. Rev Suisse Vitic Arboric Hortic 26: 123–129

Clare AS, Rittschof D, Costlow jr D (1992) Effects of the nonsteroidal ecdysone mimic RH-5849 on larval crustaceans. J Exp Zool 262: 436–440

Darvas B, Polgar L, Tag El-din MH, Eröss K, Wing KD (1992) Developmental disturbances in different insect orders caused by an ecdysteroid agonist, RH-5849. J Econ Entomol 85: 2107–2112

Dhadialla TS, Raikhel AS (1994) Endocrinology of mosquito vitellogenesis. In: Perspectives in comparative endocrinology. N Res Counc Can, pp 275–281

Feldlaufer MF (1989) Diversity of molting hormones in insects. In: Koolman J (ed) Ecdysone from chemistry to mode of action, Georg Thieme, Stuttgart, pp 308–312

Friedländer M, Brown JJ (1995) Tebufenozide (Mimic), a non-ecdysteroidal ecdysone agonist, induces spermatogenesis reinitiation in isolated abdomens of diapausing codling moth larvae (*Cydia pomonella*). J. Insect Physiol 41: 403–411

Gadenne C, Varjas L, Mauchamp B (1990) Effects of the non-steroidal mimic, RH-5849, on diapause and non-diapause of the European corn borer, *Ostrinia nubilalis* Hbn. J Insect Physiol 36: 555–559

Heller JJ, Mattioda H, Klein E, Sagenmüller A (1992) Field evaluation of RH-5992 on lepidopterous pests in Europe. Proc. Brighton Crop Prot Conf—Pests and diseases, November 1992, Brighton, vol 1, pp 59–66

Horn DHS, Bergamasco R (1985) Chemistry of ecdysteroids. In: Kerkut GA, Gilbert LI (eds) Comprehensive insect physiology biochemistry and Pergamon Press, Oxford, pp 185–248

Hsu AC-T (1991) 1, 2-Diacyl-1-alkylhydrazines: a new class of insect growth regulators. In: Baker DR, Fenyes JG, Molberg WK (eds) ACS Symp Ser 443—synthesis and chemistry of agrochemicals II. American Chemical Society, Washington DC, pp 478–490

Ishaaya I, Yablonski S, Horowitz AR (1995) Comparative toxicity of two ecdysteroid agonists, RH-2485 and RH-5992, on susceptible and pyrethroid-resistant strains of the Egyptian cotton leafworm, *Spodoptera littoralis.* Phytoparasitica 23: 139–145

Jacas J, González M, Viñuela E (1995) Influence of the application method on the toxicity of the moulting accelerating compound tebufenozide on adults of the parasitic wasp *Opius concolor* Szèpl. Meded Fac Landbouwwet Univ Gent 60: 935–939

Jindra J (1994) Gene regulation by steroid hormones: vertebrates and insects. Eur J Entomol 91: 163-187

Koolman J, Karlson P (1985) Regulation of ecdysteroid titer: degradation. In: Kerkut GA, Gilbert LI (eds) Comprehensive insect physiology, biochemistry and pharmacology, vol 7. Pergamon Press, Oxford, pp 343–361

Kreutzweiser DP, Capell SS, Wainio-Keizer KL, Eichenberg DC (1994) Toxicity of a new molt-inducing insecticide (RH-5992) to aquatic macroinvertebrates. Ecotoxicol Environ Safety 28: 14–24

Lawrence PO (1992) Egg development in *Anastrepha suspensa*: influence of the ecdysone agonist, RH-5849. In: Aluja M, Liedo P (eds) Fruit flies: recent advances in research and control programs. Springer, Berlin Heidelberg New York, pp 51–56

Mattioda H, Maigrot PH, Heller JJ, Baverez S, Chapius G (1993) Le tébufénozide: une nouvelle approche de lutte contre les lépidoptères en vigne et arboriculture fruitière. Proc 3me Conf Int sur les Ravageurs en Agriculture, January 1993, Montpellier, 1, pp 243–250

Mohammed-Ali AK, Chan TH, Thomas AW, Strunz GM, Jewett B (1995) Structure-activity relationship study of synthetic hydrazines as ecdysone agonists in the control of spruce budworm (*Choristoneura fumiferana*). Can J Chem 73: 550–557

Monthéan C, Potter DA (1992) Effects of RH-5849, a novel insect growth regulator, on Japanese beetle (Coleoptera: Scarabaeidae), and fall armyworm (Lepidoptera: Noctuidae) in turfgrass. J Econ Entomol 85: 507–513

Nakagawa Y, Shimizu B-I, Oikawa N, Akamatsu M, Nishimura K, Kurihara N, Ueno T, Fujita T (1995) Three-dimensional quantitative structure-activity analysis of steroidal and dibenzoylhydrazine-type ecdysone agonists. In: Hansch C, Fujita T, Classical and three-dimensional QSAR in agrochemistry, ACS Symp Ser, vol 606. Washington DC, pp 606, 288–301

Oberlander H, Silhacek DL, Porcheron P (1995) Non-steroidal ecdysteroid agonists: tools for the study of hormonal action. Arch Insect Biochem Physiol 28: 209–223

Oikawa N, Nakagawa Y, Soya Y, Nishimura K, Kurihara N, Ueno T, Fujita T (1993) Enhancement of N-acetylglucosamine incorporation into cultured integument of *Chilos uppressalis* by molting hormone and dibenzoylhydrazine insecticides. Pestic Biochem Physiol 47: 165–170

Oikawa N, Nakagawa Y, Nishimura K, Ueno T, Fujita T (1994a) Quantitative structure-activity analysis of larvicidal 1-(substituted benzoyl)-2-benzoyl-1-*tert*-butylhydrazines against *Chilo suppressalis.* Pestic Sci 41: 139–148

Oikawa N, Nakagawa Y, Nishimura K, Ueno T, Fujita T (1994b) Quantitative structure-activity studies of insect growth regulators. X. Substituent effects on larvicidal activity of 1-*tert*-butyl-1-(2-chlorobenzoyl)-2-(substituted benzoyl) hydrazines against *Chilo suppressalis* and design synthesis of potent derivates. Pestic Biochem Physiol 48: 135–144

Palli SR, Primavera M, Tomkins W, Lambert D, Retnakaran A (1995) Age-specific effects of a nonsteroidal ecdysteroid agonist, RH-5992, on the spruce budworm, *Choristoneura fumiferana* (Lepidoptera: Tortricidae). Eur J Entomol 92: 325–332

Park NJ, Jang KS, Cho JR, Cho KY (1992) Effects of RH-5849, an ecdysone agonist, against feeding and growth of tobacco cutworm (*Spodoptera litura* Fabricius) larvae. Korean J Appl Entomol 31: 475–479

Quack S, Fretz A, Spindler-Barth M, Spindler K (1995) Receptor affinities and biological responses of nonsteroidal ecdysteroid agonists on the epithelial cell line from *Chironomus tentans* (Diptera: Chironomidae). Eur J Entomol 92: 341–347

Rees HH (1989) Zooecdysteroids: structures and occurrence. In: Koolman J (ed) Ecdysone from Chemistry to mode of action, Georg Thieme Stuttgart, pp. 308–312

Retnakaran A, Oberlander H (1993) Control of chitin synthesis in insects. In: Mussarelli RAA (ed) Chitin enzymology. European Chitin Society, Ancona, pp 89–99

Retnakaran A, Hiruma K, Palli SR, Riddiford LM (1995) Molecular analysis of the mode of action of RH-5992, a lepidopteran-specific, non-steroidal ecdysteroid agonist. Insect Biochem Mol Biol 25: 109–117

Rohm and Haas (1989) Technical information bulletin, RH-5849, exploratory insecticide. Rohm and Haas, Spring House, Pennsylvania

Rohm and Haas (1994) Technical bulletin, Mimic-Confirm, tebufenozide (RH-5992). Rohm and Haas, Spring House, Pennsylvania

Salgado VL (1992a) Block of voltage-dependent K^+ channels in insect muscle by the diacylhydrazine insecticide RH-5849, 4-aminopyridine, and quinidine. Arch Insect Biochem Physiol 21: 239–252

Salgado VL (1992b) The neurotoxic insecticidal mechanism of the nonsteroidal ecdysone agonist RH-5849: K^+ channel block in nerve and muscle. Pestic Biochem Physiol 43: 1–13

Shimizu T (1994) Effect of RH-5849 on the emergence of *Apanteles kariyai* parasitoid from the common army worm larvae, *Leucania separata*. Int Pest Control 36: 131–133

Silhacek DL, Oberlander H, Porcheron P (1990) Action of RH-5849, a nonsteroidal ecdysteroid mimic, on *Plodia interpunctella* (Hübner) *in vivo* and *in vitro*. Arch Insect Biochem Physiol 15: 201–212

Slama K (1995) Hormonal status of RH-5849 and RH-5992 synthetic ecdysone agonists (ecdysoids) examined on several standard bioassays for ecdysteroids. Eur J Entomol 92: 317–323

Smagghe G. (1991) Toxicity and mode of action of the ecdysone agonist RH-5849 in three *Spodoptera* spp. (Lepidoptera: Noctuidae). MSc Thesis, University. Gent, Belgium

Smagghe G, Degheele D (1992a) Effect of the nonsteroidal ecdysteroid agonist RH-5849 on reproduction of *Spodoptera littoralis* (Boisd.) (*Lepidoptera: Noctuidae*). Parasitica 48: 23–29

Smagghe G, Degheele D (1992b) Effects of the first nonsteroidal ecdysteroid agonist on larvae of *Spodoptera littoralis* (Boisd.) (Lepidoptera: Noctuidae). Arch Insect Biochem Physiol 21: 119–128

Smagghe G, Degheele D (1993) Toxicity, pharmacokinetics, and metabolism of the first nonsteroidal ecdysteroid agonist RH-5849 on *Spodoptera exempta* (Walker), *Spodoptera exigua* (Hübner), and *Leptinotarsa decemlineata* (Say). Pestic Biochem Physiol 46: 149–160

Smagghe G, Degheele D (1994a) Action of a novel nonsteroidal ecdysteroid mimic, tebufenozide (RH-5992), on insects of different orders. Pestic Sci 42: 85–92

Smagghe G, Degheele D (1994b) Action of the nonsteroidal ecdysteroid mimic RH-5849 on larval development and adult reproduction of insects of different orders. Invertebr Reprod Dev 25: 227–236

Smagghe G, Degheele D (1994c) Effects of the ecdysteroid agonists RH-5849 and RH-5992, alone and in combination with a juvenile hormone analogue, pyriproxyfen, on larvae of *Spodoptera exigua*. Entomol Exp Appl 72: 115–123

Smagghe G, Degheele D (1994d) The significance of pharmacokinetics and metabolism to the biological activity of RH-5992 (tebufenozide) in *Spodoptera exempta*, *Spodoptera exigua*, and *Leptinotarsa decemlineata*. Pestic Biochem Physiol 49: 224–234

Smagghe G, Degheele D (1995a) Biological activity and receptor-binding of ecdysteroids and the ecdysteroid agonists RH-5849 and RH-5992 in imaginal wing discs of *Spodoptera exigua* (Lepidoptera: Noctuidae). Eur J Entomol 92: 333–340

Smagghe G, Degheele D (1995b) Selectivity of nonsteroidal ecdysteroid agonists RH-5849 and RH-5992 to nymphs and adults of the predatory soldier bugs, *Podisus nigrispinus* and *Podisus maculiventris* (Hemiptera: Pentatomidae). J Econ Entomol 88: 40–45

Smagghe G, Degheele D (1997) Comparative toxicity and tolerance for the ecdysteroid mimic tebufenozide in a laboratory and field strain of the cotton leafworm. J Econ Entomol (in press)

Smagghe G, Degheele D, Viñuela E, Budia F (1994) Potency and ultrastructural effects of tebufenozide on *Spodoptera exigua*. Proc Brighton Crop Prot Cont—Pests and diseases, November 1994, Brighton, vol 2, pp 339–340

Smagghe G, Böhm G-A, Richter K, Degheele D (1995) Effect of nonsteroidal ecdysteroid agonists on ecdysteroid titer in *Spodoptera exigua* and *Leptinotarsa decemlineata*. J Insect Physiol 41: 971–974

Smagghe G, Eelen H, Verschelde E, Richter K, Degheele D (1996a) Differential effects of nonsteroidal ecdysteroid agonists in coleoptera and lepidoptera: analysis of evagination and receptor binding in imaginal discs. Insect Biochem Mol Biol 26: 687–695

Smagghe G, Viñuela E, Budia F, Degheele D. (1996b) *In vivo* and *in vitro* effects of the nonsteroidal ecdysteroid agonist tebufenozide on cuticle formation in *Spodoptera exigua*: an ultrastructural approach. *Arch Insect Biochem Physiol* 32: 121–134

Sobek L, Böhm G-A, Penzlin H (1993) Ecdysteroid receptors in last-instar larvae of the wax moth *Galleria mellonella* L. Insect Biochem Mol Biol 23: 125–129

Spindler-Barth M, Turberg A, Spindler K-D (1991) On the action of RH-5849, a nonsteroidal ecdysteroid agonist, on a cell line from *Chironomus tentans*. Arch Insect Biochem Physiol 16: 11–18

Talbot WS, Swyryd EA, Hogness DS (1993) *Drosophila* tissues with different metamorphic responses to ecdysone express different ecdysone receptor isoforms. Cell 73: 1323–1337

Tateishi K, Kiuchi M, Takeda S (1993) New cuticle formation and molt inhibition by RH-5849 in the common cutworm, *Spodoptera litura* (Lepidoptera: Noctuidae). Appl Entomol Zool 28: 177–184

Wing KD (1988) RH-5849, a nonsteroidal ecdysone agonist: effects on a *Drosophila* cell line. Science 241: 467–469

Wing KD, Aller HE (1990) Ecdysteroid agonists as novel insect growth regulators. In: Casida JE (ed) Pesticides and alternatives. Elsevier Amsterdam, pp 251–257

Wing KD, Ramsay JR (1989) Other hormonal agents: ecdysone agonists. In: BCPC Monogr. No 43 Progress and prospects in insect control. BCPC, Surrey, pp 107–117

Wing KD, Slawecki RA, Carlson GR (1988) RH-5849, a nonsteroidal ecdysone agonist: effects on larval Lepidoptera. Science 241: 470–472

CHAPTER 3

Pymetrozine: A Novel Insecticide Affecting Aphids and Whiteflies

D. Fuog[1], S.J. Fergusson[2], and C. Flückiger[3]
[1]Novatris Crop Protection, Inc., Basel, Switzerland
[2]Novartis crop protection, Inc., 7145 58th Ave., Vero Baach, Florida 32967, USA
[3]Novartis Crop Protection, Inc., P.O. Box 18300, Greensboro, North Corolina 27419–8300, USA

1 Introduction

Pymetrozine (CGA 215'944) is a new insecticide, highly active and specific against sucking insect pests (Flückiger et al. 1992a). Pymetrozine is the only representative of the pyridine azomethines, a new class of insecticides, and is currently being developed worldwide for control of aphids and whiteflies in field crops, vegetables, ornamentals, cotton, hop, deciduous fruit, and citrus, and of the brown planthopper, *Nilaparvata lugens* (Staol), in rice (Flückiger et al. 1992a,b). The compound appears to have great promise in integrated pest management (IPM) programs due to its high degree of selectivity, low mammalian toxicity, and safety to birds, fish, and nontarget arthropods. Pymetrozine was discovered and is currently being registered and marketed by Ciba-Geigy with the tradenames "Fulfill", "Relay" and "Sterling" in the USA and "Plenum" or "Chess" elsewhere.

2 Names and Physicochemical Properties

Active ingredient code:	CGA 215'944
Common name:	Pymetrozine
Systematic chemical name:	4,5-dihydro-6-methyl-4-(3-pyridinyl-methylene amino)-1,2,4-triazin-3(2H)-one(CA)
Empirical formula:	$C_{10}H_{11}N_{50}$
Structural formula:	

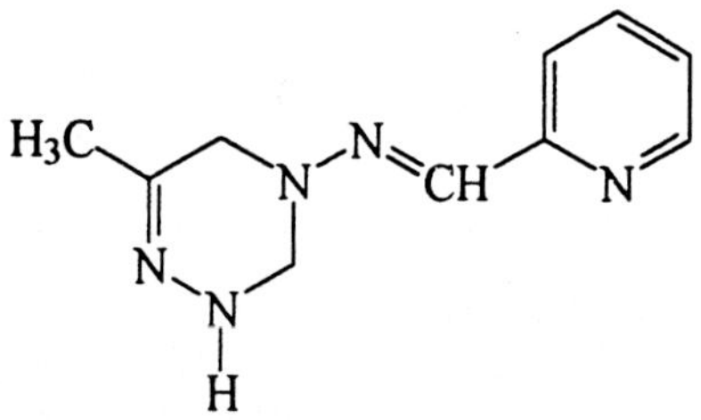

Molecular weight:	217.23
Water solubility:	290 mg/l at 25 °C
Soil mobility:	Little mobile
Partition coefficient *n-octanol/water:*	log P = − 0.18 at 25 °C
Vapor pressure:	< 4 × 10^{-6} Pa at 25 °C
Melting point:	217 °C (decomposition)

3
Mode of Action

Pymetrozine represents not only a new class of chemistry, but also a unique mode of action. Although it has no knockdown effect and is not directly toxic to insects, it does cause an immediate and irreversible cessation of feeding after exposure to the compound that is not due to a deterrent action. Rather, the mode of action is due to a blockage of stylet penetration (Kayser et al. 1994). The target mechanism of pymetrozine has not yet been identified. It would, however, be helpful to understand this mechanism to be able to use pymetrozine in an optimal way. Mortality is slow and, depending on climatic conditions, treated insects may appear normal for several days before they die from starvation (Schwinger et al. 1994). Laboratory results indicate that the feeding activity of aphids is reduced by almost 90% within 3 h postapplication although 100% mortality is not achieved until 48 h posttreatment (Flückiger et al. 1992a, Table 1).

Laboratory studies using an electrical penetration graph (EPG) technique (Tjallingii 1988) demonstrated that pymetrozine applied to aphids topically, orally, or by injection, resulted in almost immediate blockage of stylet insertion by aphids into the plant and eventual death by starvation. Topical application of pymetrozine to *Myzus persicae* (Sulz.) and *Aphis gossypii* (Glov.) resulted in the initiation of feeding being effectively blocked (Table 2). If insertion of the stylet into the phloem was achieved, it took considerably more time than normal, and the aphids ingested sap only for a very short time. Injection of pymetrozine into *M. persicae* indicated that the lowest active dose was 1.2 ng per aphid. Doses above this level resulted in immediate inhibition of stylet probing. In diet studies,

Table 1. Antifeedant activity and mortality of *Aphis craccivora* (Koch) following pymetrozine application at one g ai/l

	Hours after application			
	0–3	3–6	6–24	24–48
Reduction in feeding[a] (%)	88	85	85	–
Control of aphids[b] (%)	10	27	92	100

[1]Feeding reduction evaluated by comparing honeydew production per living individual with untreated check. Honeydew collected on filter paper, dyed with ninhydrine spray, and measured electronically.
[2]Mortality of aphids compared with untreated check.

Table 2. Effect of pymetrozine following topical application to two aphid species

Effect on	*Myzus persicae* Dose (ng per aphid)			*Aphis gossypii* Dose (ng per aphid)		
	0	50	100	0	50	100
Time (min) to first penetration	1.5	280	~	1.8	220	~
Longest feeding pattern E2[a] (min)	> 600	36	0	> 600	15	0
Mortality by 24 h (%)	2	12	90	3	14	85
Offspring per aphid in 24 h	2.0	0.8	0	2.3	1.0	0

~ not reached.
[a]E2, passive phloem ingestion waveform (Harrewijn and Kayser 1997).

a concentration of 300 μg pymetrozine ml of diet disrupted food uptake in 5–10 min of feeding (Harrewijn and Kayser 1967).

Pymetrozine has both contact and systemic activity on plants. It is distributed in the plant in both acropetal and basipetal directions via the xylem and the phloem system (Wyss and Bolsinger 1997). The compound has translaminar activity and effective control has been documented from both foliar and soil applications. Due to its translocation properties, new growth is effectively protected also after foliar application.

4 Toxicology

Pymetrozine has one of the most attractive toxicity and safety profiles for an insect control agent and is unlikely to pose any problems in normal use. Current data indicate the compound has a very low mammalian toxicity and an excellent safety profile for most non-target arthropods, birds, and fish (Table 3). Test results show a half-life in the soil of only 2–29 days, indicating rapid degradation in the environment. Pymetrozine and its major metabolites exhibited only a low leaching potential and remained generally in the upper soil layers. Under recommended use conditions, contamination of groundwater is unlikely.

5 Spectrum of Activity

As indicated in section 1, pymetrozine is selectively active against a number of agronomically important homopterous insects (Table 4). Both juvenile and adult stages of aphids are susceptible to the compound. In whiteflies the mobile first instar nymphs and adults are most susceptible to the compound. The whiteflies *Trialeurodes vaporariorum* Westw., *Bemisia tabaci* Genn., *Bemisia argentifolii* Bellows et al. and *Aleyrodes proletella* L. are only half as sensitive as aphids and require twice the dose rate. The second, third and fourth larval instars of aleyrodid species are much less sensitive than adults. Pymetrozine has proved to be extremely effective in glasshouses where reinfestation of whiteflies is limited (unpubl. data). In rice, pymetrozine has shown a good activity against leafhoppers.

Table 3. Toxicology profile of pymetrozin

	Acute Toxicity
Mammals	
Acute oral LD_{50} rat	5820 mg/kg
Acute dermal LD_{50} (24 h)	> 2000 mg/kg
Acute inhalation LC_{50} (4 h)	> 1800 mg/m^3 air
Mutagenicity (Ames, mammalian cells)	Negative
Eye irritation (rabbit)	None
Skin irritation (rabbit)	None
Skin sensitization (guinea pig)	None
Wildlife	
Daphnids EC_{50} *(Daphnia magna Straus)*	87 mg/l
Fish LC_{50} (for trout *Salmo gairdneri* Rich., catfish *Ictalurus punctatus* Raf., bluegill *Lepomis macrochirus* Raf., and carp *Cyprinus carpio* L., practically nontoxic)	> 100 mg/l
Mallard duck LD_{50} oral (*Anas platyrhynchos* L.)	> 2000 mg a.i./kg
Bees LD_{50} oral (*Apis mellifera* L.)	> 117 (µg/per bee (Phosphamidon = 0.56)
Bees LD_{50} contact (*Apis mellifera* L.)	> 200 µg per bee (Phosphamidon = 0.15)
Earthworm (*Eisenia foetida Michaelsen*)	1098 mg/kg (nontoxic)
In soil	
Little mobile	
Rapid degradation	$T_{1/2}$ = 2–29 days

Table 4. Pymetrozine activity for a range of insect species and mites

Pest[a]	Order	Method[b]	LC_{50} (mg/l)
Myzus persicae (N1) (Green peach aphid)	Homoptera	Contact, curative spray	0.2
Bemisia tabaci (N1) (Sweet potato whitefly)	Homoptera	Contact, curative spray	0.9
Nilaparvata lugens (N2) (Brown planthopper)	Homoptera	Contact, preventive spray	2.8
Musca domestica (L1) (House fly)	Diptera	Contact, preventive spray	> 1000
Diabrotica balteata (L1) (Cucumber beetle)	Coleoptera	Feeding contact, pipetting	> 1000
Heliothis virescens (L1) (Tobacco budworm)	Lepidoptera	Feeding contact, curative spray	> 1000
Spodoptera littoralis (L1) (Egyptian cotton leafworm)	Lepidoptera	Feeding contact, curative spray	> 1000
Tetranychus urticae (L1) (Two spotted spidermite)	Acari	Feeding contact, curative spray	> 1000

[a]L1, First instar larval stage; N1 first nymphal stage; N2 second nymphal stage.
[b]Standard Novartis laboratory methods were used.

Because of its selective activity pymetrozine is very safe to beneficial arthropods. Laboratory and field trials in many crops indicate that pymetrozine is highly unlikely to interfere with the action of most beneficial insects (Flückiger et al. 1992a, Table 5), in a way that clearly distinguishes it from other insecticides.

Table 5. Selectivity of pymetrozine compared with pirimicarb and dimethoate on beneficial arthropods

Beneficial insect, instar[a]		LC_{50} (g ai/100 l)	
	Pymetrozine	Pirimicarb	Dimethoate
Orius majusculus Reut. (N1, 2) (Het., Anthocoridae)	> 810	16	3
Chrysoperla carnea Steph. (L2, 3) (Neur, Chrysopidae)	> 810	> 270	30
Coccinella septempunctata L. (L2) (Col, Coccinellidae)	> 810	45	3
Amblyseius fallacis Garman. (A) (Acari Typhlodromidae)	> 1000	> 1000	3

Method: exposure to treated surfaces, feeding with untreated prey.
[a]N1, 2, First and second nymphal stage; L2, 3, second and third instar larval stage; A adult.

The development of resistance by sucking insect pests is an increasing problem in many crops (Georghiou 1986; Voss 1988), and products with a new mode of action such as pymetrozine are needed. Of particular importance is the activity of pymetrozine on sucking insects which have developed resistance to organophosphate and carbamate insecticides. Pymetrozine was evaluated on a strongly resistant and two susceptible strains of *M. persicae* (Bolsinger et al. 1993). The resistant strain used was an R3 strain capable of detoxifying a range of insecticides by overproduction of esterase E4 (Field and Devonshire 1991). Pymetrozine was compared with four organophosphates, heptenophos, dimethoate, fenthion, and parathion, and two carbamates, pirimicarb and triazamate. Among the four organophosphates tested, heptenophos was the least resisted (Table 6).

Table 6. Resistance factors determined in several independent bioassays with *Myzus* persicae (spray application and potter tower treatment on first instar and mixed populations)

Pesticide	RFR3a	RFRb	RFSc
Heptenophos[d] (most active OP)	73–83	16–18	0.3–2
OTHER OP's[d]	> 200	> 200	NA
Pirimicarb[d]	95–> 200	10–16	1–2
Triazamated	> 150	30	6.5
Pymetrozine	0.5–2	0.5–2	0.5–2

RF, resistance factor; NA, not applicable; OP, organophosphorous.
[a]R3, highly OP and carbamate resistant strain 749J (Institute of Arable Crop Research, Rothamstead, UK)
[b]R, Novartis strain, exhibiting increased levels of esterase activity.
[c]S, normal susceptible strain US1L (Institute of Arable Crop Research, Rothamsteod, UK)
[d]All data are significantly different from pymetrozine at $p \leq 0.05$ (ANOVA.)

The results show clearly that aphid strains which are resistant to other groups of chemistry show no cross-resistance to pymetrozine. The resistance factor of pymetrozine was very low on all three strains compared with the other products tested and pymetrozine was by far the least affected by the strong resistance to established insecticides of the R3 strain.

In areas where sucking insect pests may already have developed resistance or in areas where preventative resistance management programs can be established, pymetrozine has the potential to play an important role.

6 Use Recommendations

Pymetrozine will be recommended in vegetables and ornamentals against various aphid species at 10 g a.i./hl and against whiteflies at the higher dosage rate of 20 g a.i./hl (Table 7). In tobacco, cotton, and potato, the major target pests will be *M. persicae* and *A. gossypii* at dosage rates of 100–200 g a.i./ha. In the USA rates for aphid control in vegetables and cotton will be 50–100 g a.i./ha. In rice, pymetrozine is active on the brown planthopper at rates of 100–150 g a.i./ha applied as a foliar spray or at 1.5 g a.i. in seedling box application. For the control of aphids in citrus and deciduous fruit the recommended dosage rate is 5-20 g a.i./hl.

7 Effects on Virus Transmission

The control of sucking insect pests in many crops is extremely important in order to minimize the direct damage caused by feeding and to reduce the transmission of disease causing viruses. In potatoes, pymetrozine is active against all important aphid pests. In a field trial conducted in Brazil (Novartis in-house trial) against *M. persicae* to compare the initial and residual activity of pymetrozine with those of standard commercial insecticides, pymetrozine applied at 100 g

Table 7. Major use recommendations for pymetrozine in several crops

Crop	Pest	Dosage rate
Vegetables and ornamentals	*Aphis gossypii*	10–20 ga. i./hl
	Myzus persicae	10–20 ga. i./hl
	Trialeurodes vaporariorum	20–30 ga. i./hl
	Bemisia tabaci	20–30 ga. i./hl
Potato	*Myzus persicae*	100–200 ga. i./ha
Tobacco	*Myzus persicae*	200 ga. i./ha
	Aphis gossypii	200 ga. i./ha
Cotton	*Aphis gossypii*	250 ga. i./ha
Rice	*Nilaparvata lugens*	100–150 ga. i./ha
		1.5–2.0 ga. i./box[a]
Citrus	*Toxoptera spp.*	10–20 ga. i./hl

[a]Seedling box, 1624 cm^2, sufficient to plant 50 m^2.

a.i./ha gave better initial activity than the standard and was superior in residual activity 21 days postapplication (Fig.1). The data also show the delayed action of pymetrozine. Aphid control increased from day 2 to day 7.

A laboratory study was conducted to evaluate the efficacy of pymetrozine in reducing the transmission of a persistent virus, potato leaf roll luteovirus (PLRV), and a nonpersistent necrotic strain of potato Y potyvirus (PVY^N) (Harrewijn and Piron 1994). Individual aphids, *M. persicae*, were given access to potato plants infected with PLRV for 24 h or PVYN for 1 h. Aphids were then transferred to virus-free potato plants which had been sprayed 24 h before with pymetrozine (100 mg/l). Results indicated that pymetrozine reduced the transmission of PLRV and PVY^N by 97% and 75%, respectively. Also in this study, the behavioral component of virus transmission was studied using the electrical penetration graph (EPG) technique which indicated that pymetrozine does not effectively inhibit acquisition of PVY^N, a nonpersistent virus, but did reduce the subsequent transmission to healthy plants by 65%. The study showed that on treated plants, aphids begin normal stylet penetration, but there is no sustained feeding and significantly less phloem feeding explaining why inoculation of PLRV, which must be transmitted via phloem feeding, is significantly reduced.

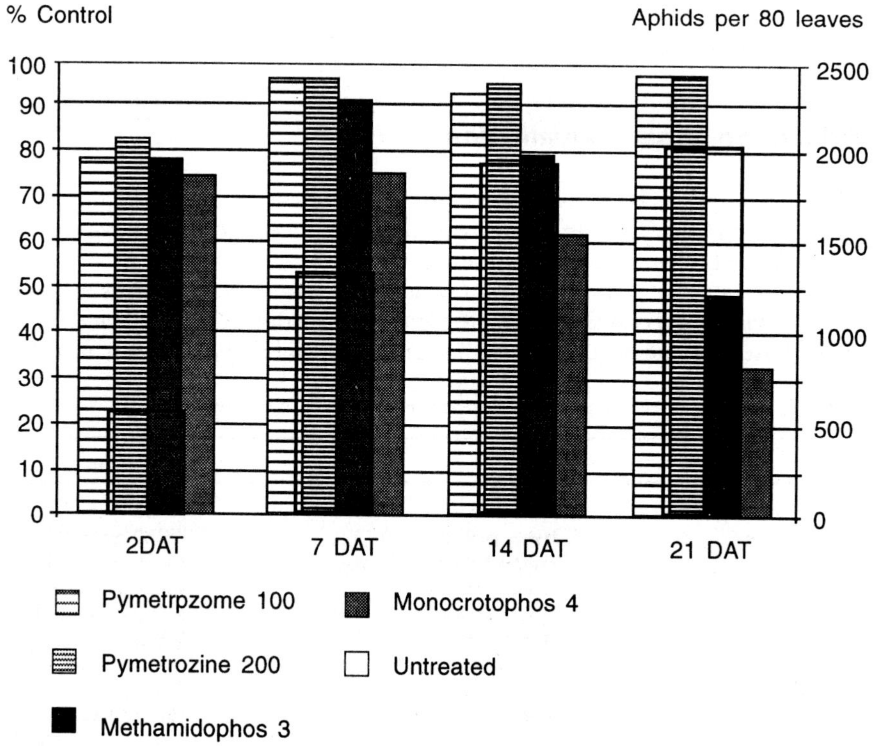

Fig. 1. Efficacy of pymetrozine against Myzus persicae on potatoes, percent control. (Brazil, 1993)

Subsequent studies in both greenhouse and field trials have been conducted in the USA. In a greenhouse trial (Novartis in-house trial by S.J.E.) with the nonpersistent bean common mosaic virus (BCMV) in peppers, previously starved apterous adults of *M. persicae* were allowed a 5-min acquisition period on BCMV-infected pepper leaves. Aphids were then placed on plants sprayed with 100 g a.i./ha of pymetrozine 24 h before test initiation. All combinations of treated and untreated source and test plants were used: untreated source and test (U-U), untreated source to treated test (U-T), treated source to untreated test (T-U), and treated source and treated test (T-T). When the source was untreated (U-U or U-T), the virus was efficiently transmitted to test plants whether treated or untreated with 87.5% infection (Fig. 2). The transmission from treated test plants (T-U or T-T) was significantly lower, with 64.6% and 38.3% for treated source to untreated test and treated source to treated test, respectively. These results indicate that under light potyvirus pressure pymetrozine may reduce transmission of a nonpersistent virus.

In another greenhouse trial (Novartis in-house trial by S.J.F.) with persistent potato leaf roll virus (PLRV) in potatoes, apterous adult *M. persicae* were given a 3-day acquisition period on infected potato plants. Three to five adults were then placed on plants treated with pymetrozine at 1 g a.i. per plant. PLRV infection was determined 4 weeks later using ELISA. Results indicated that pymetrozine completely prevented the transmission of PLRV. Subsequent field trials (Novartis in-house trial) also indicated that pymetrozine was effective in controlling transmission of the virus, particularly at the rate of 200 g a.i./ha (Fig. 3). However, other field trials conducted with peppers and tobacco indicated that in spite of providing highly effective control of *M. persicae*, transmission of nonpersistent viruses was not prevented.

These results indicate that pymetrozine can afford a valid protection against the transmission of persistent viruses. However, the practical value of this benefit is best realized in situations where infection pressure by nonpersistent viruses is not significant.

8
Summary

Pymetrozine represents a novel class of chemistry with a unique mode of action. It is specific to sucking insect pests and is highly active against aphids, whiteflies, and the brown planthopper, even if these pests have developed resistance to organophosphates and carbamates. Pymetrozine has a very favorable toxicological and environmental profile; its lack of activity on beneficial insects will increase its usefulness and inclusion into IPM and integrated resistance management programs. Finally, the reduction in transmission of persistent viruses shown in potatoes is an additional benefit that should help to make pymetrozine a very useful insecticide for the agricultural industry.

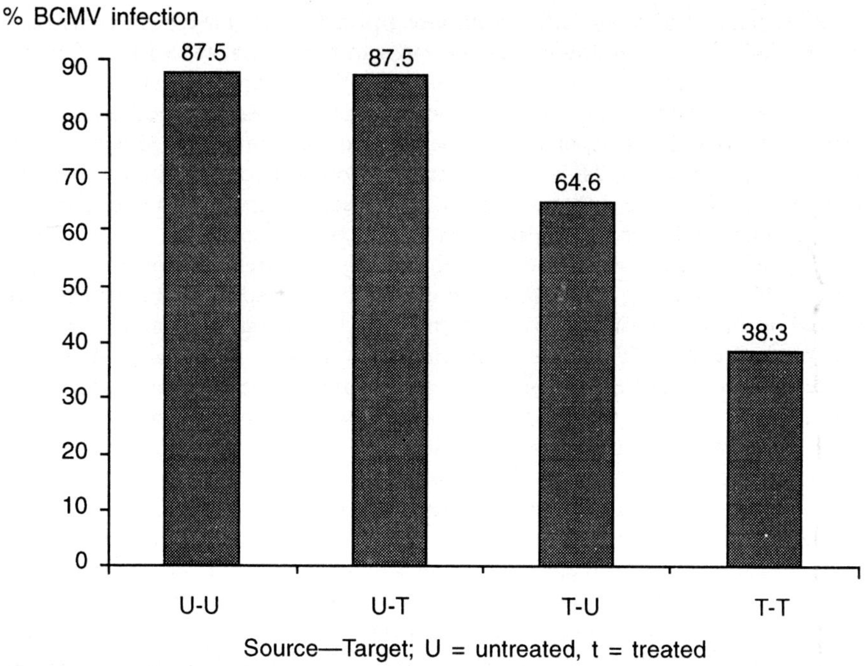

Fig. 2. Pymetrozine against BCMV in peppers in greenhouse tnals (USA, 1994). *U–U* Untreated source and test; *U–T* untreated source to treated test; *T–U* treated source to untreated test; *T–T* treated source and treated test

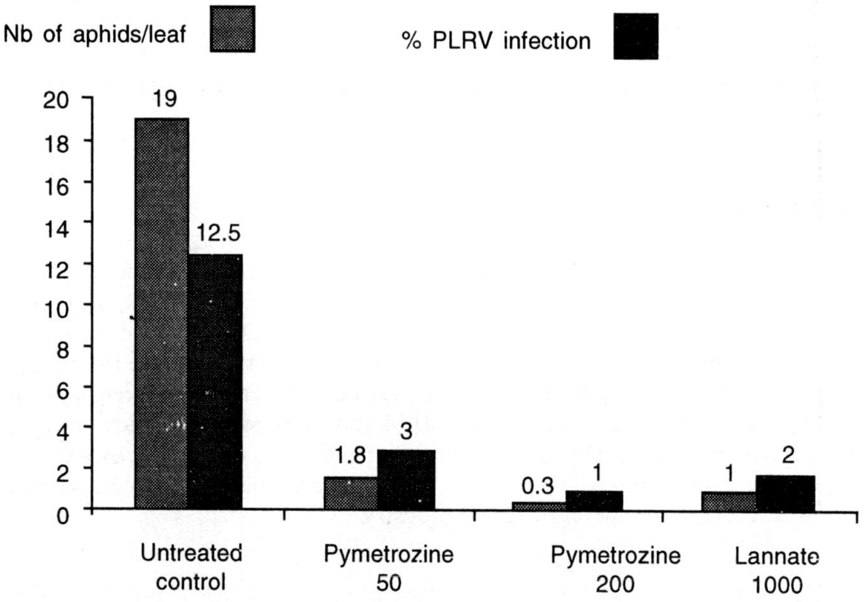

Fig. 3 Pymetrozine against PLRV in potatoes in fieldtrials (USA, 1994)

References

Bolsinger M, Ruesch O, Stieger A, Angst M (1993) Performance of pymetrozine, a novel agent to control aphids and whiteflies, on organophosphate and carbamate resistant strains of *Myzus persicae*. Proc Conf Int sur les Ravageurs en Agriculture, Montpellier, 7–9 Dec 1993, vol 2, pp 755–761

Field LM, Devonshire AL (1991) Insecticide resistance by gene amplification in *Myzus persicae*. In: Denholm I, Devonshire AL, Hollomon DW (eds) Achievements and developments in combating pesticide resistance. Proc SCI Symp 'Resistance 91'. Elsevier, Amsterdam, pp 240–250

Flückiger CR, Kristinsson H, Senn R, Rindlisbacher A, Buholzer H, Voss G (1992a) CGA 215'944—a novel agent to control aphids and whiteflies. Proc 1992 Brighton Crop Prot Conf—Pests and diseases, vol 1, pp 43–50

Flückiger CR, Senn R, Buholzer H (1992b) CGA 215'944—opportunities for use in vegetables. Proc 1992 Brighton Crop Prot Conf—Pests and diseases, col 3, pp 1187–1192

Georghiou GP (1986) The magnitude of the resistance problem. In: Georghion GP, Saito T(eds) Pesticide resistance: strategies and tactics for management. National Academy Press, Washington DC pp 14–43

Harrewijn P, Kayser H (1997) Pymetrozine, a fast acting and selective inhibitor of aphid feeding. *In situ* studies with electronic monitoring of feeding behaviour. Pestic Sci 49: 130–140

Harrewijn P, Piron PGM (1994) Pymetrozine, a novel agent for reducing virus transmission by *Myzus persicae*. Proc 1994 Brighton Crop Prot Conf—Pests and diseases, vol 2, pp 923–928

Kayser H, Kaufmann L, Schürmann F (1994) Pymetrozine (CGA 215'944): a novel compound for aphid and whitefly control. An overview of its mode of action Proc 1994 Brighton Crop Prot Conf—Pests and diseases, vol 2, pp 737–742

Schwinger M, Harrewijn P, Kayser H (1994) Effects of pymetrozine (CGA 215'944), a novel aphicide, on feeding behaviour of aphids. Proc 8th Int Union Pure and Applied Chemistry, Int Conf Pesticide Chemistry, Washington, vol 1, p 230

Tjallingii WF (1988) Electrical recording of stylet penetration activities. In: Minks AK, Harrewijn P (eds.) Aphids: their biology, natural enemies and control, 2B. Elsevier, Amsterdam, pp 95–108

Voss G (1988) Insecticide/acaracide resistance: industry's effort and plans to cope. Pestic Sci 23 149–156

Wyss P, Bolsinger M (1997) Translocation of pymetrozine in plants. Pestic Sci (in press)

CHAPTER 4

Imidacloprid, a Novel Chloronicotinyl Insecticide: Biological Activity and Agricultural Importance

A. Elbert, R. Nauen and W. Leicht
Bayer AG, Institute for Insect Control, Agrochemical Division,
Centre Monheim, 51368 Leverkusen, Germany

1 Introduction

Following the discovery of the insecticidal properties of the heterocyclic nitromethylenes (Soloway et al. 1978), chemists of Nihon Bayer Agrochem started in 1979 to optimize these structures. In 1985 the coupling of the chloropyridyl moiety to the N-nitro substituted imidazolidine ring system enabled the synthesis of the highly active insecticide imidacloprid (Fig. 1). Imidacloprid is the first commercial example of the chloronicotinyl insecticides acting on nicotinic acetylcholine receptors (Leicht 1993). It is now registered in more than 60 countries as a compound with a new or non-conventional mode of action to combat highly resistant insect pests (Elbert et al. 1991; Elbert et al. 1996; Nauen et al. 1996a). Chloronicotinyl insecticides will grow in importance in the coming years because other close analogues of imidacloprid, such as Takeda's and Nippon Soda's open chain derivatives nitenpyram and acetamiprid, respectively, have been described (Tomizawa et al. 1995; Yamamoto et al. 1995). During recent years several studies have demonstrated the excellent activity of imidacloprid on pest species of different orders. The present chapter gives an overview of the biological activity of imidacloprid on different target pests, its selectivity even at the molecular level, its physicochemical properties which led to good systemicity and its agricultural importance.

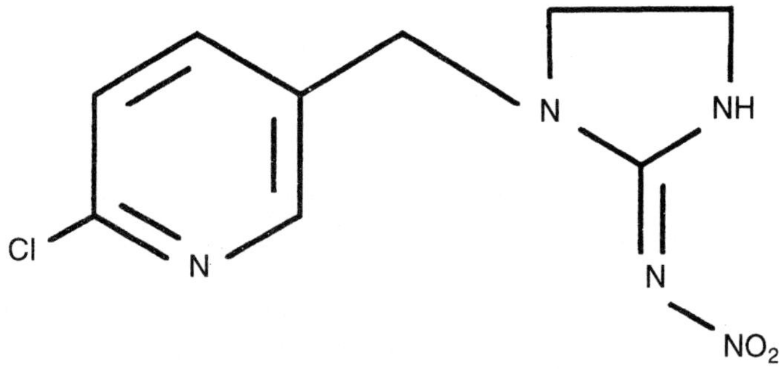

Fig. 1. Chemical structure of the chloronicotinyl insecticide imidacloprid

2
Biological Activity

2.1
Efficacy on Target Pests

Imidacloprid is highly effective for the control of homopteran pests, i.e. aphids, leafhoppers, planthoppers, thrips and whiteflies (Elbert et al. 1991). The compound is also active against some species of the orders Coleoptera, Diptera and Lepidoptera. No activity against nematodes and spider mites has been found. Imidacloprid has a good xylem-mobility which makes it especially useful for seed treatment and soil application, but it is equally effective after foliar application (Elbert et al. 1991). The trade names for soil application/seed treatment are Admire and Gaucho, the trade names for foliar application are Confidor and Provado and Premise for termite control.

2.1.1
Foliar Application

The spray application is especially targeted against pests attacking crops such as cereals, maize, rice, potatoes, vegetables, sugar beet, cotton and deciduous fruits. Table 1 shows the acute activity (estimated LC_{95} in ppm a.i.) of imidacloprid against a variety of pests following foliar application (dip and spray treatment) of host plants underlaboratory and greenhouse conditions. Imidacloprid was very active on a wide range of aphids. Most susceptible was the damson hop aphid *Phorodon humuli* (LC_{95} = 0.32 ppm) which is often highly resistant against conventional insecticides, i.e. organophosphates, carbamates and pyrethroids (Lewis and Madge 1984). Imidacloprid was highly effective against some of the most important rice pests, such as leafhoppers and planthoppers, rice leaf beetle *Lema oryzae* and rice water weevil *Lissorhoptrus oryzophilus.* Although imidacloprid is generally less effective against biting insects, the efficacy against Colorado potato beetle *Leptinotarsa decemlineata* is relatively high (LC_{95} = 40 ppm). The LC_{95}-values for the second or third instar larvae of some of the most deleterious pest species of the order Lepidoptera, the noctuids *Helicoverpa armigera, Spodoptera frugiperda* and *Plutella xylostella were* approx. 200 ppm and higher than those for the above-mentioned species (Elbert et al. 1991).

2.1.2
Soil Application and Seed Treatment

Typical soil insect pests such as *Agriotes* sp., *Diabrotica balteata* or *Hylemyia antiqua* were controlled by incorporation of 2.5–5 ppm a.i. into the soil. Higher concentrations of imidacloprid are necessary to control *Reticulitermes flavipes* (7 ppm) and *Agrotis segetum* (20 ppm). However, much more pronounced is the activity of imidacloprid against sucking early season pests which attack the aerial parts of a wide range of crops. Soil concentrations as low as 0.15 ppm a.i. gave

Table 1. Efficacy (estimated LC_{95} in ppm a.i.) of imidacloprid after foliar application against a variety of pests under laboratory conditions

Pest species	Developmental stage	Estimated LC_{95} (ppm)
Homoptera		
Aphis fabae	Mixed	8
Aphis gossypii	Mixed	1.6
Aphis craccivara	Mixed	1.6
Aphis pomi	Mixed	8
Brevicoryne brassicae	Mixed	40
Myzus nicotianae	Mixed	8
Myzuz persicae	Mixed	1.6
Phorodon humuli	Mixed	0.32
Laodelphax striatellus	Larvae third instar	1.6
Nephotettix cincticeps	Larvae third instar	0.32
Nilaparvata lugens	Larvae third instar	1.6
Sogatella furcifera	Larvae third instar	1.6
Pseudococcus comstocki	Larvae	1.6
Bemisia tabaci	Larvae second instar	8
Hercinothrips femoralis	Mixed	1.6
Lepidoptera		
Chilo suppressalis	Larvae first instar	8
Helicoverpa armigera	Larvae second instar	200
Plutella xylostella	Egg	200
	Larvae second instar	200
Spodoptera frugiperda	Egg	200
	Larvae second instar	40
Heliothis virescens	Egg	40
Coleoptera		
Leptinotarsa decemlineata	Larvae second instar	40
	Adult	40
Lema oryzae	Adult	8
Lissorhoptrus oryzophilus	Adult	40
Phaedon cochleariae	Larvae second instar	40

excellent control of *Myzus persicae* and *Aphis fabae* in greenhouse experiments (Elbert et al. 1991). A good residual activity is essential for the protection of young plants; Table 2 shows the residual activity of imidacloprid and some other standard insecticides following soil treatment or seed dressing against aphids.

2.2 Systemicity

The water solubility of imidacloprid is 0.51 g/l at 20 °C, thus providing the molecule a considerable mobility in the xylem of plants. Due to its lack of any acidic hydrogen, the compound's pKa value is > 14 and therefore transport of imidacloprid within the phloem is unlikely, as been shown in several studies (Stein-Dönecke et al. 1992; Tröltzsch et al. 1994). The translaminar transport of imidacloprid from the treated upper side of a leaf to the lower surface is excellent, as shown in some investigations using cabbage leaves (Elbert et al.

Table 2. Residual activity (weeks after which mortality is still higher than 95%) of various insecticides against *M. persicae* on cabbage and against *A. fabae* on beans following soil treatment or seed dressing

Insecticide	Concentration (ppm a.i. soil)	*Myzus persicae* (weeks)	*Aphis fabae* (weeks)	Dose (ga. i./100 kg seed)	*Aphis fabae* (weeks)
Imidacloprid	2.5	> 8	> 5	100	> 5
	1.25	> 8	5	25	5
	0.625	> 8	4	6	3
Ethiofencarb	2.5	4	2	100	> 5
	1.25	1	1	25	2
	0.625	< 1	< 1	6	< 2
Carbofuran	2.5	2	–	–	–
	1.25	< 1	–	–	–
	0.625	< 1	–	–	–
Aldicarb	2.5	5	–	–	–
	1.25	3	–	–	–
	0.625	3	–	–	–

1991). Of the numerous studies performed on the translocation of imidacloprid in monocotyledons and dicotyledons, two examples were chosen to demonstrate in more detail the uptake of the compound from the seed shell into the plant.

2.2.1
Translocation in Winter Wheat

Studies on seeds of winter wheat treated with (pyridinyl-^{14}C-methylene) imidacloprid and unlabelled compound at doses of 100 g a.i./100 kg seeds (approx. 51 μg a.i. per seed) revealed a continuous increase of the applied radioactivity into the sprouting wheat from 1% at the first leaf stage to approximately 19% at full maturity, depending on soil moisture. The distribution of the radioactivity within the plant was typical for a xylem-mobile compound, i.e. apical accumulation and an obvious concentration gradient between the oldest and the youngest leaf (Fig. 2). Concentrations of imidacloprid as low as 0.12 mg/kg fresh mass (younger leaves at the end of the shooting phase) showed an insecticidal efficacy of 98% against *Rhopalosiphum padi* (Stein-Dönecke et al. 1992).

2.2.2
Translocation in Cotton

The uptake and translocation of (pyridinyl-^{14}C-methylene) imidacloprid from treated cotton seeds (250 g a.i./100 kg seed) into the growing parts of the cotton plant were totally different from that described for winter wheat or sugar beet (Tröltzsch et al. 1994). In greenhouse experiments using the cotton variety Delta Pine 50, only 5–6% of the radioactivity applied to the seed had been taken up by the growing cotton plant; most of the remaining radioactivity could be extracted from the surrounding soil or the seed-coats as unchanged parent

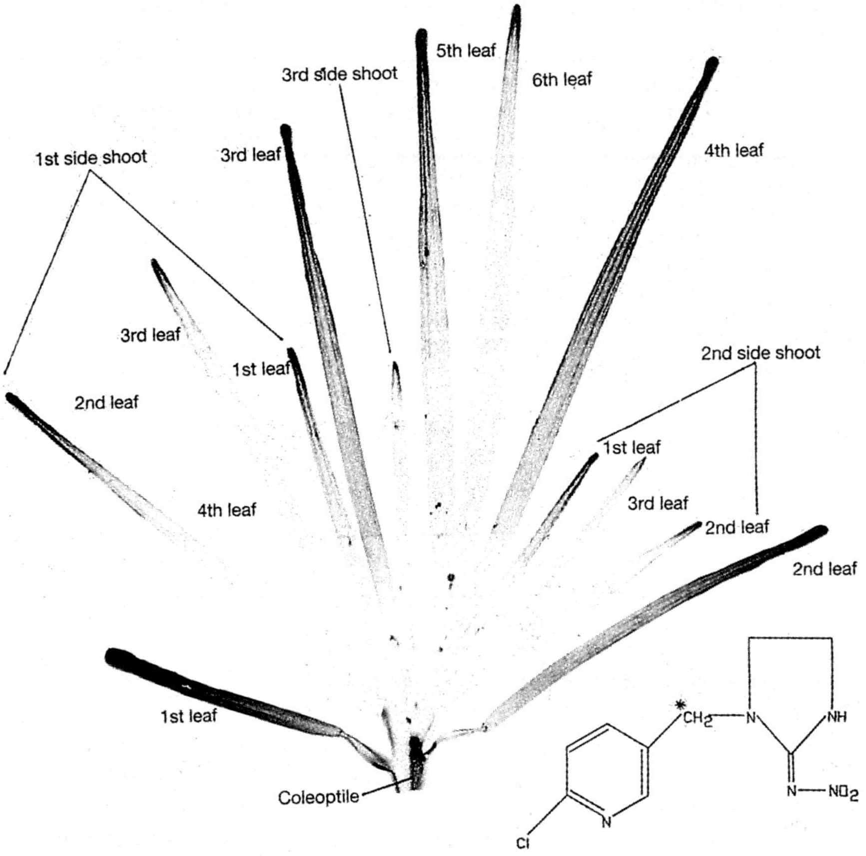

Fig. 2. Translocation of radiolabelled imidacloprid in winter wheat after seed treatment

compound. Most of the translocated radioactivity was found in the cotyledons, and as early as 27 days after sowing only 6% of the extracted radioactivity from the green plant parts represented unchanged parent compound; however, 52 days after planting no imidacloprid could be detected in the green parts of the cotton plant (Tröltzsch 1995). Only small amounts of radioactivity are translocated into the true leaves, but these amounts are sufficient to protect the young cotton plant from early season pests like *Aphis gossypii*, jassids and thrips for 4-6 weeks. That is a shorter period than those reported for other crops like sugar beet or winter wheat (Stein-Dönecke et al. 1992; Westwood et al. 1995).

Macroautoradiographs of young cotton plants revealed a compartmentalized translocation of radioactivity as tiny spots in the first two true leaves (Fig. 3) as well as in the younger leaves, thus providing evidence for an accumulation of the parent compound and/or its metabolites in the inner glands of cotton (Fig. 4, Nauen and Elbert 1994). This type of compartmentalization of active ingredients in the lysigenic glands of cotton has also been described for herbicides

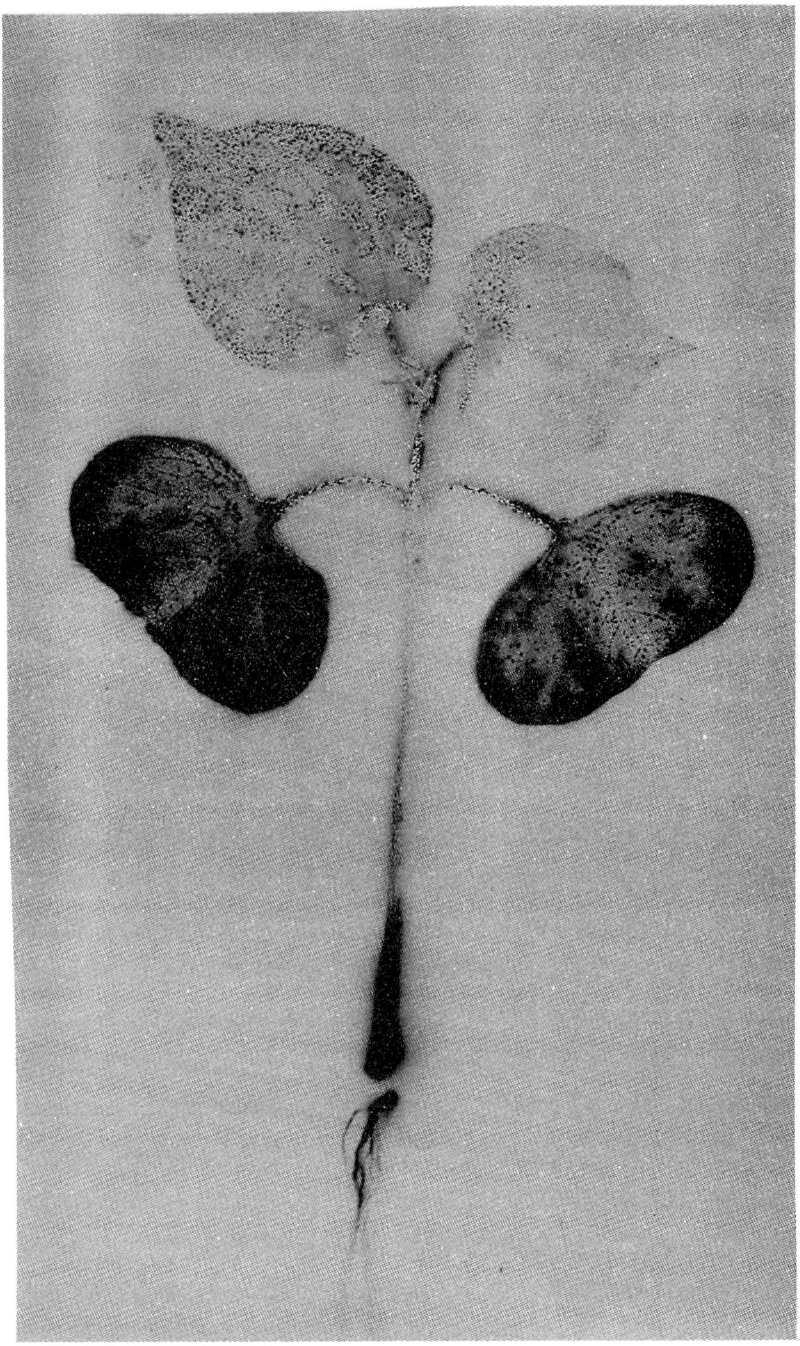

Fig. 3. Translocation of radiolabelled imidacloprid in cotton seedling

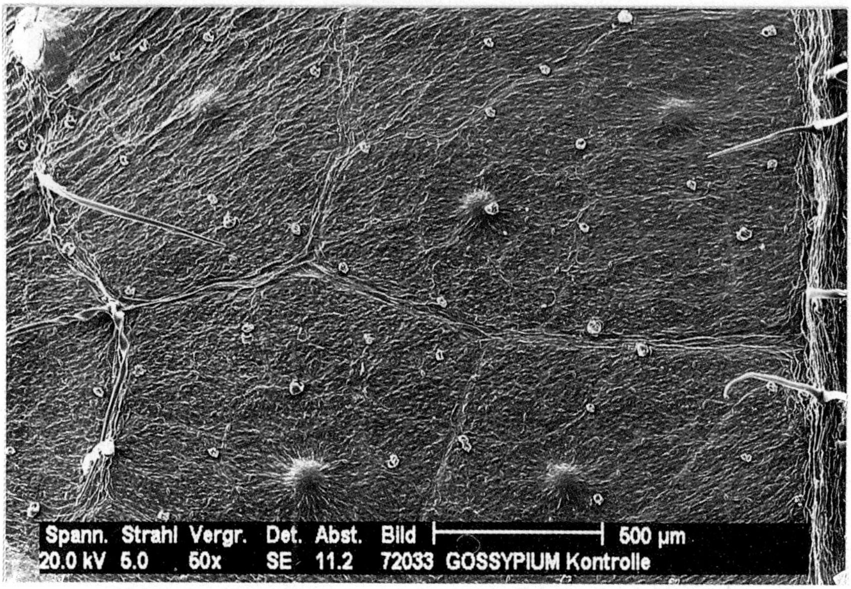

Fig. 4. Electron micrograph displaying inner and outer glands on the underside of cotton leaves

and other insecticides (Strang and Rogers 1971a, b; Phillips et al. 1978). Nauen and Elbert (1994) speculated that *A. gossypii* avoided these internal glands when penetrating the leaf surface, thus explaining why cotton aphids are not controlled that quickly in comparison to aphid populations on other crops. Experiments with other aphid species such as *M. persicae* which did not possess an evolutionary adaption to cotton and which were controlled for a longer period after sowing gave further evidence for this assumption.

2.3 Sublethal Effects

Understanding how low concentrations of systemic insecticides like imidacloprid affect especially aphids and some other pests could be useful to assess their real potential, because even when the concentration in treated crops declines as a result of plant growth or metabolism, such compounds may still be able to control certain species by virtue of their effects on behaviour.

2.3.1 *Antifeedant Effect*

When applied at recommended field doses imidacloprid induced effects typical of compounds interfering with the insect nervous system, namely uncoordinated movement, tremor and later on paralysis (Leicht 1993). Apart from these findings, effects on the behaviour of certain insects, especially on aphids, have been described. Aphids walked off treated leaves or probed for shorter periods on systemically treated leaves (Dewar and Read 1990; Woodford and Mann 1992). Nauen (1995) described the behaviour-modifying effects of low systemic concentrations of imidacloprid on *M. persicae* with special reference to an antifeeding response. This response had previously been observed in the form of decreased honeydew excretion in the cotton aphid *A. gossypii* (Nauen and Elbert 1994). Earlier studies on *M. persicae, A. fabae, Acyrthosiphon pisum and Macrosiphum avenae* did not reveal antifeedant effects (Elbert et al. 1991; Grothe 1992; Kaust and Poehling 1992), possibly due to the high concentrations of imidacloprid used.

Sublethal concentrations (< 10 ppb) have dramatic effects on the feeding behaviour of aphids, resulting in the depression of honeydew excretion, wandering and, subsequently, death due to starvation (Nauen 1995; Devine et al. 1996). Such a depression of honeydew excretion has also been observed in declining field populations of *A. gossypii* on seed-treated cotton plants, demonstrating the possible relevance of such sublethal effects under practical conditions (W. Mullins, pers comm). The starvation response is reversible, since aphids transferred from treated to untreated leaves recovered, began feeding on the untreated leaves and showed a steady increase in weight and honeydew production (Nauen 1995). Furthermore, choice tests with *M. persicae* and *M. nicotianae* revealed migration from the leaves treated with sublethal concentrations of imidacloprid to untreated leaves (Nauen 1995). Lösel and Goodman (1993) described a similar antifeedant effect in the brown planthopper *Nilaparvata lugens* when applying to rice leaves sublethal concentrations of the structurally related 2-nitromethylene-1,3-thiazinan-3-yl-carbamaldehyde. The antifeedant effects of imidacloprid have not only been described for aphids, but also for black maize beetles *Heteronychus arator*, as well as false wireworms *Somaticus* sp., when feeding on stems of seed-treated maize plants (Drinkwater 1994; Drinkwater and Groenewald 1994). Other species showing this antifeeding response were *Anthonomus grandis* (B. Monke, pers. comm.) and *Heliothis virescens* (R. Tiemann, pers. comm.), though other studies did not reveal any antifeedant potential in the latter (Lagadic et al. 1993).

2.3.2
Reduction of Aphid Viability

The often observed reduction in viability of aphids exposed to imidacloprid-treated plants may be a result of a reduction in feeding activity and therefore fewer larvae are produced. The viability of the susceptible *M. persicae* clone US1L is reduced by more than 50% at a systemically applied concentration of imidacloprid of just 0.2 ppb (Devine et al. 1996). This concentration was more than 20 times lower than the actual EC_{50} (feeding prevention) for imidacloprid in this clone. The same authors did not find similar effects in *M. nicotianae* at such low concentrations, suggesting that some species are not as susceptible as others. Often, a number of the larvae produced by adult *M. persicae*, feeding on leaves systemically treated with sublethal concentrations of imidacloprid, were non-viable (Devine et al. 1996). Similar effects could be seen when applying sublethal doses of imidacloprid to *Aphis craccivora, A. fabae and P. humuli* (Nauen, unpubl. results). Starving aphids of these species did not deposit non-viable embryos, thus suggesting direct effects of imidacloprid on the fertility of aphids. Considerable reduction in fertility and deposition of non-viable larvae were also reported for some grain aphids such as *R. padi, Sitobion avanae and Metopolophium dirhodum*, feeding on seed-treated winter barley and oat plants (Knaust and Poehling 1992). The effects of such sublethal concentrations of imidacloprid on the fertility of aphids may prevent rapid buildup of populations in the field, even later in the season when the concentration of imidacloprid in seed-treated plants has declined.

2.4 Action on Resistant Pest Species

2.4.1 *Aphids*

Studies on carbamate, organophosphate and pyrethoid resistant *M. persicae* from a German greenhouse revealed no cross resistance against imidacloprid (Fig. 5; Elbert et al. 1991; Nauen et al. 1996a). A monitoring programme, based at Integrated Approach to Crops Rexarch (IACR)-Rothamstead, which was started in 1991 with detection of baseline susceptibility of imidacloprid against standard susceptible strains, including collections of *M. persicae, M. nicotianae, P. humuli and A. gossypii* from all over Europe, revealed no resistance to imidacloprid in the investigated populations (Elbert et al. 1996). It was observed that strains of *M. nicotianae* and tobacco-associated *M. persicae* are somewhat less susceptible (resistance factors [RF] of 2–10, depending on the bioassay used) to imidacloprid (Devine et al. 1996; Nauen et al. 1996a; Nauen and Elbert 1997). It has been speculated that tobacco-associated strains of *Myzus* sp. could have a reduced receptor affinity, due to their adaption to nicotine-containing tobacco plants (Devine et al. 1996). This assumption could not be confirmed when the binding of radiolabelled imidacloprid to the nicotinic acetylcholine receptor was determined in homogenates of a red Japanese strain of *M. persicae* closely related to *M. nicotianae*, slightly resistant to cartap and nicotine, but highly resistant to organophosphates and carbamates (Nauen et al. 1996a). Other strains of *M. nicotianae* have also been investigated in this manner and did not reveal any reduced receptor affinity (Nauen et al. 1996b). Some cases of reduced susceptibility to imidacloprid have been shown to correlate with the hardiness of field strains as compared with the susceptible reference clones used in resistance studies (Nauen and Elbert 1997).

2.4.2 *Whiteflies*

Some of the most serious pests of steadily growing importance on cotton, vegetables and ornamentals are the different biotypes of the cotton whitefly *Bemisia tabaci* (Costa and Brown 1991; Cahill et al. 1995). Whiteflies resistant

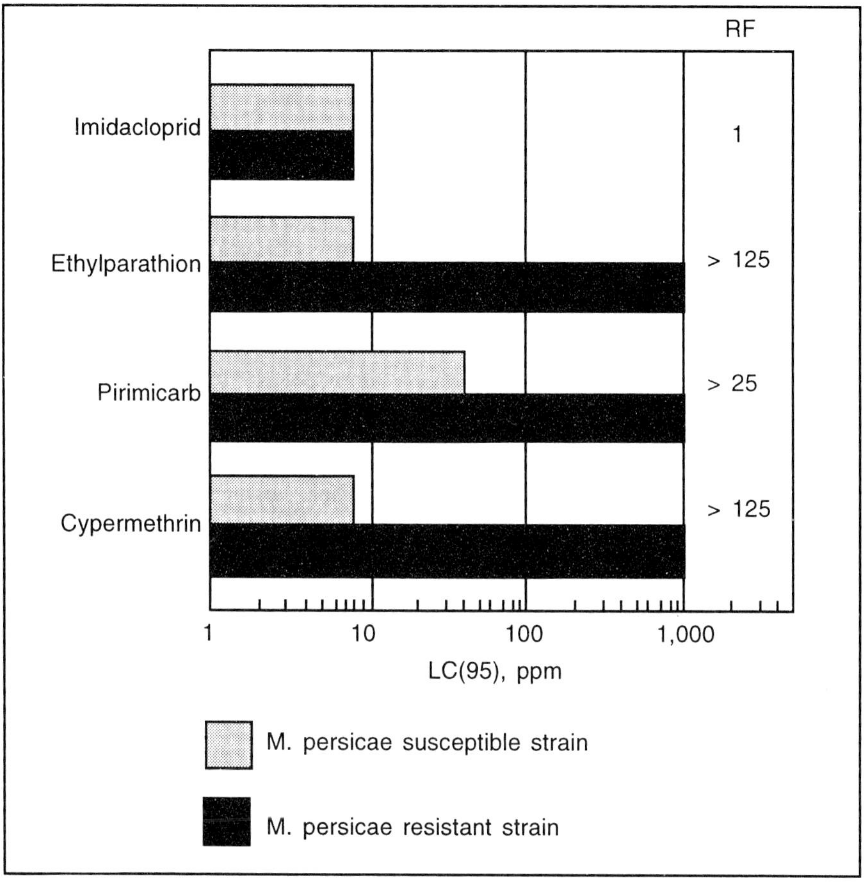

Fig. 5. Efficacy of imidacloprid on a highly resistant strain of *M. persicae*. *RF* = Resistance factor

to organophosphates, carbamates, pyrethroids and endosulfan from various regions of the world, including the recently described B-type, showed no cross-resistance to imidacloprid (Cahill et al. 1996). The first stage of this study carried out at IACR-Rothamsted was to develop an appropriate bioassay system, which resulted in a systemic test using cotton leaf discs of leaves formerly immersed with their petioles into solutions of imidacloprid (Cahill et al. 1996). A baseline susceptibility towards imidacloprid was defined on adults using different strains, which showed different degrees of resistance against all conventional insecticides but apparently no cross resistance to imidacloprid. The only case of reduced susceptibility to imidacloprid of B. tabaci has been reported from the Almeria-region in Spain, where the product has been used intensively since 1992 (Cahill et al. 1996). However, no loss of field efficacy has been observed in this region between 1988 and 1996 (Elbert and Nauen 1996).

2.4.3
Leafhoppers and Planthoppers

All field strains of *Nephotettix cincticeps, Laodelphax striatellus, Sogatella furcifera* and *Nilaparvata lugens* investigated so far gave the same response as susceptible laboratory populations (Elbert et al. 1996). Field populations of *L. striatellus* resistant to organophosphates and carbamates respond highly susceptibly towards imidacloprid using the topical application method (Sone et al. 1995). Only in the laboratory has a strain of L. striatellus has been selected for imidacloprid resistance, and this strain showed a resistance factor of 18 compared with a susceptible strain (Sone et al. 1995). This resistant laboratory strain was fully susceptible when imidacloprid was orally ingested, but moderately resistant to the topically applied compound (Elbert et al. 1996). The investigations performed so far do not implicate reduced insecticide penetration with the observed resistance (S. Sone, pers. comm.).

2.4.4
Colorado Potato Beetle

The Colorado potato beetle *L. decemlineata* is one of the chewing pests which can be effectively controlled by imidacloprid. Larvae as well as adults irrespective of their resistance factors against conventional insecticides are susceptible to this compound. An extensive monitoring programme (Dively, pers. comm.) in the USA revealed certain variation in the response of many field populations of *L. decemlineata*, with different degrees of resistance to azinphos-methyl and/or fenvalerate. No apparent cross resistance between imidacloprid and other insecticides could be detected. As imidacloprid has never been used at these locations, the observed variation in baseline response is regarded as natural variability of different populations (Elbert et al. 1996).

3
Mode of Action and Selectivity

3.1
Mode of Action on Insects and Vertebrates

In studies using the ganglia of the American cockroach *Periplaneta americana*, imidacloprid displaces radiolabelled α-bungarotoxin, a specific ligand of the nicotinic acetylcholine receptor (nAChR) from its binding site. This experiment demonstrated that imidacloprid acts directly on the nAChR causing its toxic effects in insects. Electrophysiological studies with imidacloprid on the cholinergic motor neuron Df of *P. americana* revealed a depolarization of the cell membrane (at concentrations below 1 μM), which is comparable with the activity of acetylcholine (Bai et al. 1991). In addition, the mode of action of imidacloprid was studied using the CNS of the stable fly *Stomoxys calcitrans.* An almost complete and virtually irreversible blockage of postsynaptic nicotinic acetylcholine receptors was found (Abbink 1991). Tomizawa and Yamamoto (1992) described the inhibition of α-bungarotoxin binding to nAChR from heads of *Apis mellifera* by imidacloprid and nicotinoids, and Liu and Casida (1993) reported high affinity binding sites for radiolabelled imidacloprid in housefly head membrane preparations, with 95% specific binding and a dissociation constant of 1.2 nM. At least the same high affinity binding to nAChR has also been described for one of the homopteran target organisms, the green peach aphid *M. persicae* (Nauen et al. 1996a). Since the structure of vertebrate and insect nAChR is similar but not identical, the action of imidacloprid on the nAChR in vertebrates has been examined. The influence of imidacloprid on rat muscle nAChR was investigated in *Xenopus* oocytes which had expressed this receptor following the injection of specific messenger-RNA. Activation of the nAChR by imidacloprid was determined in whole-cell voltage clamp studies. As a result, it has been shown that imidacloprid activates also nAChR from rat muscle. However, this effect is more than 1000 times weaker than the action on insect nAChR, and therefore a high selectivity of imidacloprid can be demonstrated at the molecular level (Methfessel 1992). Zwart et al. (1994) confirmed this selectivity when measuring the binding of some nitromethylene heterocycles including imidacloprid to nAChR from locust neurones and mammalian cell lines.

Comparing the acute activity of imidacloprid in *in vivo* experiments on aphids and rats, the previous results are confirmed. The LD_{50} for *M. persicae* in a topical application bioassay was determined with 0.062 mg a.i./kg body weight in comparison with a LD_{50} for rats with 450 mg a.i./kg after oral administration. The resulting factor of 7300, as compared with 300 and 51 for pirimicarb and oxydemeton-methyl, respectively, demonstrates the high selectivity of imidacloprid (Table 3).

3.2 Selectivity on Arthropods

As shown in Section 2, imidacloprid controls a wide range of sucking pests and a few species of chewing insects. In extensive laboratory and field studies the ecobiological profile of imidacloprid has been established by Pflüger and Schmuck (1991). The activity of soil microorganisms is not impaired even at very high dose rates of 2000 g a.i./ha. Spray applications, four times overdosed, had only a transient effect on earthworm populations, which had been equalized until autumn at the year of application. In laboratory examinations imidacloprid-coated sugar beets (150 g a.i. per unit) had no effect on *Eisenia fetida*. The

Table 3. Contact activity of various insecticides on aphids compared with the oral toxicity on rats

Active ingredient	Myzus persicae LD_{50} mg/kg	Rat LD_{50} mg/kg	Factor LD_{50} rat/aphid
Imidacloprid	0.062	450	7300
Cyfluthrin	0.024	400	17000
Oxydemeton-methyl	0.98	50	51
Pirimicarb	0.50	150	300

product is toxic to birds in the acute test but only moderately toxic in subacute tests and in a bird reproduction study. Imidacloprid has a pronounced deterrent effect on birds (Avery et al. 1994); therefore, in a good agricultural practice, after seed treatment and granular application there is a negligible risk that birds will ingest lethal quantities of the active ingredient. Safety margins for algae Daphnia and fish are so large, that harmful effects against these organisms can be excluded.

Detailed field studies have been performed in Germany, Italy and Spain between 1987 and 1991 to evaluate possible adverse effects of imidacloprid against beneficial organisms (H.W. Schmidt, pers. comm.). Figure 6 summarizes the effects on important beneficial organism in typical single field trials. Imidacloprid applied as foliar spray (Confidor 200SL, soluble liquid) does not affect diplopoda and spiders such as Linyphiidae, Araneidae, *Typhlodromus piri* and the carabid beetle *Platynus dorsalis*. Anthocoridae, the staphilinid beetle *Philonthus* sp., *Chrysopa carnea* and Braconidae are affected to various degrees by imidacloprid sprays, although populations of Anthocoridae recovered quickly after exposure to imidacloprid. In addition to these results, it has been demonstrated that parasitic stages developing within their host as well as pupal stages of many beneficial insects are not affected by the product. Adult *Coccinella septempunctata* are mainly affected by the lack of prey; once hit by recommended field dosages of imidacloprid, a knock-down effect followed by recovery has been observed (H.W. Schmidt, pers. comm.). The product is harmful to bees and should not be applied during the flowering period (Pflüger and Schmuck 1991).

Virtually no apparent impact on beneficials has been observed after seed dressing (Gaucho 70 WS, water dispersible powder; Fig. 6). Therefore, all types of systemic applications such as seed dressing, soil treatment or stem painting/injection are preferred whenever possible. Hence, imidacloprid as a systemic agent can be used in integrated pest management (IPM) systems without reservation.

4 Agricultural Importance

Due to its unique properties—high intrinsic acute and residual activity against sucking and some chewing insect species, high efficacy against suceptible and resistant populations such as aphids, whiteflies, leafhoppers and planthoppers, Colorado potato beetles and others, and excellent acropetal translocation — imidacloprid can be used in a broadvariety of crops. These include e.g. aphids in vegetables, sugar beet, cotton, pome fruit, cereals and tobacco; leafhoppers, planthoppers and water weevil in rice; whiteflies in vegetables, ornamentals, cotton and citrus; lepidopteran leafminer in pome fruit and citrus; and wireworms in sugar beet and corn. Termites and turf pests such as white grubs are also covered by imidacloprid (Elbert et al. 1990, 1991).

As the product controls important vectors of virus diseases, the secondary spread of viruses in various crops is impaired. This has been shown for the persistent barley yellow dwarf virus (BYDV) transmitted by *R. padi* and *S.*

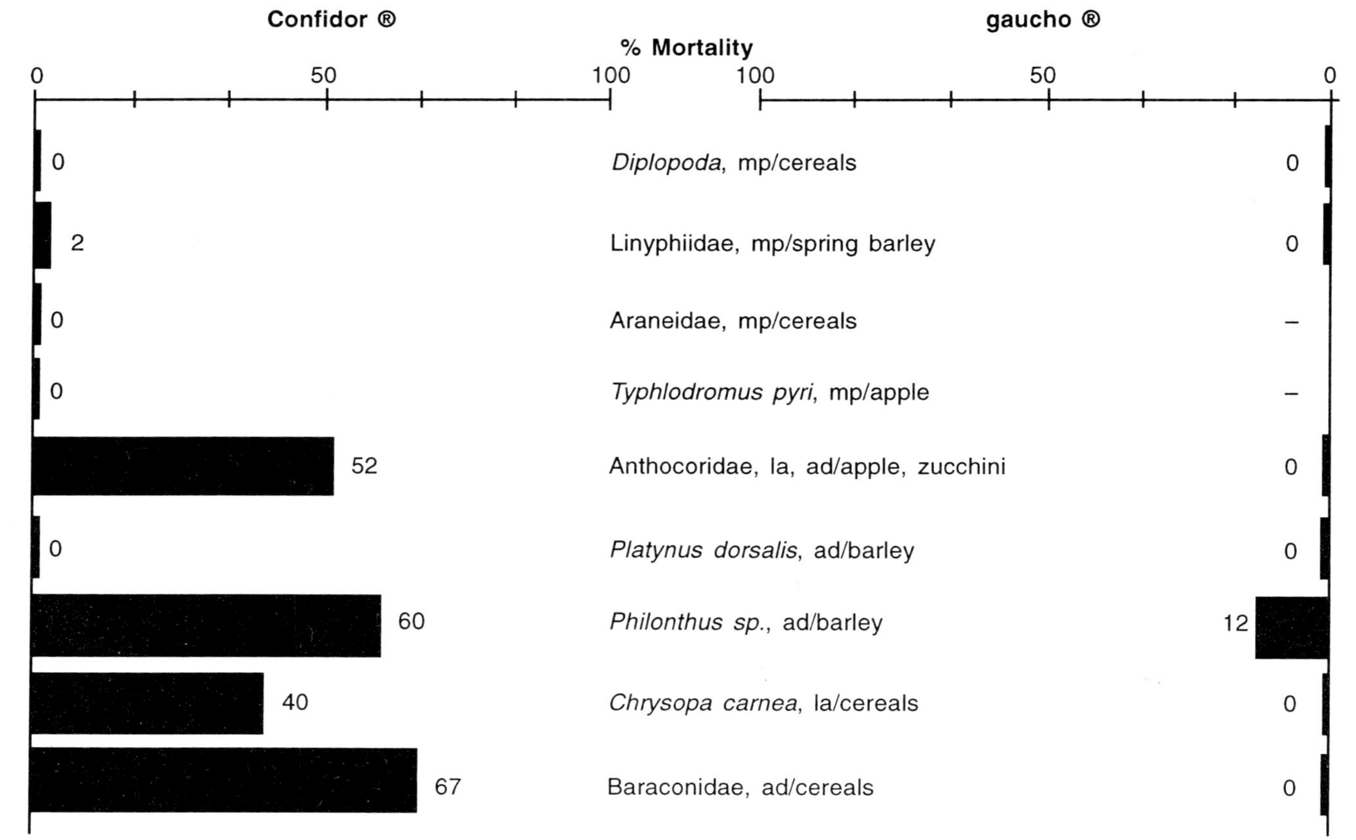

Fig. 6. Effect of imidacloprid on beneficials in field trials in Germany, Italy and Spain. Confidor was used as 25 WP or 200 SL at 120–250 ga. i./ha for spray application, Gaucho was used as 70 WS at 70 ga. i./kg seed for seed treatment. *La* Larvae; *ad* adults; *mp* mixed population

avenae (Knaust and Poehling 1992). Seed treatments proved highly effective in controlling BYDV vectors and the subsequent infection in a series of field trials in southern England (Bluett and Birch 1992). Sugar beet seed pelleted with imidacloprid was well protected especially against infections with beet mild yellow virus (BMYV) transmitted by M. persicae (Dewar 1992).

Due to its high systemicity, diverse ways of application are feasible and have been introduced into practice. Soil treatments can be done by incorporation of granules, injection, application with irrigation water, spraying, use of tablets, etc. Plants or parts of them are treated by seed dressing, pelleting, implantation, dipping, injection, and painting. The mentioned methods have led to a more economic and environmentally friendly use of the product that fits well into various IPM programmes. In Sections 4.1 to 4.4 a few examples of the multiple purpose of imidacloprid in various crops are given. These are rice in Japan, cotton in the mid-south USA, vegetables in Mexico and cereals in France. In these crops the product is targeted mainly against early season pests.

As imidacloprid is an excellent tool for the management of resistant pests, an overuse could lead to the development of new types of resistance against this valuable insecticide. For example, Prabhaker et al. (in press) demonstrated that the selection of an imidacloprid resistant whitefly population is possible at least under artificial laboratory conditions. Therefore, general resistance management guidelines for chloronicotinyl insecticides have been set up in order to assure a rational use and to prolong the lifespan of products from this important chemical class (Elbert et al. 1996). Important principles of the guidelines are:

1. Long-term rotation acts against rapid selection of resistant populations.
2. Use effective doses of the individual components when applying tank mixes and the full recommended rate of in–can mixtures.
3. Control should not be carried out with products of just one a.i. class.
4. The use of non-specific products helps prevent the development of resistance.
5. All possible cultivation techniques should be used alongside physical and biological pest control methods.
6. Crop protection products should be used in such a way as to reduce the risk to beneficial organisms.
7. Preparations should be used at the recommended doses and spray intervals.
8. Ensure that uniform spray coverage is achieved.
9. Where resistance reduces effectiveness, do not carry out a follow-up treatment with an a.i. of the same class.
10. Monitor the situation wherever possible so as to detect the first signs of resistance.

4.1 Rice

In Japan, nursery box application of insecticides has developed to become a standard procedure and it is currently used in 50% of the rice growing area. This type of treatment is directed against early season pests like *L. oryzophilus* (rice water weevil) and *L. oryzae* (rice leaf beetle). In contrast to commercially available products like carbosulfan, benfuracarb, propoxur and others, which control mainly rice water weevil and rice leaf beetle, imidacloprid covers the whole spectrum of early season pests including *N. cincticeps* (green rice leafhopper), *L. striatellus* (smaller brown planthopper), and also mid and late season hopper species like *S. furcifera* (white-backed planthopper) and *N. lugens* (brown planthopper). Benefits of nursery box application are: reduced labour input, longer residual effect and reduced side-effects against non-targetorganisms in comparison with the whole surface treatment.

Based on 5-year field trials, the use pattern for imidacloprid against main rice pests has been established. Applied as a granule, 2 GR with 1 g a.i. per nursery box (= 200 g a.i./ha) 0-3 days before transplanting, the product gives full protection between 50 and 90 days against hoppers depending on the species. By controlling *L. striatellus*, a 91% reduction on rice stripe virus was observed. An excellent residual effect against *L. oryzophilus* was detected up to 50 days and against *L. oryzae* up to 60 days after transplanting.

Apart from the seedling box application the product can be used as a granule for water surface application, as wettable powder (WP) or as dust formulation for foliar treatment and also as seed dressing for direct seeded rice to control the above-mentioned pests. Under field conditions, no apparent effects of imidacloprid treatments on populations of the spiders *Pardosa pseudoannulata* and *Tetragnatha vermiformis* were observed. Therefore, imidacloprid can be regarded as very effective against rice insects, flexible to use but most efficient in nursery box application. It outperforms conventional standards due to broader efficacy and lower dose rates. Virus diseases are suppressed through effective vector control and low toxicity has been observed against important beneficials in rice (Iwaya and Tsuboi 1992).

4.2
Cotton

Main cotton pests in the mid-south USA are *Thrips tabaci* (potato thrips), *Lygus lineolaris* (tarnished plant bug), *Neurocolpus nubilis* (clouded plant bug), *A. grandis* (boll weevil), *H. virescens* (tobacco budworm) and *H. zea* (bollworm). Other pests such as *Pseudatomoscelis seriatus* (cotton fleahopper), *A. gossypii* (cotton aphid), *Spodoptera exigua* (beet armyworm) and *S. frugiperda* (fall armyworm) are of lesser importance in that region.

Thrips tabaci attacks the terminal bud and two to four true leaves. The plant is damaged by reduced stand, retarded growth, killed buds and delayed fruiting. Significant influence on the yield has been demonstrated. *Lygus lineolaris* damages by feeding on pinhead squares, which causes young squares to shed. Terminal bud injury leads to multiple branched plants("crazy cotton"). Symptoms caused by *N. nubilis* are similar to those of *L. lineolaris.*

It is evident that damage caused by early season pests such as thrips and bugs can be very important. Loss of first position squares can have detrimental effects on the yield potential of the plant. If normal fruiting of cotton plants is prevented this can cause a delay in maturity, which in turn causes additional late season applications to control bollworms, tobacco budworms and boll weevils.

The flowable concentrate Gaucho 480 FS is applied as seed dressing against early infestations of thrips and aphids at 250 g a.i./100 kg cotton seed (37 g a.i./ha). Against bugs and later attack of aphids, three foliar split applications at low dose rates are recommended with Provado 1.6 F (flowable concentrate, 3 × 15 g a.i./ha; Fig. 7). These should be done as band sprays at intervals of 7–10 days. The first application starts 30 days after emergence in the five-

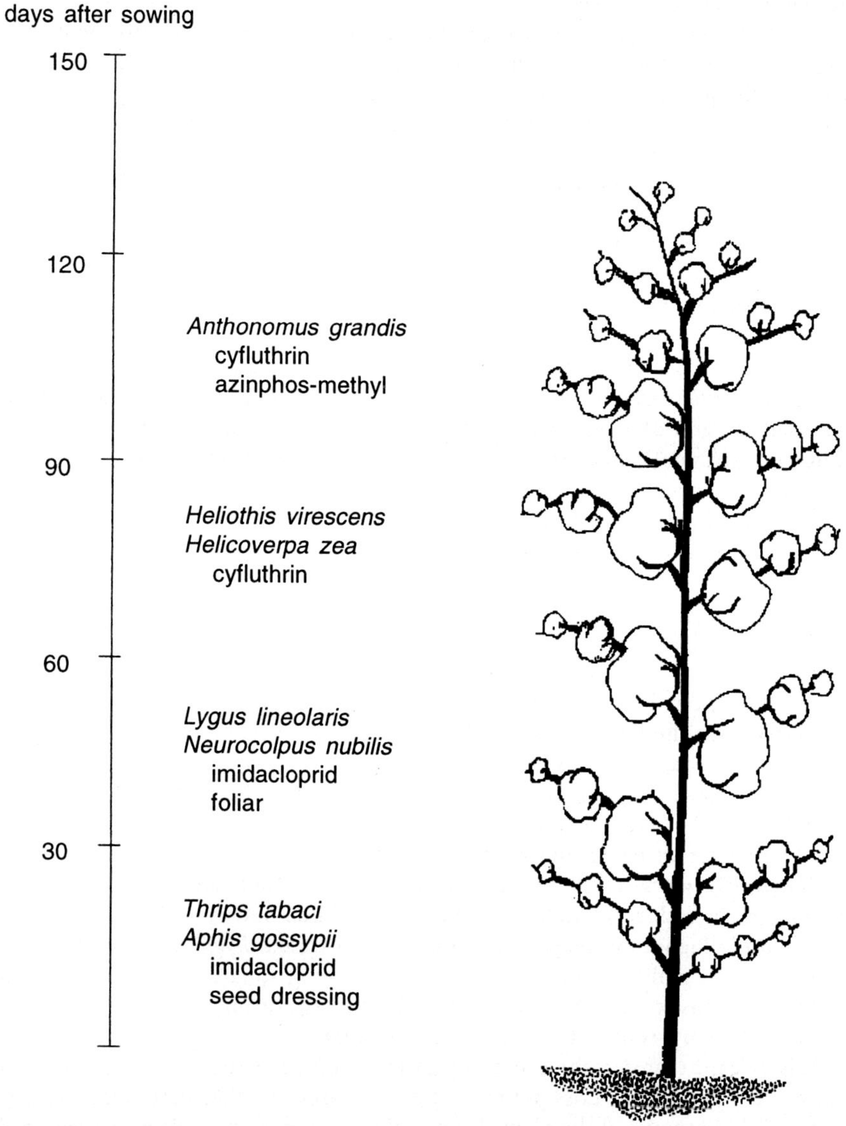

Fig. 7. Insect management in cotton in mid-south USA

leafstage subsequent to the Gaucho protection period. The programme corresponds to the long-term rotation system for resistance management for glasshouse pests recommended by Sanderson and Roush (1995). Imidacloprid will be restricted to the first 2 months of the crop cycle and other products/systems than chloronicotinyl insecticides should be used afterwards. The *Heliothis/Helicoverpa* complex is controlled by conventional products in mid season, whereas boll weevil and armyworms will be controlled by azinphos-methyl or cyfluthrin, respectively. The subsequent use of Gaucho and Provado ensures a full protection of cotton plants against early season pests for up to 60 days. Early fruit setting and maturity are distinctly enhanced and ensure the basis for a high yield.

4.3 Vegetables

Virus diseases have been the main factors affecting tomato and chilli production in Mexico during the past 20 years and have led to significant yield reductions in many regions. These crops are grown on approximately 160,000 ha with a constant virus infestation in 30–35% of the field. Persistent viruses are the most problematic. In tomato crops, viruses can infest up to 100% of the crop. The damage depends on the type of virus, the population density of vectors and the developmental stage of the crop. Protection of the crop by conventional insecticide treatments under plastic-covered greenhouses has not given satisfactory results.

A programme based on imidacloprid applications, which protect the plant during the critical period of virus transmission, was suggested by Bayer de Mexico. This period starts at emergence and ends at commence of flowering; later on, the losses provoked by viruses are of lesser importance. Imidacloprid is capable of restricting the spread of viruses by efficient control of vectors such as *B. tabaci, Trialeurodes vaporariorum, M. persicae, Paratrioza cockerelli* and some *Thrips species* (Table 4). During the critical period of up to 60 days from sowing, a permanent protection of the young plant by imidacloprid has to be assured.

The procedure described in Fig. 8 ensures the prevention of virus diseases in tomato through efficient vector control. Similar procedures have been developed

Table 4. Important virus diseases in Mexican tomato production

Disease	Abbreviation	Type of virus	Vector
Serrano golden mosaic virus	SGMV	Persistent	*Bemisia tabaci/Trialeurodes vaporariorum*
Tomato spotted wilt virus	TSWV	Persistent	*Caliothrips phaseoli/ frankliniella spp.*
Tobacco etch virus	TEV	Non-persistent	*Myzus persicae*
Tomato mosaic virus	TOMV	Non-persistent	*Myzus persicae*
Tobacco ringspot virus	TRSV	Non-persistent	*Mysuz persicae*
Cucumber mosaic virus	CMV	Non-persistent	*Myzus persicae*
Potato leaf roll virus	PLRV	Persistent	*Myzuz persicae*
Tomato leaf roll virus	TLRV	Persistent	*Bemisia tabaci/ Trialeurodes vaporariorum*

Fig. 8. Tomato plants protected by imidacloprid in comparison with a regional standard treatment A seed treatment is recommended with Gaucho 70 WS, 54 ga. i./kg seed. Three to 5 days before transplantation, 0.7 ga. i. (Confidor 350 SC, suspension concentrate) is applied into the substrate of 1000 young plants. Three to 5 days after transplantation, 350 ga. i./ha is applied to 20000–30000 plants into the soil

for chilli, tobacco and potato. The use of imidacloprid should be restricted to three applications as described. According to the resistance management guidelines, insecticides other than chloronicotinyls should be used afterwards.

4.4 Cereals

The traditional method of controlling virus vectors in Europe is the regular monitoring of aphid attack of young wheat and barley plants after emergence in autumn. If the threshold of 10% infested plants with at least one aphid per plant has been surpassed, usually a pyrethroid will be sprayed. The monitoring has to be continued because the residual effect of the treatment may be insufficient. Due to climatic conditions and duration of aphid attack, a second or a third application may be required. This rather time-consuming and difficult procedure can be avoided by a seed treatment with imidacloprid. BYDV vectors such as *R. padi* and *S. avenae* are controlled and the secondary spread of the disease is inhibited. In contrast to one or several sprays, the seed dressing as the only treatment assures prolonged protection during the critical period, where virus transmission is of importance. Furthermore, pyrethroid sprays with broad spectrum activity can be substituted by only one imidacloprid application, which has virtually no effects on beneficials and other non-target organisms.

High yielding varieties achieve their optimum yield potential only if they are sown early, which implies an enhanced risk of infestation with aphids and, consequently with BYDV. As Gaucho controls aphids and suppresses the infection with BYDV, early sown cereals are especially well protected against insects and diseases. Therefore, a better tillering is achieved and 15-25% of the seed rate can be saved.

Figure 9 shows the results of a typical field trial in winter wheat in France. From 100 untreated plants 26 were attacked by aphids 41 days after sowing and of 100 leaves 23 showed BYDV-symptoms 191 days later. The yield in untreated wheat was 4070 kg. Both imidacloprid seed treatment and pyrethroid spray

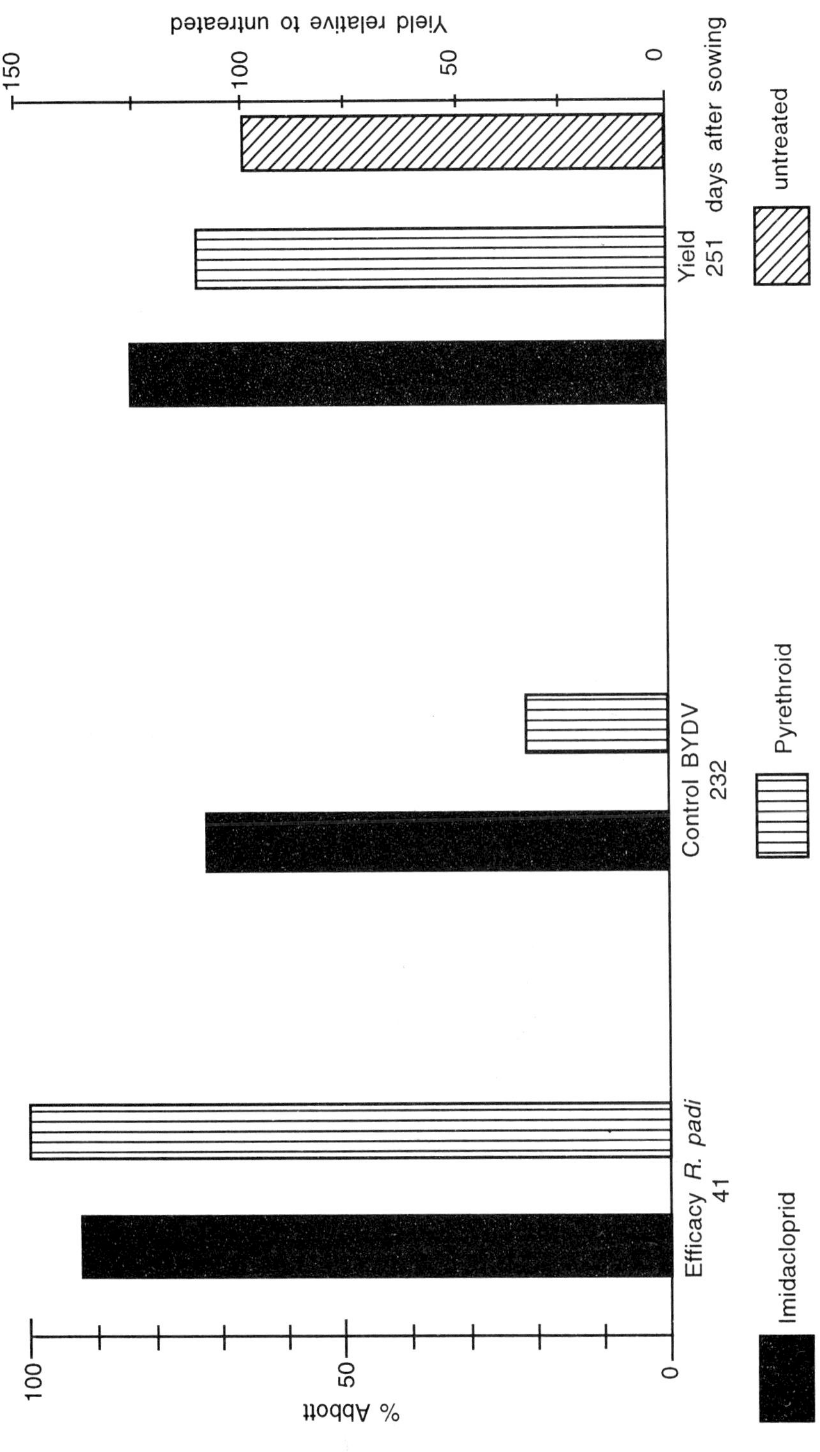

Fig. 9. Effect of imidacloprid on aphid control, virus symptoms and yield increase in a field trial in winter wheat in France. Imidacloprid was applied as seed dressing (Gaucho blé) at 70 ga. i./100 kg seed. A standard pyrethroid spray was applied at 7.5 ga. i./ha on 21 November, 1989

application showed a very-good-to-excellent effect against the vector *R. padi*. The acute effect of the pyrethroid was even better due to its quick knock-down activity. Imidacloprid, on the other hand, acts slower, therefore it took more time to kill invading aphids, but its residual effect was much more pronounced. This fact results in long lasting control of aphids and, as a consequence, a considerably better prevention of virus symptoms. A yield increase of 26% over untreated was achieved in comparison with 10% of the pyrethroid treatment.

Field trials over 4 years in France confirmed that both wheat and barley are well protected against the mentioned pests and diseases. In an average of 76 trials in wheat (66 trials in barley), a yield increase of 3.1 dt/ha (4.4 dt/ha in barley, 1 dt = 100 kg) in comparison with untreated was achieved by the Gaucho treatment.

5 Summary

Imidacloprid, a new chloronicotinyl insecticide, was synthesized in 1985. Its unique biological properties have been detected in laboratory and greenhouse trials. It is highly active against sucking and some biting insects, a wide range of aphid, leaf- and planthopper species, whiteflies and some beetles; micro lepidopteran and dipteran species are controlled by foliar spray. Due to its systemic properties, soil treatment, seed dressing, use in irrigation water, stem application, etc. are possible. Detailed examinations revealed the fate and biological activity of the insecticide after seed dressing in wheat, cotton and other crops.

Sublethal doses of imidacloprid led to pronounced effects on aphid behaviour and on reproduction rates. Honeydew production of *M. persicae* is reduced and aphids left treated leaves. They may die due to starvation if no alternative food source is available. Also in *M. persicae*, more than 50% reduction of viability at imidacloprid concentration as low as 0.2 ppb, applied systemically, was observed. Both effects, the antifeedant effect and the reduction of fertility, add to the overall activity of imidacloprid.

Due to its mode of action it has been demonstrated that imidacloprid controls a wide range of pests resistant to conventional insecticides. Therefore, the product fits well in resistance management strategies. A general resistance management guideline for chloronicotinyl insecticides has been defined.

Imidacloprid's site of action in insects is the nicotinic acetylcholine receptor (nAChR), which has not been targeted so far by economically important insecticides. It has been demonstrated that imidacloprid shows a high selectivity at the molecular level, because its binding at the nAChR in vertebrates is more than 1000 times lower than on the insect receptor. In field tests no effects on soil microorganisms, algae, water fleas and fish were found, and only transient effects on earthworms have been detected. Though toxic to birds, due to its pronounced repellent effect, no toxic quantities of the active ingredient are ingested.

Imidacloprid does not affect diplopoda, spiders and predatory mites after foliar application. Other beneficial insects might be affected depending on species, developmental stage and exposure to the insecticide, and the product is classified as toxic to bees. However, after soil treatment and seed dressing no adverse effects on beneficial organisms have been found. Therefore, imidacloprid can be regarded as an important component of integrated pest management, if applied by various methods on or into the soil.

The described biological, ecobiological and toxicological properties make way for diverse applications of imidacloprid in various crops against many insect pests. Examples are given in rice for the control of *L. oryzophilus, L. oryzae, L. striatellus, N. cincticeps, S. furcifera*, and *N. lugens*, in cotton for *T. tabaci, A. gossypii, L. lineolaris* and *N. nubilis*, in vegetables for *B. tabaci, T. vaporariorum, M. persicae, Paratrioza cockerelli* and *Thrips* sp. and in cereals for *R. padi*.

Acknowledgements

We thank our colleagues R. Altmann, W. Mullins, K. Nevermann, H.W. Schmidt and K. Sturm for their contributions to this chapter.

References

Abbink J (1991) The biochemistry of imidacloprid. Pflanzenschutz-Nachr Bayer 44: 183–194

Avery ML, Decker GD, Fischer DL (1994) Cage and flight pen evaluation of avian repellency and hazard associated with imidacloprid treated rice seed. Crop Prot 13: 535–540

Bai D, Lummis SCR, Leicht W, Breer H, Satelle DB (1991) Actions of imidacloprid and related nitromethylene on cholinergic receptors of an identified insect motor neurone. Pestic Sci 33: 197–204

Bluett DJ, Birch PA (1992) Barley yellow dwarf virus (BYDV) control with imidacloprid seed treatment in the United Kingdom. Pflanzenschutz-Nach Bayer 45: 455–490

Cahill M, Byrne FJ, Gorman K, Denholm I, Devonshire AL (1995) Pyrethroid and organophosphate resistance in the tobacco whitefly *Bemisia tabaci* (Homoptera: Aleyrodidae). Bull Entomol Res 85: 181–187

Cahill M, Gorman K, Day S, Denholm I, Elbert A, Nauen R (1996) Baseline determination and detection of resistance to imidacloprid in *Bemisia tabaci* (Homoptera: Aleyrodidae). Bull Entomol Res 86: 343–349

Costa HS, Brown JK (1991) Variation in biological characteristics and esterase patterns among populations of *Bemisia tabaci*, and the association of one population with silverleaf symptom induction. Entomol Exp Appl 61: 211–219

Devine GJ, Harling ZK, Scarr AW, Devonshire AL (1996) Lethal and sublethal effects of imidacloprid on nicotine-tolerant *Myzus nicotianae and Myzus persicae*. Pestic Sci 48: 57–62

Dewar AM (1992) The effects of imidacloprid on aphids and virus yellows in sugar beet. Pflanzenschutz-Nachr Bayer 45: 423–442

Dewar AM, Read LA (1990) Evaluation of an insecticidal seed treatment, imidacloprid, for controlling aphids on sugar beet. Proc Brigthon Crop Prot Conf, Pests and diseases, pp 721–726

Drinkwater TW (1994) Comparison of imidacloprid with carbamate insecticides, and the role of planting depth in the control of false wireworms, *Somaticus* species, in maize. Crop Prot 13: 341–345

Drinkwater TW, Groenewald LH (1994) Comparison of imidacloprid and furathiocarb seed dressing insecticides for the control of the black maize beetle, *Heteronychus arator* Fabricius (Coleoptera: Scarabaeidae), in maize. Crop Prot 13: 421–424

Elbert A, Nauen R (1996) Bioassays for imidacloprid for a resistance monitoring against the whitefly *Bemisia tabaci*. Proc. Brighton Crop Prot Conf—Pests and diseases, November 1996, pp 731–738

Elbert A, Overbeck H, Iwaya K, Tsuboi S (1990) Imidacloprid, a novel systemic nitromethylene analogue insecticide for crop protection. Proc Brighton Crop Prot Conf, Pests and diseases, pp 21–28

Elbert A, Becker B, Hartwig J, Erdelen C (1991) Imidacloprid—a new systemic insecticide. Pflanzenschutz-Nachr Bayer 44: 113–136

Elbert A, Nauen R, Cahill M, Devonshire A, Scarr A, Sone S, Steffens R. (1996) Resistance management for chloronicotinyl insecticides using imidacloprid as an example, Pflanzenschutz-Nachr Bayer 49: 5–54

Grothe H (1992) Investigations into the effect of the new synthetic compound imidacloprid ("Confidor"), in comparison with other insecticides, on behavioural changes in various aphid species (Homoptera: Aphididae). Thesis from the Department of Biology, carried out at the Institut für Phytopathologie und Angewandte Zoologie der Justus-Liebig-Universität Gießen

Iwaya K, Tsuboi S (1992) Imidacloprid—a new substance for the control of rice pests in Japan. Pflanzenschutz-Nachr Bayer 45: 197–227

Knaust HJ, Poehling HM (1992) Studies of the action of imidacloprid on grain aphids and their efficiency to transmit BYD-Virus. Pflanzenschutz-Nachr Bayer 45: 381–408

Lagadic L, Bernard L, Leicht W (1993) Topical and oral activities of imidacloprid and cyfluthrin against susceptible laboratory strains of *Heliothis virescens* and *Spodoptera littoralis* (Lepidoptera: Noctuidae). Pestic Sci 38: 323–328

Leicht W (1993) Imidacloprid—a chloronicotinyl insecticide. Pestic Outlook 4: 17–21

Lewis GA, Madge DS (1984) Esterase activity and associated insecticide resistance in the damson-hop aphid, *Phorodon humuli* (Schrank) (Homoptera: Aphididae). Bull Entomol Res 74: 227–238

Liu M-Y, Casida JE (1993) High affinity binding of [^{3}H]-imidacloprid in the insect acetylcholine receptor. Pestic Biochem Physiol 46: 40–46

Lösel PM, Goodman LJ (1993) Effects on the feeding behaviour of *Nilaparvata lugens* (Stål) of sublethal concentrations of the foliarly applied nitromethylene heterocycle 2-nitromethylene-1, 3-thiazinan-3-yl-carbamaldehyde. Physiol Entomol 18: 67–74

Mellis JM, Kirkwood RC (1980) The uptake, translocation and metabolism of epronaz in selected species. Pestic Sci 11: 324–330

Methfessel C (1992) Action of imidacloprid on the nicotinic acetylcholine receptors in rat muscle. Pflanzenschutz-Nachr Bayer 45: 369–380

Nauen R (1995) Behaviour modifying effects of low systemic concentrations of imidacloprid on *Myzus persicae* with special reference to an antifeeding response. Pestic Sci 44: 145–153

Nauen R, Elbert A (1994) Effect of imidacloprid on aphids after seed treatment of cotton in laboratory and greenhouse experiments. Pflanzenschutz-Nachr Bayer 47: 181–215

Nauen R, Elbert A (1997) Apparent tolerance of a field-collected strain of *Myzus nicotianae* to imidacloprid due to strong antifeeding response, Pestic Sci 49: 252–258

Nauen R, Strobel J, Otsu K, Tietjen K, Erdelen C, Elbert A (1996a) Aphicidal activity of imidacloprid against a carbamate and organophosphate resistant Japanese strain of the tobacco feeding form of *Myzus persicae* (Homoptera: Aphididae) closely related to *Myzus nicotianae*. Bull Entomol Res 86: 165–171

Nauen R, Hungenberg H, Tollo B, Tietjen K, Elbert A (1996b) Antifeedant effect, biological efficacy and high affinity binding of imidacloprid to acetylcholine receptors in tobacco-associated *Myzus persicae* (Sulzer) and *Myzus nicotianae* Blackman (Homoptera: Aphididae). Proc 20th Int Congr Entomology 25–31 Aug 1996, Firenze, Italy, p 597

Pflüger W, Schmuck R (1991) Ecotoxicological profile of imidacloprid. Pflanzenschutz-Nachr Bayer 44: 145–158

Phillips FT, Etheridge P, Kavadia VS, Sethi GR, Sparrow PE (1978) Translocation of carbon-14 dieldrin from small droplets on cotton leaves. Ann Appl Biol 89: 51–60

Prabhaker N, Toscano NC, Castle SJ, Henneberry TJ (1997) Selection for resistance to imidacloprid in silverleaf whiteflies in imperial valley and development of a hydroponic bioassay method for monitoring of imidacloprid resistance. Pestic. Sci (in press)

Sanderson JP, Roush RT (1995) Management of insecticide resistance in the greenhouse. In: Bishop A, Hausbeck M, Lindquist R (eds) Proc 11th Conf on Insect and disease management on ornamentals, Fort muers, Florida

Soloway SB, Henry AC, Kollmeyer WD, Padgett WM, Powell JE, Roman SA, Tiemann CH, Corey RA, Horne CA (1978) Nitromethylene insecticides. In: Geissbühler H, Brooks GT, Kearney C (eds) Advances in pesticide science, part 2. Pergamon Press, pp 206–227

Sone S, Hattori Y, Tsuboi S, Otsu Y (1995) Difference in susceptibility to imidacloprid of the populations of the small brown planthopper, *Laodelphax striatellus* Fallen, from various localities in Japan. J Pestic Sci 20: 541–543

Stein-Dönecke U (1992) Beizhofausbildung, Aufnahme, Translokation und Wirkung von [14C]Imidacloprid bei Winterweizen und Zuckerrüben nach Saatgutbehandlung und unter dem Einfluß verschiedener Bodenfeuchten. PhD Thesis, Universität Bonn

Stein-Dönecke U, Führ F, Wieneke J, Hartwig J, Leicht W (1992) Influence of soil moisture on the formulation of dressing zones and uptake of imidacloprid after seed treatment of winter wheat. Pflanzenschutz-Nachr Bayer 45: 327–368

Strang RH, Rogers RL (1971a) A microautoradiographic study of 14C-Trifluralin absorbtion. Weed Sci 19: 363–369

Strang RH, Rogers RL (1971b) A microautoradiographic study of 14C-Diuron absorbtion by cotton. Weed Sci 19: 355–362

Tomizawa M, Yamamoto I (1992) Binding of nicotinoids and the related compounds to the insect nicotinic acetylcholine receptor. J Pestic Sci 17: 231

Tomizawa M, Otsuka H, Miyamoto T., Eldefrawi ME, Yamamoto I (1995) Pharmacological characteristics of insect nicotinic acetylcholine receptor with its ion channel and the comparison of the effect of nicotinoids and neonicotinoids. J Pestic Sci 20: 57–65

Tröltzsch C-M (1995) Einfluß unterschiedlicher Bewässerungsverfahren und Wirkstoffaufwandmengen von [^{14}C]Imidacloprid auf die Beizhofausbildung, die Aufnahme und Wirkung in Baumwollpflanzen nach Saatgutbeizung. PhD Thesis, Universität Bonn

Tröltzsch C-M, Führ F, Wieneke J, Elbert A (1994) Influence of various irrigation procedures on the uptake of imidacloprid by cotton after treatment. Pflanzenschutz-Nachr Bayer 74: 249–303

Westwood F, Dewar AM, Bromilow RH, Bean KM (1995) Study of uptake of [^{14}C]imidacloprid in sugar beet. Proc IIRB Conf, Dijon

Woodford JAT, Mann JA (1992) Systemic effects of imidacloprid on aphid feeding behaviour and virus transmission. Proc Brighton Crop Prot Conf—Pests and diseases, pp 557–562

Yamamoto I, Yabuta G, Tomizawa M, Saito T, Miyamoto T, Kagabu S (1995) Molecular mechanism for selective toxicity of nicotinoids and neonicotinoids. J Pestic Sci 20: 33–41

Zwaart R, Oortigiesen M, Vijverberg HPM (1994) Nitromethylene heterocycles: selective agonists of nicotinic receptors in locust neurons compared to mouse N1E-115 and BC3H1 cells. Pestic Biochem Physiol 48: 202

CHAPTER 5

Buprofezin: A Novel Chitin Synthesis Inhibitor Affecting Specifically Planthoppers, Whiteflies and Scale Insects

A. De Cock and D. Degheele
Faculty of Agricultural and Applied Biological Sciences, University of Gent. Belgium

1 Introduction

Insect growth regulators (IGRs) are selective insecticides interfering with normal growth and development. Buprofezin, 2-*tert*-butylimino-3-isopropyl-5-phenylperhydro-1,3,5-thiadiazin-4-one, developed by Nihon Nohyaku in 1981, is one of the first IGRs mainly acting against sucking insects such as whiteflies and scale insects. As a chitin synthesis inhibitor, it expresses its action at the time of moulting; the affected insects are not able to shed their cuticle and die during this process. These symptoms resemble those induced by the benzoylphenylureas, although the chemical structure of buprofezin (Fig. 1) is not analogous to that of the benzoylphenylureas (Retnakaran and Wright 1987).

Buprofezin is a thiodiazin derivative with a molecular weight of 305.5 and a melting point for the pure substance of 106.1 °C. The solubility at room temperature is 0.9 and 240 mg/l in water and acetone, respectively. Buprofezin is formulated as a wettable powder, emulsifiable concentrate or granules.

Its most successful applications are against various homopterous pests in paddy fields and citrus groves, i.e. crops in which the effectiveness of many of the hydrochlorides, organophosphates, carbamates and even plant substances like pyrethroids is seriously affected due to the development of resistance.

2 Mode of Action

The insect integument generally consists of an epidermal cell layer with the overlying cuticle which it has secreted (Fig. 2). In the cuticle, chitin (a polymer of acetylglucosamine) always appears in association with protein (the ratio of

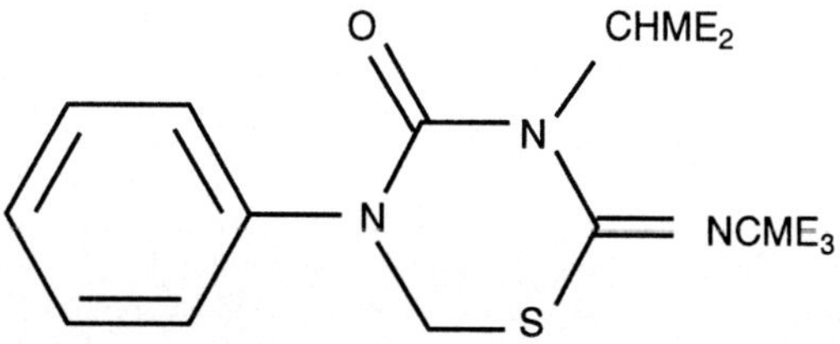

Fig. 1. Structure of buprofezin

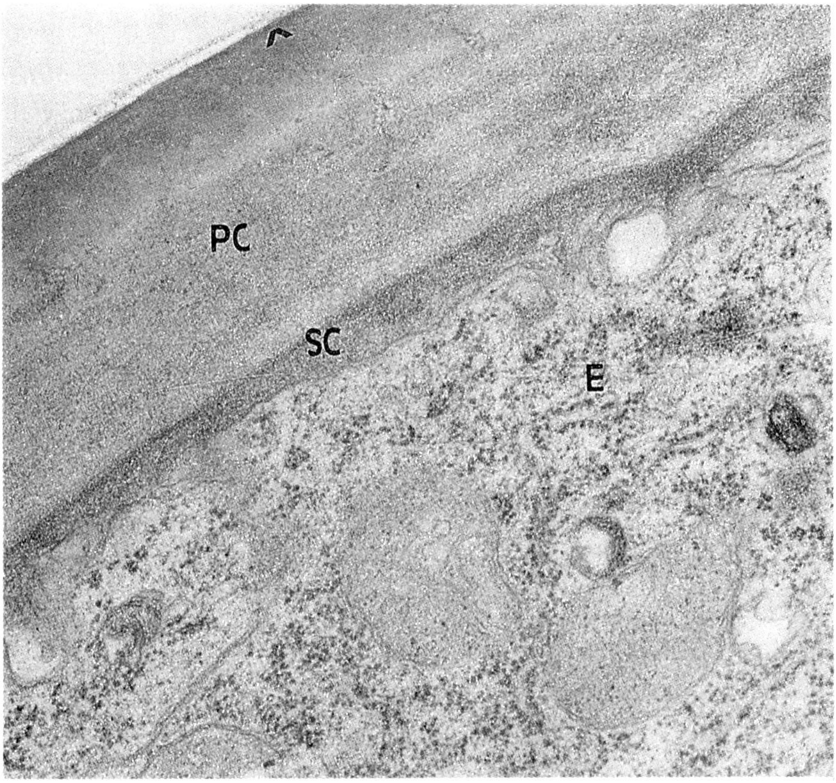

Fig. 2. Fine structure of the integument of a third instar nymph of *T. vaporatiorum* indicating four main layers of the integument: epicuticle (*arrow*), procuticle (*PC*), subcuticle (*SC*) and epidermis (*E*) (× 35 500)

protein to chitin differing in cuticles of different species). The cuticle can be divided into a thin outermost epicuticle and a thicker lamellated procuticle. Growth of each juvenile stage is limited, since the cuticle is only capable of limited stretching, so a moulting process is necessary. At the beginning of each new instar, the old cuticle is shed (ecdysis) after detaching the procuticle from the epidermis (apolysis) and dissolving of a part of it. Reynolds (1987) described the regulation of moulting processes by ecdysteroids: after an initial period of growth, the first event of the moulting cycle is the separation of the old cuticle of the epidermis. Apolysis is then promoted by the rising of the 20-hydroxyecdysone level. The space between the epidermis and the cuticle becomes filled with a moulting fluid, containing inactive precursors of the enzymes that will ultimately break down the old cuticle and allow its materials to be recovered. The cell proliferation of the epidermis and the secretion of a new cuticle can follow apolysis under a high level of the moulting hormone. Prior to ecdysis, however, the level of 20-hydroxyecdysone begins to decline. Its declining phase initiates enzyme activation in the moulting fluid and consequently digestion of the old cuticle.

The benzoylphenylureas were the first compounds reported to interfere with chitin synthesis and the deposition of integumentary cuticle of insects. The characteristic symptoms caused by a buprofezin treatment are similar to those of the benzoylphenylureas (De Cock and Degheele 1991). Poisoning symptoms are only seen around the time of moulting. As a result of chitin deficiency, the procuticle disrupts under the pressure of the exuvial fluid and the insect becomes unable to moult. Immature stages are incapable of casting their exuviae and die in a typical moulting position accompanied by blackening and a loss of moisture. Uchida et al. (1985) studied the cuticle deposition after treatment with buprofezin in the nymphs of *Nilaparvata lugens* Stål (Hom.: Delphacidae) by means of microscopic observations of stained sections. No histological differences were observed until 71 h after treatment. A staining difference occurred between treated and untreated insects 84 h after treatment, but even at this time no abnormality was seen. The failure of new cuticle deposition occurred just before death of the nymphs. This phenomenon was further studied in whiteflies by Hegazy et al. (1990) and De Cock and Degheele (1991). Treatment with buprofezin at the beginning of the third instar of *Trialeurodes vaporariorum* Westwood (Hom.: Aleyrodidae) larvae resulted in the death of the larvae at the time of moulting. Electron microscopic observations, after treatment of the larvae with 20 mg a.i./l buprofezin, did not reveal any differences in ultrastructural profile before apolysis. Post-apolysial formation of a normal lamellated procuticle was disturbed (Figs. 2, 3). The pharate procuticle was amorphous, varying in thickness and the mean thickness was greatly reduced as compared with a non-treated one. The subcuticle, a layer between the epidermis cells and the procuticle, was interrupted several times at locations corresponding to the thinnest places of pharate cuticle. In the same studies, unusual structures such as myelin figures and hypertrophied mitochondria were observed, indicating that buprofezin may have an additional toxic effect on epidermal cells.

Izawa et al. (1985) indicated that the biosynthesis of protein from ^{3}H-labelled glucose of [^{3}H]tyrosine was not affected by buprofezin treatment. ^{3}H-labelled nucleic acid synthesis from ^{3}H-glucose was slightly reduced by buprofezin. The chitin synthesis from N-acetyl-D (^{3}H)-glucosamine was strongly inhibited by a treatment of buprofezin. Similar results have been obtained when using [U-^{14}C] glucose as the precursor (Uchida et al. 1985). Since no inhibition was observed in the synthesis of the ultimate precursor of chitin, (UDP-N-acetyl-D-^{3}H) glucosamine, from N-acetyl-D-[^{3}H]glucosamine, buprofezin seemed to block the polymerization of UDP-N-acetyl-D-glucosamine to chitin. Partial inhibition of chitin biosynthesis seems enough to kill the nymphs. Ten mg a.i./l kills 100% of N. *lugens nymphs*, but only partially (35-70%) inhibits chitin biosynthesis. The I_{25} of buprofezin against chitin biosynthesis, 8.5×10^{-6} M (2.6 mg/l), was close to the LC_{50} (Izawa et al. 1985). Izawa et al. (1986) injected the chitin precursor N-acetyl-D[1,6-3H] glucosamine into fourth instar larvae of *Henosepilachna vigintioctopunctata* Fabricius (Col.: Coccinellidae) 2 days after a treatment with 4 μg per insect of buprofezin. In treated larvae, chitin biosynthesis was inhibited by 74.4% relative to the untreated control. Although no lethal effect of buprofezin on adult insects is reported, many authors report on the capability of this IGR to affect reproduction. Asai et al. (1985) reported that female adult

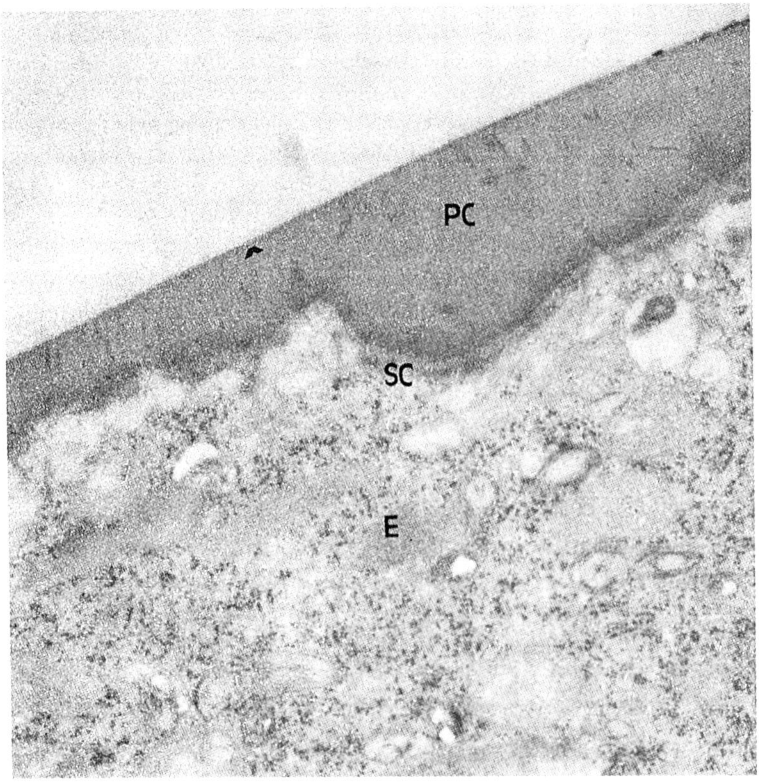

Fig. 3. Fine structure of the integument of a third instar nymph of *T. vaporariorum* treated with buprofezin, showing a newly formed epicuticle and an abnormal procuticle which lacks lamellae. Epicuticle (*arrow*), procuticle (*PC*), subcuticle (*SC*) and epidermis (*E*) (× 50 400)

life span of *N. lugens* was 25% reduced if treated within the first 48 h of emergence. The life span of males was little affected by buprofezin. If treated within 24 h of emergence, buprofezin considerably inhibited oviposition without interrupting ovarian development. Uchida et al. (1987) attributed the suppression of egg laying to the inhibition of prostaglandin E_2 biosynthesis. Histological observation showed that the eggs accumulated in ovaries of the buprofezin-treated *N. lugens* were deposited after a 10-pg prostaglandin E_2 injection per female. This kind of injection prompted the buprofezin-treated adults to reach a normal level of oviposition. Izawa et al. (1986) could reverse the suppression of egg laying in *H. vigintioctopunctata* in the same way. Injection of 1 ng prostaglandin in the abdomen after 5 days' feeding on tomato leaves treated with 1000 mg/l buprofezin accelerated the egg laying of treated females to a normal level. Interfering with prostaglandin synthesis has never been previously known from IGRs.

Uchida et al. (1986) stated that due to the low probability that a compound can exhibit two intrinsic activities, there might be a relationship between the two types of effect (inhibition of chitin synthesis and inhibition of prostaglandin synthesis). The authors briefly indicated that both effects were cancelled by 20-hydroxyecdysone. In-depth studies to clarify this antagonism to 20-hydroxyecdysone have been carried out by Kobayashi et al. (1989). Light microscopic observation indicated that apolysis and the subsequent cell proliferation of epidermis were unaffected, but that the digestion of old cuticle and the deposition of new cuticle were severely inhibited in the buprofezin-treated nymphs during 48–66 h after the last ecdysis. 20-Hydroxyecdysone applied topically 48 h after ecdysis exhibited a similar effect. The level of 20-hydroxyecdysone in untreated fourth instar nymphs showed two peaks at 24 and 33 h after ecdysis, but then declined rapidly. Buprofezin remarkably elevated the level of the moulting hormone and extremely retarded the fall of 20-hydroxyecdysone level in the buprofezin-treated nymphs, in which its precursor, ecdysone, was also observed to accumulate during 33–42 h. Thus, buprofezin seems to inhibit the fall of the 20-hydroxyecdysone level resulting in the inhibition of events following apolysis and cell proliferation, such as old cuticle digestion and new cuticle deposition.

De Cock and Degheele (1993) revealed, by means of lectin gold labelling techniques, that the malformation of the pharate cuticle was due to a lack of chitin incorporation. In untreated nymphs, the epicuticle and epidermis cells were devoid of label, while the procuticle was heavily labelled. In buprofezin-treated nymphs, only the pre-apolysial cuticle remained heavily labelled, independently of the concentration used. The post-apolysial cuticle, however, was hardly labelled, the intensity of labelling depending on the applied concentration. These results lead to the conclusion that buprofezin is only active after apolysis (initiated by the rise of the 20-hydroxyecdysone level) and can be related to the previously mentioned results of Kobayashi et al. (1989) that buprofezin inhibits the fall of the 20-hydroxyecdysone level and the initiation of subsequent events.

3 Fields of Application

Buprofezin is primarily active against hemipterous insects, but its larvicidal activity extends to some Coleoptera and Acarina (Kanno et al. 1981). In this respect, it has a complementary action to most earlier-developed IGRs such as diflubenzuron, chlorfluazuron, hexaflumuron and triflumuron.

In general, the compound has an excellent nymphicidal activity, declining with the nymphal stage. Its adulticidal and ovicidal activities are weak. However, suppression of egg viability in target species, after females came in contact with the material, is frequently reported. Vapour phase and contact toxicity seem to be major factors contributing to the potency of buprofezin. In some cases a weak translaminar activity was observed, suggesting that buprofezin acts in part by ingestion (De Cock et al. 1990).

3.1 Planthoppers and Leafhoppers

The activity of buprofezin against rice pests is described by several authors. Kanno et al. (1981) first reported on the activity on this group of insects. An application of 1 and 0.5 mg/l proved to be effective against *N. lugens, Sogatella furcifera* (Horvath) (Hom.: Delphacidae), *Laodelphax striatella* Fallen and *Nephotettix cincticeps* Uhler (Hom.: Ciccadellidae). For first-to-fifth instar of *N. lugens*, LC_{50}s range from 0.13 to 0.58 mg/l (Asai et al. 1983; Konno 1990). Nagata (1986) conducted field experiments during 3 years and found the optimal application time (with respect to maximum reduction of the post-treatment density of the nymphs) to be the mid-nymphal stages of the second generation. Applications at the early nymphal stages were relatively ineffective and caused a considerable post-treatment increase of nymphal density. Although buprofezin exerts an excellent control on rice pests, problems may occur when virus diseases need to be controlled. Virus diseases such as tungro and stripe are transmitted by the green rice planthopper *Nephotettix virescens* (Distant) (Hom.: Ciccadellidae) and the smaller brown planthopper *L. striatella*, respectively. Due to its slow action and its lack of repellent activity, buprofezin does not contribute to the reduction of the incidence of virus disease (Umeda and Hirano 1990). Tiongco et al. (1990) suggested in these cases the use of a combination of cypermethrin and buprofezin.

3.2 Whiteflies

The control of whiteflies (Hom.: Aleyrodidae) has become a major problem in recent years due to the widespread occurrence of strains with high levels of resistance against classical insecticides. The greenhouse whitefly *T. vaporariorum* and the sweetpotato whitefly *Bemisia tabaci* Gennadius are causing problems worldwide. Buprofezin has been extensively tested against both species and efficiently kills nymphs at recommended field rates between 100 and 250 mg/l. Adulticidal and ovicidal activity is generally non-existent.

For *T. vaporariorum* the LC_{50}s range from 0.6 mg a.i./l for the first instar to 15.7 mg a.i./l for the fourth instar, the estimated concentration for 50% cumulative larval toxicity over the four instars, was 2.1 mg a.i./l. For B. *tabaci*, the LC_{50}s range from 0.6 mg a.i./l to 47 mg a.i./l for the first and fourth instar, respectively, whereas the LC_{50} for cumulative larval toxicity was 6-7 mg/l (Yasui et al. 1985; De Cock et al. 1990; Gerling and Sinai 1994).

Hatchability of eggs from *T. vaporariorum* and *B. tabaci* females treated for some time with buprofezin was extremely low. The estimated concentration for 50% inhibition of egg hatch applied through females was ca. 15 mg/l; this effect on embryogenesis correlated well with the length of female exposure to buprofezin. A total suppression of egg hatch was achieved after 24 h exposure time of the adults and effects lasted for 72 h in adults free of buprofezin (Yasui et al. 1987; Ishaaya et al. 1988; De Cock et al. 1990).

Persistence of a buprofezin treatment was studied by Yasui et al. (1990) and De Cock et al. (1990). Fifty percent loss of foliar buprofezin varied from a short period (2.3 days) at 62.5 mg/l to a relatively long period (13 days) at 250 mg/l. At the latter concentration a complete suppression of progeny formation was observed on cotton up to 14 days after treatment. On tomatoes no viable eggs or larvae were found up to 28 days after treatment and double spraying with a 14-day interval extended this period to 42 days. Field trials were conducted on beans, cucumbers, tomatoes and *Poinsettias* in the Netherlands and the UK during 1983-1988. A good control of *T. vaporariorum* after one or two applications with 75 mg a.i./l was obtained. Control of *B. tabaci* was approximately 80% after a single application in the same series of trials. This finding was confirmed on *Poinsettia* where control of *B. tabaci* was also insufficient (maximum 56%) (Koch 1989). Excellent control of *T. vaporariorum* in bean trials could be achieved in former Czechoslovakia (Laska 1986). Positive experiences in azalea culture have been reported by Heungens and Buysse (1991).

Combinations with other insecticides such as fenpropathrin, fluvalinate and carbofuran are frequently mentioned to overcome buprofezin's lack of activity against adult whiteflies (Ishaaya et al. 1988; Rat-Morris 1990; Boukadido and Michelakes 1993; Bae et al. 1994). Results of the study of Castañer et al. (1989) on *Aleurothrixus floccosus* (Mask.) were less promising. Buprofezin never reached a 100% control level for all nymphal instars. A combination with piperonyl butoxide was more efficient, but fourth-instar larvae were partly unaffected (maximum 88% mortality). Dhouibi (1992) mentioned that buprofezin was able to control *Parabemisia myricae* Kuw. and *A. floccosus* at a dose of 125 mg a.i./l. The greenhouse whitefly is a serious pest in tamarillo (*Cyphomandra betacea*) crops in New Zealand. Field tests with buprofezin in this crop revealed that the compound gave similar control to the usually used deltamethrin/oil combination. However, ten applications were necessary and the control of the insects was not complete (Blank et al. 1991).

3.3
Scales and Mealybugs

The Californian red scale *Aonidiella aurantii* (Mask.) (Hom.: Diaspididae) and the Mediterranean black scale *Saissetia oleae* (Olivier) (Hom.: Coccidae) are major pests in most citrus-growing areas of the world. LC_{50}s for *A. aurantii* were 0.13 and 0.14 mg a.i./l and for *S. oleae* 0.008 and 0.031 mg a.i./l, for first and second instars, respectively. Buprofezin had a moderate effect on male pupae of *A. aurantii*, with an LC_{50} of 93 mg a.i./l. Females of both scales were not significantly affected by buprofezin. At 100 mg a.i./l, buprofezin suppressed embryogenesis of both scale species, resulting in a significant reduction in production of *A. aurantii* crawlers and in a strong suppression of egg viability of *S. oleae*. The effect on *S. oleae* embryogenesis was associated with length of exposure of females to buprofezin before oviposition. Buprofezin was quite persistent, with the RLT_{50}s (residual lethal time of which potency is reduced to 50% of the original value) of 125 mg a.i./l under indoor and outdoor conditions, 87.9 and 33.5 days, respectively (Yarom et al. 1988; Ishaaya et al. 1989).

Wettable powder (WP) formulation applied as a single spray of 125 mg a.i./l or as two sprays of 62.5 mg a.i./l each gave insufficient control of *A. aurantii*; addition of 0.5% mineral oil considerably increased the potency of this formulation. The EC (emulsifiable concentrate) formulation was more potent and gave a good control of the scales; addition of 0.5% mineral oil further increased the performance of the EC (Ishaaya et al. 1992). A combination of buprofezin (WP 50%) and citrex, an oil with a narrow distillation range, was used by Grout and Richard (1991) to determine whether variation in treatment efficacy was a direct result of timing of application. Treatments were applied at crawler emergence and midway between two peaks of crawler emergence (interemergence) in populations with either two or three cohorts. Crawler emergence treatments were significantly more efficacious than interemergence treatments only in dual-cohort populations during winter, when the intercohort periods were longest. In populations with three cohorts, application timing did not significantly influence treatment efficacy, probably because the time between crawler emergence and interemergence was shorter and within the effective residual period for buprofezin plus oil.

Research of Mendel et al. (1991) proved that buprofezin effectively reduced egg hatch and development of larval stages of the cottony cushion scale *Icerya purchasi* (Maskell) (Hom.: Margarodidae) and the citrus mealybug *Planococcus citri* (Risso) (Hom.: Pseudococcidae) at 0.05% a.i./l. The overall effectiveness against *I. purchasi* was better as compared with *P. citri*. Du Toit and de Villiers (1990) found that buprofezin had activity on second-instar larvae of *Protopulvinaria pyriformis* (Cockerell) (Hom.: Coccidae) and on first and second instars of *Cornuaspis beckii* (Newman) and *Carulaspis japonicus* Green (Hom.: Coccidae) at 500 mg a.i./l. Difficulties in controlling resistant mealybugs *Pseudococcus affinis* (Hom.: Pseudococcidae) with chlorpyrifos in apple and pear orchards in New Zealand could be solved with buprofezin (Walker et al. 1993). Another pest in New Zealand (and elsewhere) is the Latania scale *Hemiberlesia lataniae* (Signoret) (Hom.: Diaspididae) on avocados. The extra complication in avocado rearing is the presence of mature fruit on the tree while the new season's crop is set. Protection during this period requires bee-safe material to prevent scale settlement on mature fruit. Buprofezin seemed to be a suitable candidate in this respect. The compound did not prevent settlement of the scales but appeared to act as a synergist when combined with fluvalinate, increasing the protection to 97% as compared with 47% when buprofezin is used on its own after 19 days (Blank et al. 1993).

Kanno et al. (1981) reported an excellent activity (100% mortality in second instars) against the arrowhead scale *Unaspis yanonensis* (Kuwana) (Hom.: Diaspididae) and the Comstock mealybug *Pseudococcus comstocki* (Kuwana), (Hom.: Pseudococcidae) until 69 and 14 days after treatment, respectively. Buprofezin is further recommended against *Pseudococcus maritimus* (Ehrhorn) (Hom.: Pseudococcidae), *Ceroplastes ceriferus* (Fabricius) (Hom.: Coccidae) and *C. rubens* (Maskell), *Chloropulvinaria aurantii* (Cockerell) (Hom.: Coccidae), *Pseudaulacaspis pentagona* (Targioni Tozzetti) (Hom.: Diaspididae) and *Unaspis citri* (Comstock) (Hom.: Diaspididae) (Anonymous 1985, 1987).

In general, the effect of buprofezin on scales and mealybugs can be summarized as follows: in the case of treated larvae, a delayed larvicidal action occurs, and in the case of treated adults there is no direct toxicity but a decrease in progeny formation; in the case of treated eggs a direct ovicidal action occurs.

3.4 Psyllidae

Activity has also been reported against Psyllidae. Buprofezin was compared with flufenoxuron, triflumuron and hexaflumuron by registering mortality in the progeny of treated citrus psylla *Trioza erytreae* (Del Guercia) (Hom.: Psyllidae), a vector of the organism causing greening disease in citrus in South Africa. Flufenoxuron was the most active compound; the activity of buprofezin decreased after 12 days. The effect on Psyllidae was similar to that on other insects: no adult mortality, high larval mortality and failure of egg hatch (Claassens 1993).

3.5 Others

Forty percent of second-instar nymphs of *Dysdercus cingulatus* F. (Het.: Pyrrhocoridae) were killed within 4 days and the others almost totally retarded in development, leading to death during moult (Fonagy and Darvas 1988). The technical bulletin of Nihon Nohyaku mentions activity against Eriophyidae and Tarsonemidae (Acarina) (Anonymous 1985, 1987).

4 Residues

Very few reports exist on buprofezin residues in crops. The limits of detection of buprofezin with high performance liquid chromatography (HPLC) in water and in rice plants are 0.002 mg/kg, and 0.005 mg/kg, respectively. Average recoveries varied between 72 and 91%. Rice plants gradually took buprofezin from the water and raised their concentration, the bioaccumulation coefficient being 2.4–3.4, indicating a limited transport in the plants (Bae er al. 1994; Shuzhao et al. 1994).

Valverde-Garcia et al. (1993) assessed three different multiresidue methods (the Mills, Lucke and Leary methods) for the extraction of buprofezin residues from peppers, bean and egg plants. Recoveries of >81 ± 11% were obtained in all instances, but the Leary method was the most efficient. The degradation rate constant and the half-life period obtained for buprofezin in egg plants grown in a commercial greenhouse in Almeria, Spain were, respectively, 0.15 day^{-1} and 4.6 days based on the assumption of a pseudo first order degradation behaviour. In a later study, Valverde-Garcia and Fernandez-Alba (1994) used a gas chromatographic/mass selective detection method with an average recovery of 78 ± 10% on cucumber, pepper, tomato, squash and egg plant. The mean recoveries were less than those obtained with the previously mentioned methods. The limit of quantitation of the analytical methods was 0.01 mg/kg.

Chemical residues on cotton leaves decayed at different rates and apparently with different kinetics at 62.5 and 250 mg a.i./l; the decay of the highest concentration followed a first order kinetics, whereas that of the low concentration seemed to agree with biphasic first order kinetics. Treatment with the highest concentration initially left a deposit that was 2.8 times higher and decayed 5.6 times slower than the lower concentration. According to the half-life term, which usually characterizes pesticide longevity, the low concentration possesses a value ($T_{1/2}$ = 2.27 days) typical of relatively short-lived insecticides whereas the high concentration value ($T_{1/2}$ = 12.7 days) is within the range common for the most persistent types (Willis and McDowel 1987). The applied foliar buprofezin was divided into several fractions. Of the initial deposit, 60–70% is surface residue (dislodgeable) and is susceptible to weathering processes. Apart from washing by rain, volatility is the main route for surface decay. Volatilization is strongly influenced by the deposit density and the particle size which in turn are affected by the concentration used. The other fractions are resistant to washing by rain and consist of a small proportion of penetrated residue (± 4%) and a large fraction (25–30%), the exact nature of which is not known. However, there is some evidence that this fraction is absorbed to the leaf wax layer or to the residual inert ingredient of the formulation. The non-washable fraction is probably a major factor responsible for the persistence of buprofezin, whereas the dislodgeable fraction may contribute to the vapour phase toxicity of buprofezin observed in this study. The small amount that penetrates into the leaf accounts for the moderate translaminar effect of buprofezin (De Cock et al. 1990).

5 Environmental Fate

The degradation ability of buprofezin in soils has been studied by Funayama et al. (1986) in flooded and upland soils under laboratory conditions. Buprofezin gradually decomposed with half-life times of 104 and 80 days for flooded and upland soils, respectively. After 150 days, five degradation products were identified by thin-layer chromatography: 2-*tert*-butylimino-5-(4-hydroxyphenyl)-3-isopropyl-perhydro-1,3,5-thiadiazin-4-one; 3-isopropyl-5-phenyl-phenyl-perhydro-1,3,5-thiadiazin-2,4-dione; 1-*tert*-butyl-3-isopropyl-5-phenyl-biuret; 1-isopropyl-3-phenylurea and phenylurea. As minor products, 2-*tert*-butylimino-5-phenyl-perhydro-1,3,5-thiadiazin-4-one or buprofezin sulfoxide were found in the soils. [^{14}C]Carbondioxide and bound ^{14}C residue accounted for 23-24% and 13–21% of the applied radioactivity, respectively. The degradation of buprofezin was markedly delayed in sterile soils. Since neither formation of $^{14}CO_2$ nor ring hydroxylation was observed in the sterile soils under both conditions, the authors assume that the degradation appears mostly through biological transformation by soil microorganisms.

6 Resistance

In the case of buprofezin, data concerning resistance have been reported since

1994. A buprofezin-resistant strain of *T. vaporariorum* was found in a commercial tomato crop in northwestern Belgium where it was detected after 24 treatments with buprofezin during 4 years. LC_{50} values could not be determined, with concentrations varying between 5 and 333 mg a.i. buprofezin/l (De Cock et al. 1995). A repetition of the tests after 6 months without insecticide treatment did not change these results. The strain was used in a study for cross-resistance between buprofezin and diafenthiuron or pyriproxyfen. No such cross-resistance could be demonstrated. First instars of both the buprofezin resistant (R) and susceptible strain had an equal susceptibility to diafenthiuron, whereas adults of the R strain may have had a slightly higher tolerance. For pyriproxyfen, four- to sixfold tolerance was indicated (based on LC_{50} and LC_{90} values) for egg hatch suppression (De Cock et al. 1995).

A decrease in susceptibility of whiteflies to buprofezin after multiple sprayings was described also by Horowitz and Ishaaya (1992), who reported a threefold increase in LC_{90} of buprofezin for *B. tabaci* through the course of a cotton season. The same authors reported a four- to fivefold increase in tolerance to buprofezin after two applications of buprofezin in greenhouses (Horowitz and Ishaaya 1994). Cahill et al. (1996) calculated baseline data for *B. tabaci* using a leaf-dip assay. The baseline LC_{50} (0.53 mg/kg buprofezin) was compared with field collections from Japan, Pakistan, Spain, Sudan, Mexico, the Netherlands and the UK. Populations known or suspected to have been intensively selected with buprofezin showed substantial tolerance to the insecticide, both at the LC_{50} and at specific diagnostic doses. The authors underline the threat that arises when resistance genes are transported via the ornamental plant trade into remote areas.

7 Toxicity to Non-Target Organisms

Buprofezin shows low toxicity to mammals (acute oral LD_{50} for rats: 8140 mg/kg)) and fish (LC_{50} (48 h) carps: 2–10 mg/l). Because the compound is used in integrated pest management (IPM) systems, a great deal of literature is devoted to its effect on beneficial organisms. Stolz (1990) and Blumel and Stolz (1993) evaluated the side effects of buprofezin on the predatory mite *Phytoseiulus persimilis* Athias-Henriot (Acarina: Phytoseidae) following the International Organization for Biological and Integrated Control of Noxious Animals and Plants (IOBC) guidelines (Hassan 1992). These guidelines utilize four evaluation categories for laboratory trials: (1) harmless (< 50% mortality); (2) slightly harmful (50–79% mortality); (3) moderately harmful (80–99% mortality); (4) harmful (>99%). Since the pesticide did not reveal any adverse effect on the mites and their reproductive capacity, it was ranked 'harmless'. Kanno et al. (1981) reported that buprofezin was safe for *Amblyseius longispinosus* Evans (Acarina: Phytoseidae).

Garrido et al. (1984) followed the IOBC guidelines to evaluate the effect of buprofezin on *Encarsia formosa* Gahan (Hym.: Aphelinidae) and *Cales noacki* Howard (Hym.: Aphelinidae) two aphelinid parasites of the whiteflies *T. vaporariorum* and *A. floccosus.* Although 10% mortality occurred at 125 and 250 mg a.i./l for *C. noacki*, the compound was still classified 'harmless'. No mortality occurred for *E. formosa*. Castañer et al. (1989) reported the same outcome for the combination buprofezin-*C. noacki*. They added that a combination of buprofezin and piperonyl butoxide caused 38% mortality (Castañer et al. 1989). Martin and Workman (1986) also executed field trials to evaluate the effect of buprofezin on *E. formosa*. A low concentration (25 mg a.i./l; 10% of the recommended field rate) was sprayed in a greenhouse with a high population of whiteflies (*T. vaporariorum*) and *Encarsia* wasps. It substantially reduced whitefly population but allowed some to survive and support the parasite population. The same idea was successfully applied in glasshouse crops and Bouvardia culture (Wilson and Anema 1988; Blümel 1990). Dhouibi (1992) extended this result to *Encarsia transvena* Timberlake (Hym.: Aphelinidae). Gerling and Sinai (1994) studied the susceptibility of *Encarsia luteola* Howard (Hym.: Aphelinidae) and *Eretmocerus* sp., two parasitoids of *B. tabaci*. They observed a 40% decrease in emergence success in *Eretmocerus* sp., this in contrast to *E. luteola* where such a response was not observed. However, *Eretmocerus* was unaffected when emerged from *B. tabaci* treated as pupa, while emergence of *E. luteola* was reduced by 77% after this kind of treatment. The rate of parasite oviposition was unaffected by buprofezin. In ornamentals, insect-free plants are required, which hampers the implementation of biological control. However, some examples exist of the combination of biological and selective chemical compounds. Stenseth (1993) promoted a system in which buprofezin and *E. formosa* complemented each other to control whiteflies. The wasp was used to keep mother plants clean and the cuttings were dipped in buprofezin before planting.

Buprofezin severely affected larvae of *Rodolia cardinalis* Mulsant (Col.: Coccinellidae) and prevented egg hatch of *Chilocorus bipustulatus* L. (Col.: Coccinellidae) (Mendel et al. 1994). Smith and Papacek (1990) could not prove adverse effects on adults, but observed an increased mortality of juveniles in *Cryptolaemus montrouzieri* Mulsant, another coccinellid. Detection of the effect on the predation of the ladybird *Orcus chalybeus* (*Halmus chalybeus*) (Boisduval) (Col.: Coccinellidae) on soft wax scales *Ceroplastes destructor* Newstead revealed that the predation of the adults was not affected. The latter study, however, did not include effects on larvae, which is most important when studying IGRs such as buprofezin (Lo and Blank 1992). Gravena et al. (1992) proved that buprofezin was only moderately toxic to *Pentilia egena* (Muls.) (Col.: Coccinellidae); a 36% larval reduction occurred. A recent study in southern Africa proved that the extensive use of IGRs for the control of *A. aurantii* led to extensive disruption of the biocontrol of *I. purchasi* provided by the coccinellids *R. cardinalis* and other indigenous *Rodolia* sp. The adverse effects of field-weathered residues of insect growth regulators were also studied, and revealed a nearly complete failure of egg hatch when adults were exposed to 3-week-old residues of buprofezin, but 7-week-old residues no longer significantly reduced egg hatch (Hattingh and Tate 1995). This reveals an often neglected part of side effect studies. A harmless reduction of a beneficial, in the sense that the population cannot be built up soon enough again after treatment, might have devastating effects on a biological control programme that relies on these beneficials.

Direct and residual toxicity of buprofezin tested at commercial concentrations on adults of *Oligota* sp. (Col.: Staphylinidae), important predators of the citrus red mite *Panonychus citri* McGregor (Acarina: Tetranychidae), showed no toxic effects (Gyoutoku and Kasio 1990). Carvalho et al. (1994) investigated the effect of buprofezin on parasitism and development of *Trichogramma pretiosum* Riley (Hym.: Trichogrammatidae) without detecting negative effects.

No adverse effects were reported on adult survival and daily oviposition of *Ceraeochrysa cubana* (Hagen) (Neur.: Chrysopidae) (Carvalho et al. 1994). The same was true for *Compariella bifasciata* (Howard) (Hym.: Encyrtidae), *Encyrtus infelix* Embleton (Hym.: Encyrtidae) and *Elatophilus hebraicus* Pericart (Het.: Anthocoridae). A slight effect was observed on immature stages of *Cryptochaetum iceryae* Williston (Dip.: Cryptochaetidae) and none of the larvae of *R. cardinalis* (Col.: Coccinellidae) developed into adults after application of buprofezin. Buprofezin is considered safe for the carabid beetle *Poecilus cupreus* (Col.: Carabidae) (Abdelgader and Heimbach 1992).

Kanaoka et al. (1994) stated little (maximum 20% mortality) or no effect after laboratory studies on *Microvelia horvathi* (Het.: Veliidae), *M. atrolineata* Bergroth, *Gerris paludum insularis* Fabricius (Het.: Gerridae), *Cyrtorhinus lividipennis* Reuter (Het.: Miridae) and *Lycosa pseudoannulata* Boes et Str. (Aranea: Lycosidae), all predators, and *Paracentrobia andoi* (Hym.: Trichogrammatidae) and *Hapalogonatopus* sp., both parasitoids of *N. lugens* and *N. cincticeps* in rice fields. They obtained the same results for *Sigara substriata* (Fieb.) (Het.: Corixidae), *Luciola cruciata* Motschutsky (Col.: Lampyridae) and *Hydaticus grammicus* Germ. (Col.: Dytiscidae).

Van de Veire and Vacante (1988) monitored parasitism of the leafminer *Liriomyza trifolii* (Burgess) (Dip.: Agromyzidae) by the ectoparasites *Diglyphus isaea* (Walker) (Hym.: Eulophidae) and *Dacnusa sibirica* Telenga (Hym.: Braconidae) and reported a complete compatibility with a buprofezin treatment. Beitia et al. (1991) generally confirmed these results but mentioned a 23% mortality in eggs of *D. isaea*. Gravena et al. (1992) showed that buprofezin was harmless for chrysopids.

Percentage parasitism of *A. aurantii* by *Compariella bifasciata* Howard (Hym.: Encyrtidae) and *Aphytis* sp. (Hym.: Aphelinidae) remained the same after buprofezin treatment (Ishaaya et al. 1992). Buprofezin had only slightly adverse effects on the parasitoids *Aphytis mytilaspidis* (Le Baron) and *Aphytis lepidosaphes* Compere (Darvas et al. 1994). Nagai (1991) proved the compound to be safe for *Orius* spp. (Het.: Anthocoridae) in an IPM programme for eggplants (*Solanum melogena* L.)

Although chitin synthesis inhibitors interfere with a specific metabolic pathway and are selective in this respect, the effect of these compounds on natural enemies is often not neglectable. Results of toxicity tests are highly dependent on test procedures, life stage and species tested. In the case of buprofezin, most known beneficials seem to be spared. The major exception to be concerned are the Coleoptera, especially the Coccinellidae, of which the juveniles are highly vulnerable to buprofezin.

8 Conclusions

Buprofezin combines excellent properties to control its target insects (scales, whiteflies, planthoppers and leafhoppers) with a favourable selectivity towards beneficials. Unfortunately, its target insects are those known for quick development of resistance. To prevent such an evolution it is recommended to use the compound as a support for biological control in integrated pest management programmes or to develop insect resistance management programmes to promote alternation of insecticides and exposure of only one pest generation per year to a specific insecticide. These factors may help in preventing resistance development and/ or diluting the (possible) resistant genotypes. The compound manifests its effect on Homoptera, Coleoptera and some Acarina. Buprofezin is considered a relatively safe insecticide to most parasitoids currently used on a commercial basis belonging to Heteroptera and Hymenoptera. Buprofezin should be used on young immature stages, especially first and Second instars; this will ensure a reduction of the pest population with little effect on the parasitoids because adult parasitoids will survive and probably will be able to keep the remaining pest population under control with little need for further intervention.

References

Abdelgader H, Heimbach U (1992) The effect of some insect growth regulators (IGRs) on first.instar larvae of the carabid beetle *Poecilus cupreus* (Coleoptera: Carabidae) using different application methods. Aspects Appl Biol 31: 171–177

Anonymus (1985) Technical information. Applaud 25 WP on citrus scales and mealybugs. Nihon Nohyaku, Tokyo.

Anonymus (1987) Applaud, new pesticide (insect growth regulator), technical information. Nihon Nohyaku, Tokyo

Asai A, Fukada M, Maekawa S, Ikeda K, Kanno H (1983) Studies on the mode of action of buprofezin. I. Nymphicidal and ovicidal activities on the brown rice planthopper *Nilaparvata lugens* Stål. (Homoptera: Delphacidae). Appl Entomol Zool 18: 550–552

Asai A, Kajihara O, Fukada M, Maekawa S (1985) Studies on the mode of action of buprofezin. II. Effects on reproduction of the brown planthopper *Nilaparvata lugens* Stål. (Homoptera: Delphacidae). Appl Entonol Zool 20: 111–117

Bae YH, Lee JH, Hyun JS (1994) A systematic application of insecticides to manage early season insect pests and migratory planthoppers on rice. Korean J Appl Entomol 33: 270–280

Beitia F, Garrido A, Castañer M (1991) Mortality produced by various pesticides applied to eggs of *Diglyphus isaea* (Walker) (Hym: Eulophidae) in laboratory tests. Ann Appl Biol 118: 16–17

Blank RH, Dawson TE, Richardson AC (1991) A comparison of buprofezin and deltamethrin/ oil for control of greenhouse whitefly on Tamarillo. Proc 44th NZ Weed and Pest Control Conf 1991, pp 237–241

Blank RH, Olson MH, Clark JB, Gill GSC (1993) Investigating two bee-safe materials for controlling Latania scale on avocados during pollination. Proc 46th NZ Plant Prot Conf 1993, pp 80–85

Blümel S (1990) Nützlingseinsatz zur integretierten Schädlingsbekämpfung in Bouvardien und Rosen unter Glas. Pflanzenschutzberichte 51: 25–35

Blümel S, Stolz M (1993) Investigations on the effect of insect growth regulators and inhibitors on the predatory mite *Phytoseiulus persimilis* A.H. with particular emphasis on cyromazine. J. Plant Dis Prot 100: 150–154

Boukadido R, Michelakis S (1993) Biological and integrated methods in the control of the whitefly *Trialeurodes vaporariorum* on greenhouse tomato. IOBC WPRS Bull 16(2): 17–18

Cahill M, Jarvis W, Gorman K, Denholm I (1996) Resolution of baseline responses and documentation of resistance to buprofezin in *Bemisia tabaci* (Homoptera: Aleyrodidae). Bull Entonol Res 86: 117–122

Carvalho GA, Salgado LO, Rigitano RLO, Velloso AHP (1994) Efeitos de compostos reguladores de crescimento de insetos sobre adultos de *Ceraeochrysa cubana* (Hagen) (Neuroptera: Chrysopidae). An. Soc Entomol Bras pp 335–339

Castañer M, Garrido A, Del Busto T, Malagon J (1989) Efecto de diversos insecticidas en laboratorio sobre la mortalidad de los estados inmaduros de la mosca blanca algodonosa *Aleurothrixus floccosus* (Mask.) e incidencia sobre el insecto util *Cales noacki* How. Invest Agraria: Prod. Prot. Veg. 4: 413–427

Claassens VE, Van Den Berg MA, Sutherland B (1993) Effect of four chitin synthesis inhibitors on adult citrus psylla, *Trioza erytreae* (Del Guercio) (Hemiptera:Triozidae). J S Afr Hortic Sci 3: 43–45

Darvas B, Abd EL-Kareim AI, Camporese P, Farag AI, Matolcsy G, Ujvary I (1994) Effects of some proinsecticide type, fenoxycarb derivates and related compounds on some scale insects and their hymenopterous parasitoids. J. Appl. Entomol. 118: 51–58

De Cock A, Degheele D (1991) Effects of buprofezin on the ultrastructure of the third instar cuticule of *Trialeurodes vaporariorum*. Tissue Cell 23: 755–762.

De Cock A, Degheele D (1993) Cytochemical demonstration of chitin incorporation in the cuticle of *Trialeurodes vaporariorum* after buprofezin treatment. Int J. Insect Morphol Embryol. 22: 119–125

De Cock A, Ishaaya I, Degheele D, Veierov D (1990) Vapor pressure toxicity and persistence of buprofezin under greenhouse conditions for controlling the sweetpotato whitefly *Bemisia tabaci*. J. Econ. Entomol. 83: 1254–1260.

De Cock A, Ishaaya I, Van de Veire M, Degheele D (1995) Response of buprofezin resistant-strains of the greenhouse whitefly *Trialeurodes vaporariorum* to pyriproxyfen and diafenthiuron. J. Econ. Entomol. 88: 763–767

Dhouibi MH (1992) Efficacité de certains produits chimiques à l'égard des mouches blanches: *Aleurothrixus floccosus* Mask. et *Parabemisia myricae* Kuw (Hom. Aleyrodidae) et leur impact sur les auxiliaires *Cales noacki* Howard et *Encarsia transvena* Timberlake (Hym. Aphelinidae). Meded. Fac. Landbouwwet (Rijksuniv Gent) 57: 493–504

Du Toit WJ, De Villiers EA (1990) Effects of insect growth regulators on the development of heart-shaped scale (Hemiptera: Coccidae) on avocados in South Africa. Ann. Appl. Biol. 11: 1–2

Fonagy A, Darvas B (1988) Moulting disturbances in *Dysdercus cingulatus* larvae following diflubenzuron or buprofezin treatment. In: Sehnal F, Zabza A, Denlinger D (eds). Endocrinological frontiers in physiological insect ecology. Wroclaw Technical University Press, Wroclaw, pp. 587–591

Funayama S, Uchida M, Kanno H, Tsuchiya K (1986) Degradation of buprofezin in flooded and upland soils under laboratory conditions. J. Pestic. Sci. 11: 605–610

Garrido A, Beita F, Gruenholz P (1984) Effects of PP618 on immature stages of *Encarsia formosa* and *Cales noacki* (Hymenoptera: Aphelinidae). Proc Brit Crop Prot Conf–Pests and Diseases, Brighton, 1984, vol 4A-12, pp 305–310

Gerling D, Sinai P (1994) Buprofezin effects on two parasitoid species of whitefly (Homoptera: Aleyrodiae). J. Econ. Entomol. 87: 842–846

Gravena S, Fernandes CD, Santos AC (1992) Efeito de buprofezin e abamectin sobre *Pentilia egena* (Muls) e crispidos em citros. Anais de Soc Entomol Bras 21: 215–222

Grout TG, Richards GI (1991) Effect of buprofezin applications at different phenological times on California red scale (Homoptera: Diaspididae). J. Econ. Entomol. 84: 1802–1805

Gyoutoku Y, Kasio T (1990) Toxicity of pesticides on the Oligota spp. (Coleoptera: Staphylinidae). Proc Assoc Plant Prot Kyushu 36: 155–159

Hassan SA (1992) Guidelines for testing the effects of pesticides on beneficial organisms: description of test methods. IOBC/WPRS Bull 15(3): 18–39

Hattingh V, Tate B (1995) Effects of field weathered residues of insect growth regulators on some Coccinellidae (Coleoptera) of economic importance as biocontrol agents. Bull. Entomol. Res. 85: 489–493

Hegazy G, De Cock A, Degheele D (1990) Ultrastructural changes in the cuticle of the greenhouse whitefly *Trialeurodes vaporariorum* induced by the insect growth regulator, buprofezin. Entomol Exp. Appl. 57: 299–302

Heungens A, Buysse G (1991) Control of the greenhouse whitefly *Trialeurodes vaporariorum* (Westw.) in azalea culture with buprofezin and methamidophos. Meded Fac. Landbouwwet Rijksuniv Gent 56: 1211–1215

Horowitz AR, Ishaaya I (1992) Susceptibility of the sweetpotato whitefly (Homoptera: Aleyrodidae) to buprofezin during the cotton season. J. Econ. Entomol. 85: 318–324

Horowitz AR, Ishaaya I (1994) Managing resistance to insect growth regulators in the sweetpotato whitefly (Homoptera: Aleyrodidae). J. Econ. Entomol. 87: 866–871

Ishaaya I, Mendelson Z, Melamed-Madjar V (1988) Effect of buprofezin on embryogenesis and progeny formation of the sweetpotato whitefly (Homoptera: Aleyrodidae). J. Econ. Entomol. 81: 781–784

Ishaaya I, Blumberg D, Yarom I (1989.) Buprofezin a novel IGR for controlling whiteflies and scale insects. Meded Fac. Landbouwwet Rijrsuniv Gent 54: 1003–1008

Ishaaya I, Mendel Z, Blumberg D (1992) Effect of buprofezin on California red scale *Aonidiella aurantii* (Maskell), in a citrus orchard. Isr J. Entomol. XXV-XXVI: 67–71

Izawa Y, Uchida M, Sugimoto T, Asai T (1985) Inhibition of chitin synthesis by buprofezin analogs in relation to their activity controlling *Nilaparvata lugens* Stål. Pestic. Biochem. Physiol. 24: 343–347

Izawa Y, Uchida M, Yasui M (1986) Mode of action of buprofezin on the twenty-eight spotted ladybird *Henosepilachna vigintioctopunctata* Fabricus. Agric. Biol. Chem. 50: 1369–1371

Kanaoka A, Kodama H, Yamguchi R, Konno T, Kajihara O, Maekawa S (1994) Influence of buprofezin on natural enemies and non-target insects in the paddy field. J. Pestic. Sci. 19: 309–312

Kanno H, Ikeda I, Asai T, Kaekawa S. (1981) 2-tert-butylimino-3-isopropylperhydro 1,3,5-thiadiazin-4-one (NNI-750), a new insecticide. Proc. 1981 Crop Prot. Conf. Pesticides and Diseases, pp 59–67

Kobayashi M, Uchida M, Kuriyama K (1989) Elevation of 20-hydroxy-ecdysone level by buprofezin in Nilaparvata lugens Stål nymphs. Pestic. Biochem. Physiol. 34: 9–16

Koch W (1989) Eine "neue" Weisse Fliege im Gewächshaus, Beschreibung und Bekämpfungsmöglichkeiten von *Bemisia tabaci*. Deutscher Gartenbau 14: 892–894

Konno T (1990) Buprofezin: a reliable IGR for the control of rice pests. In: Grayson BT, Green, MB, Copping LG (eds.) Pest management in rice. Elsevier Barking, pp 210–222

Laska P (1986) Test of buprofezin, cyhalothrin and other chemicals used against glasshouse whitefly. Test of Agrochemicals and Cultivars 7: 2–3

Lo PL, Blank RH (1992) Effect of pesticides on predation of soft wax scale by the steel-blue ladybird. Proc 45th NZ Plant Prot Conf, pp 99–102

Lu XK, Gong CC (1990) Experiment on the control of *Nilaparvata lugens* Stål by Applaud. Insect Knowledge 27: 269–271

Martin NA, Workman P (1986) Buprofezin: a selective pesticide for greenhouse whitefly control. Proc. 39th NZ Weed and Pest Control Conf, pp 234–236

Mendel Z, Blumberg D, Ishaaya I (1991) Effect of buprofezin on *Icerya purchasi* and *Planococcus citri*. Phytoparasitica 19: 103–112

Mendel Z, Blumberg D, Ishaaya I (1994) Effects of some insect growth regulators on natural enemies of scale insects (Hom.: Coccidae). Entomophaga 39: 199–209

Nagai K (1991) Integrated control programs for *Thrips palmi* Karny on eggplants (*Solanum melongena* L.) in an open field. Jpn. J. Appl. Entonol Zool. 35: 283–289

Nagata T (1986) Timing of buprofezin application for control of the brown planthopper, Nilaparvata lugens Stål (Homoptera: Delphacidae). Appl. Entomol Zool. 21: 357–362

Rat-Morris E (1990) Contribution à la lutte contre l'aleurode Bemisia tabaci sur poinsettia. Phytoma 421: 46–47

Retnakaran A, Wright JE (1987) Control of insect pests with benzoylphenylureas. In: Wright JE, Retnakaran A (eds), Chitin and benzoylphenylureas. Junk, Dordrecht, pp 205–282

Reynolds SE (1987) The cuticle, growth and moulting in insects: the essential background to the action of acylurea insecticides. Pestic. Sci. 20: 131–146

Shuzhao L, Yingwei Q, Zhongxin W, Zhiyi D, Yizhong Y (1994) The residue dynamics and the transportation of buprofezin in water and rice plants. Acta Phytophylacica Sin 21: 187–192

Smith D, Papacek DF (1990) Buprofezin: an effective and selective insect growth regulator against *Unaspis citri* (Hemiptera: Diaspididae) on citrus in south-east Queensland. Gen Appl. Entomol. 22: 25–29

Stenseth C (1993) Biological control of cotton whitefly *Bemisia tabaci* (Genn.) (Homoptera: Aleyrodidae) by *Encarsia formosa* (Eulophidae: Hymenoptera) on *Euphorbiae pulcherrima* and *Hypoestes phullostachya*. IOBC/WPRS Bull 16(8): 135–138

Stolz M (1990) Testing side effects of various pesticides on the predatory mite *Phytoseiulus persimilis* Athias-Henriot (Acarina: Phytoseidae) in laboratory. Pflanzenschutzberichte 51: 127–138

Tiongco ET, Cabunagan RC, Flores ZM, Hibino H, Koganezawa H (1990) Timing of insecticide treatment for rice tungro (RTV) control. Int Rice Res. Newsl. 15: 23–24

Uchida M, Asai T, Sugimoto T (1985) Inhibition of cuticle deposition and chitin biosynthesis by a new insect growth regulator buprofezin in *Nilaparvata lugens* Stål. Agric. Biol. Chem. 49: 1233–1334

Uchida M, Izawa Y, Sugimoto T (1986) Antagonistic effect of 20-hydroxy-ecdysone to an insect growth regulator, buprofezin, in *Nilaparvata lugens* Stål. Agric. Biol. Chem. 50: 1913–1915

Uchida M, Izawa Y, Sugimoto T (1987) Inhibition of prostaglandin biosynthesis and oviposition by an insect growth regulator, buprofezin, in *Nilaparvata lugens* Stål. Pestic Biochem. Physiol. 27: 71–75

Umeda K, Hirano M (1990) Inhibition of rice virus transmission by esfenvalerate and its mechanisms. Appl. Entomol Zoolonol 25: 59–65

Valverde-Garcia A, Fernandez-Alba A (1994) Analysis of buprofezin residues in vegetable crops by gas chromatography with mass selective detection in selected ion monitoring mode. J. AOAC Int 77: 1041–1046

Valverde-Garcia A, Gonzales-Pradas E, Aguilera-Del Real A (1993) Analysis of buprofezin residues in vegetables. Application to the degradation study on eggplant grown in a greenhouse. J. Agric Food. Chem. 41: 2319–2323

Van de Veire M, Vacante V (1988) Buprofezin: a powerful help to integrated control in greenhouse vegetables and ornamentals. Bol San Veg Proc Scientific Congr, Spain, Parasitis, pp 425–435

Walker JTS, White V, Charles JG (1993) Field control of chlorpyrifos-resistant mealybugs (*Pseudococcus affinis*) in a Hawkes Bay orchard. Proc 46th NZ Plant Prot Conf pp 1993, 126–128

Willis GH, McDowell LL (1987) Pesticides' persistence on foliage. Rev. Environ. Contam. Tox. 100: 23–73

Wilson D, Anema BP (1988) Development of buprofezin for control of whitefly *Trialeurodes vaporariorum* and *Bemisia tabaci* on glasshouse crops in the Netherlands and the UK. Crop Prot Conf – Pests and diseases, Brighton, 1988, vol. 3C-2, pp 175–180

Yarom I, Blumberg D, Ishaaya I (1988) Effects of buprofezin on California red scale (Homoptera: Diaspididae) and Mediterranean black scale (Homoptera: Coccidae). J. Econ. Entomol. 81: 1581–1585

Yasui M, Fukada M, Maekawa S (1985) Effects of buprofezin on different developmental stages of the greenhouse whitefly, *Trialeurodes vaporariorum* (Westwood)—(Homoptera: Aleyrodidae). Appl. Entomol Zool. 20: 340–347

Yasui M, Fukada M, Maekawa S (1987) Effects of buprofezin on reproduction of the greenhouse whitefly, Trialeurodes *vaporariorum* (West.) (Homoptera: Aleyrodidae). Appl. Entomol Zool. 22: 266–271

Yasui M, Nishimatu T, FukadaM, Maekawa S (1990) Long-term suppressive effect of buprofezin on population growth of the greenhouse whitefly *Trialeurodes vaporariorum* (Westwood) (Homoptera: Aleyrodidae). Appl. Entomol Zool. 20: 271–274

CHAPTER 6

New Perspectives on the Mode of Action of Benzoylphenyl Urea Insecticides

H. Oberlander and D.L. Silhacek
Center for Medical, Agricultural and Veterinary Entomology,
Agricultural Research Service, United States Department of Agriculture,
P.O. Box 14565, Gainesville, Florida 32604, USA

1 Introduction

Chitin is a major constituent of insect cuticle that contributes to the structural integrity of the exoskeleton. It is synthesized as part of a chitin-protein complex during the molt/intermolt cycle. The current view of chitin synthesis (see review by Retnakaran and Oberlander 1993) is that oligosaccharides are transported across the cell membrane with assistance from a dolichol carrier, while assembly involves a lectin protein. Generally, 20-hydroxyecdysone (20-OHE) is needed to program chitin synthesis which commences in the epidermis after the fall in hormone titer. Manifestly, such a system that is so critical a part of the insect exoskeleton, as well as being specific to arthropods among the animals, can provide a significant opportunity for the development of biologically rational pesticides.

Benzoylphenyl ureas (BPUs) are selective insecticides that act on insects of various orders by inhibiting chitin formation (Ishaaya and Casida 1974; Post et al. 1974), thereby affecting the elasticity and firmness of the endocuticle (Grosscurt 1978; Grosscurt and Anderson 1980). The search for more potent acylureas led to the development of new compounds such as chlorfluazuron (CFA) and teflubenzuron (TFB) which are considerably more active than the parent compound, diflubenzuron (DFB) (Fig. 1), against various agricultural pests (Ascher and Nemny 1984; Ishaaya et al. 1986, 1987). Chitin synthesis inhibitors (CSIs) have diverse actions *in vivo* on the life cycle of pest insects including ovicidal and larvicidal effects (e.g., Ascher et al., 1987). Impairment of cuticle secretion in affected embryos may be the cause of reduced hatchability as a result of treatment with DFB (Grosscurt 1978). While larvicidal effects of CSIs are most likely due to interference with formation of a new cuticle, the mechanistic basis for ovicidal effects and inhibition of hatchability of eggs is less evident. Clearly, whatever the mode of action of the CSIs may be, their potential agricultural applications are manifold (Retnakaran and Wright 1987).

Despite many studies on BPU compounds typified by DFB, their mode of action remains elusive (e.g., Binnington and Retnakaran, 1991; Cohen, 1993). Unlike antibiotics, such as the polyoxins and nikkomycins, which are *Streptomyces*-derived competitive inhibitors of chitin synthetase, BPUs do not readily inhibit chitin synthesis in cell-free systems (e.g., Cohen and Casida, 1980; Mayer et al.,

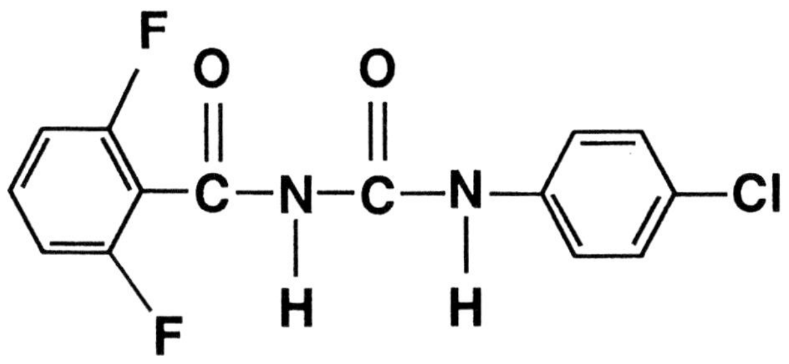

Diflubenzuron

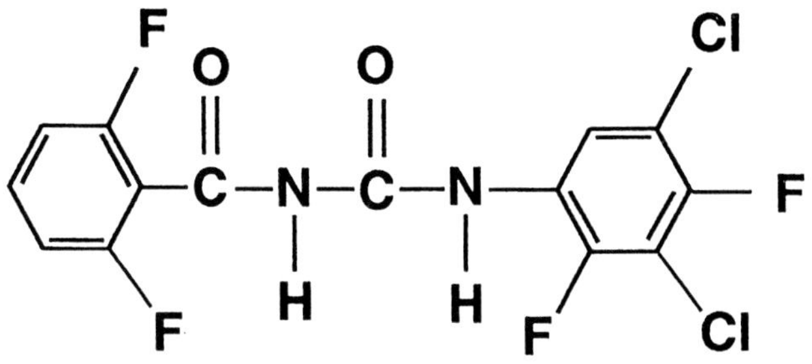

Teflubenzuron

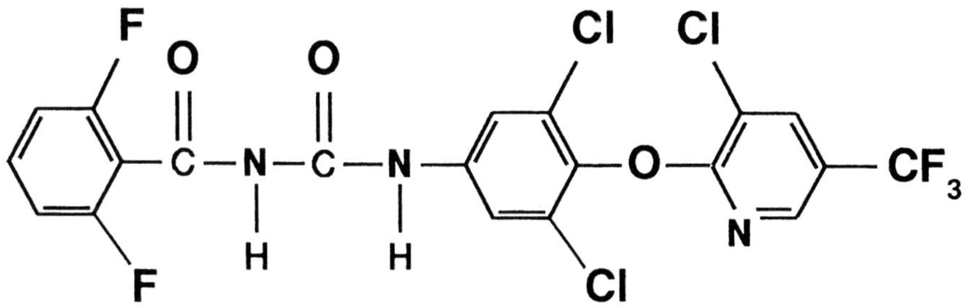

Chlorfluazuron

Fig. 1. Chemical structure of diflubenzuron (DFB), teflubenzuron (TFB) and chlorfluazuron (CFA)

1981; Ferkovich and Oberlander, unpubl. data; also reviews by Cohen, 1987; Binnington and Retnakaran, 1991), nor do they block the chitin biosynthetic pathway between glucose and UDP-GlcNAc (uridine diphosphate-N-acetyl-glucosamine) in intact larvae (Post et al., 1974). However, some authors reported limited success with cell-free assays from the sheep blowfly and the brine shrimp (Horst, 1981; Turnbull and Howells, 1983). Paradoxically, the antibiotics which inhibit chitin synthetase in both fungi and insects are ineffective for insect control, presumably because of lack of penetration through the cuticle and degradation after ingestion, while the BPUs which are promising insect control agents do not inhibit chitin synthesis in fungi *in vivo*. Manifestly, there are unique aspects to the regulation of chitin synthesis in insects that result in vulnerable points of attack by BPUs.

Some of the early hypotheses that have been put forward for the mode of action of DFB are based upon *in vivo* studies and include the following: (1) an increase in chitinase activity (Ishaaya and Casida, 1974); (2) inhibition of ecdysteroid metabolism (Yu and Terriere, 1977); and (3) proteolytic activation of a zymogenic chitin synthetase (Leighton et al., 1981). These hypotheses have been considered in detail by Cohen (1993). First of all, the rapidity of action of DFB makes it unlikely that inhibition of chitin synthesis is a secondary effect of long-term alterations of the endocrine system or of increases in chitinase levels. In addition, there is no evidence that BPUs affect the endocrine system. Finally, the rapid inhibition of chitin synthesis makes it unlikely that proteolytic activation of inactive zymogenic chitin synthetase would inhibit chitin synthesis that is already in progress and the enzyme is already in an activated state.

The general lack of activity of these CSIs in cell-free systems coupled with the apparently misleading results due to long-term effects *in vivo* make it clear that what is needed is a defined *in vitro* chitin synthesizing system that is ecdysteroid-responsive. We will consider recent developments in the use of established cell lines and organ cultures as systems for examining the action of BPUs and conclude with a discussion of recent hypotheses of inhibitor action.

2 Chitin Synthesis and Inhibition *in Vitro*

2.1 Cell Lines

Two chitin synthesizing cell lines have been described that are affected by ecdysteroids in an opposite manner. Ward et al. (1988) showed that the cockroach cell line, UMBGE spontaneously produced a base-resistant material that was degraded to chitobiose and GlcNAc. Production of this chitin-like material was stimulated by 20-OHE and inhibited by DFB. In another system, spontaneous chitin synthesis was reduced by 20-OHE in a dose-dependent manner in an epithelial cell line from *Chironomus tentans* (Spindler-Barth et al., 1989; Spindler-Barth, 1993). In this system, SIR 8514, a BPU derivative, also reduced chitin synthesis, but only in the cultured cells, not in cell-free homogenates (Londerhausen et al., 1993).

2.2 Organ Cultures

Imaginal discs of *Plodia interpunctella* have been used as a model system to study chitin synthesis and the cellular and biochemical activities necessary for its occurrence. This system was amenable to investigation with inhibitors of RNA, protein synthesis, and microtubule integrity, as well as CSIs *per se* (Oberlander et al., 1978, 1980, 1983). These experiments on isolated imaginal discs demonstrated that 20-OHE acted directly on target tissues in the absence of the endocrine system as well as other tissues. This conclusion was also reached by Ferkovich et al. (1981) and Soltani et al. (1987) based on work with cultured integument of *P. interpunctella, Galleria mellonella and Tenebrio molitor*, respectively. Based on the research on both imaginal discs and epidermis, Ferkovich et al. (1981) postulated a scheme in which 20-OHE could have its effects on chitin synthesis by stimulating the uptake of chitin precursors, the synthesis of chitin synthetase, and cuticular proteins. It became clear that these isolated hormonally responsive systems could be used to investigate the mode of action of CSIs in the absence of effects on the organism as a whole.

A detailed analysis of the dynamics of the action of BPUs on chitin synthesis was carried out with *in vitro* cultured imaginal wing discs of *Spodoptera frugiperda* (Mikólajczyk et al., 1994). In these studies, wing discs isolated from last instar larvae were exposed to CFA, DFB, or TFB at various concentrations and for different exposures during culture. Radiolabeled GlcNAc was added to the cultures for 4 h after 68 h of culture and the incorporation of GlcNAc into chitin was assessed. All three CSIs blocked ecdysteroid-dependent GlcNAc incorporation into chitin. Moreover, the effectiveness of the inhibitors was not affected by the time of application, i.e., inhibition occurred with 4 h exposures of the CSI before or at the completion of treatment with 20-OHE. Most surprisingly, culture of freshly dissected wing discs with CFA for only 15 min resulted in a 90% inhibition of chitin synthesis when measured 3 days later (Mikólajczyk et al., 1994). This is in contrast to the expectation that the effects of BPUs on chitin synthesis would be reversible (Binnington and Retnakaran, 1991). All three of these CSIs inhibited chitin synthesis in a similar dose-dependent manner *in vitro*, despite differences in effectiveness that have been observed with in vivo applications (Mikólajczyk et al., 1994).

3
Transport Hypotheses

3.1
Dolichol

Dolichol-dependent sugar transport is necessary for chitin synthesis, as demonstrated in several insect cell lines (Grosscurt and Jongsma, 1987). Tunicamycin has been used to investigate the role of dolichol-dependent transport in chitin synthesis, since it prevents the transfer of GlcNAc-1-P from UDP-GlcNAc to dolichylP, a lipid carrier. Dolichyl-pyrophosphyryl-GlcNAc then acts as an acceptor for the addition of a second GlcNAc molecule leading to the formation of a lipid-linked oligosaccharide, whose formation can be blocked by tunicamycin (Binnington and Retnakaran, 1991). Quesada-Allue (1982) showed that tunicamycin inhibited chitin synthesis in cultured epidermis of *Galleria mellonella*, and Porcheron et al. (1993) reported that it prevented incorporation of GlcNAc into a tri-chloro acetic acid precipitate from cultured IAL-PID2 cells. In addition, Londershausen et al. (1993) found that tunicamycin inhabited chitin synthesis in both a tick and *Chirnomus tentans* cell line. While there is evidence that some classes of CSIs such as polyoxin D and nikkomycin inhibit dolichol phosphate-dependent transport of chitin precursors, the BPUs have no such effect, and must inhibit chitin synthesis after the formation of lipid intermediates (Grosscurt and Jongsma, 1987). Thus, this route of inhibition cannot be postulated as an explanation for the effects of BPUs.

3.2
Precursor Uptake

Investigations of the mode of action of CSIs with established cell lines do not necessarily depend upon the ability of the cell lines to synthesize chitin, as various cellular properties can be examined directly, and then tested in a chitin synthesizing system. One such example is the suggestion that BPUs may act by promoting the transport of chitin precursors into target cells. Inhibition of precursor uptake across the cell membranes would explain why BPUs generally fail to work in cell-free systems. The precursor uptake hypothesis is supported by analogous experiments that may explain the inhibition of DNA synthesis by DFB (Mitlin et al., 1977; DeLoach et al., 1981; Soltani et al., 1984). Mayer et al. (1984) found that DFB inhibited uptake of uridine, adenosine and cytidine, but not thymidine, in murine melanoma cells. These inhibitor effects were seen within 5 min after treatment of the cells to the CSI. Similar results were obtained by Klitschka et al. (1986) with an established *Manduca sexta* cell line (CH-MRRL). Based on these results, Mayer et al. (1988) proposed that DFB and related CSIs act by affecting cellular membranes, and suggested that the observed inhibition of DNA synthesis by DFB is a result of a reduced uptake of DNA precursors by the cells.

These results on membrane effects of CSIs suggested the importance of testing directly whether BPUs would inhibit uptake of chitin precursors in target cells. This question was addressed with a cell line established from *P. interpunctella* imaginal wing discs (Lynn and Oberlander, 1983) which displayed patterns of amino sugar uptake that are similar to that of intact imaginal discs (Porcheron et al., 1988). While the effects of CSIs have been studied in other cell lines (Marks et al., 1984; Soltani et. al., 1987; Londershausen et al., 1988; Ward et al., 1988), these cell lines are not ecdysteroid-responsive as is the *P. interpunctella* imaginal disc-derived cell line (IAL-PID2). Porcheron et al. (1988) reported that the IAL-PID2 cells responded to treatment with 20-OHE with an increased uptake of GlcNAc, N-acetyl-galactosamine, and D-glucosamine, but not D-glucose or D-mannose. Moreover, the requirements for optimal hormonal stimulation of GlcNAc uptake were similar to those for uptake of this chitin precursor by intact imaginal wing discs. When this *in vitro* system was treated with TFB there was no diminution of GlcNAc uptake by the IAL-PID2 cells, while DFB had a small inhibitory effect on this process (Oberlander et al., 1991). The failure to link the inhibition of chitin synthesis by BPUs to membrane transport demonstrated the need for direct testing of this hypothesis in a more complete chitin synthesizing system.

Lepidopteran imaginal disc organ cultures have been used to test the hypothesis that inhibition of chitin synthesis by BPUs might be caused by an inhibition of uptake of GlcNAc from the culture medium. In cultured wing discs of *P. interpunctella* DFB and TFB virtually completely inhibited incorporation of radiolabeled GlcNAc into chitin. However, there was no correlation between the inhibition of chitin synthesis and the inhibition of uptake of GlcNAc or of the transported but nonmetabolized sugars, 3-O-methyl-D-glucose and 2-deoxy-D-glucose (Oberlander et al., 1991). In another organ culture system, *S. frugiperda* wing discs, DFB, and TFB, but not CFA, partially reduced uptake of GlcNAc, while all three of these CSIs completely inhibited chitin synthesis (Mikólajczyk et al., 1994; Fig. 2). Thus, there was no consistent correlation between the reduced uptake of GlcNAc and the inhibition of chitin synthesis in either *P. interpunctella* or *S. frugiperda* imaginal discs. Taken together, these experiments fail to support the hypothesis that the inhibition of uptake of chitin precursors is an adequate explanation for the action of BPUs.

3.3 Precursor Export

A new approach was taken to determining the mode of action of BPUs by Mitsui et al. (1984, 1985), who focused on a role for membrane transport in the action of DFB. These authors reported that midgut from last instar larvae of *Mamestra brassicae* could be cultured *in vitro* and tested for chitin synthesis. When the cultures were exposed to radiolabeled GlcNAc in the presence of DFB there was an accumulation of UDP-GlcNAc and an inhibition of chitin production. On the other hand, if the midguts were turned inside out and ligated at both ends and incubated with GlcNAc and UDP-GlcNAc, polyoxin D, but not DFB, inhibited chitin production. From these results the authors concluded that DFB acted by inhibiting the transport of UDP-GlcNAc across the midgut epithelium, since providing this direct chitin precursor to the exposed villi in the inverted midguts allowed chitin synthesis despite the presence of DFB. General acceptance of this interesting hypothesis would require both the isolation of the putative UDP-GlcNAc transporter, as well as demonstration of the phenomenon in an ecdysteroid-dependent chitin synthesizing system.

To provide a clear test of the transport hypothesis of Mitsui et al. (1985), an attempt was made to utilize *S. frugiperda* imaginal wing discs *in vitro* for this purpose (Mikólajczyk et al. 1995). As an analogy to the inverted midguts, it was necessary to culture imaginal wing discs in which the chitin-synthesizing surface was exposed. To do so, the wing discs were treated with a low concentration of 20-OHE which causes separation of the peripodial membrane from the wing disc and allows the peripodial sac to be physically removed. In this way, the chitin-synthesizing surface of the wing disc could be exposed to chitin precursors in the culture medium. If the wing discs operated in the same manner as the midgut, then BPUs which inhibit chitin synthesis in intact wing discs would not be effective in wing discs from which the peripodial sac had been removed and, then, the wing discs cultured in medium containing UDP-GlcNAc. These experiments resulted in the surprising finding that the imaginal wing discs would

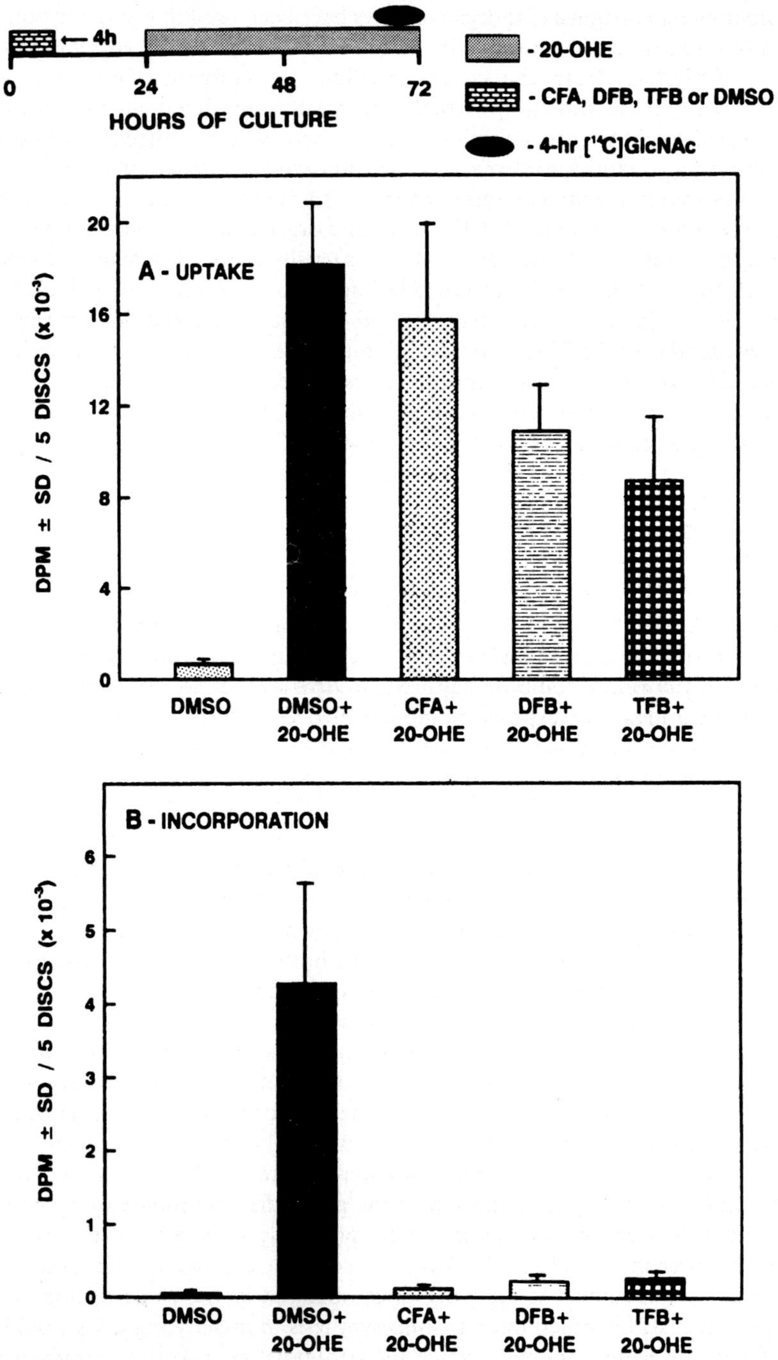

Fig. 2. Effects of a 4-h treatment of CFA, DFB, and TFB at the start of the culture period of imaginal wing discs of *Spodoptera frugiperda*. 20-OHE was added at 24 h, and uptake and incorporation of [^{14}C] GlcNAc were measured at 68 h of culture. (Mikólajczyk et al. 1994)

not synthesize chitin unless the peripodial membrane was intact (Mikólajczyk et al. 1995). Thus, it was not possible to test directly the transport hypothesis of Mitsui et al. (1985) with this system. However, the effort to do so uncovered the critical importance of the peripodial membrane for chitin synthesis in the developing imaginal wing disc.

4
Ultrastructural Analysis

The availability of a highly responsive ecdysteroid-dependent *in vitro* culture system in which the application of the inhibitor is required for only a brief period provided a unique opportunity for examining the nature of the effects of BPUs at the cellular level without any influence from the *in vivo* environment. Thus, the action of CFA was investigated *in vitro* in cultured *S. frugiperda* imaginal wing discs. The ultrastructure was examined with an electron microscope to assess the subcellular effects of the inhibitor, while cytochemical labeling with the lectin, wheat germ agglutinin, provided an opportunity to determine the localization of chitin and nonpolymerized GlcNAc (Zimowska et al., 1994). Importantly, CFA selectively inhibited 20-OHE-stimulated chitin synthesis and procuticle deposition in the wing discs. On the other hand, no differences were observed in the hormone-treated wing discs with respect to tracheole migration, evagination, exocytosis, endocytosis, or the accumulation of nonpolymerized GlcNAc in the extracellular matrix between the peripodial membrane and the wing pouch of CFA-treated wing discs as compared with control wing discs (Figs. 3, 4). Thus, CFA prevented the appearance of wheat germ agglutinin-labeled chitin in newly formed procuticle, without any effect on the deposition of proteinaceous cuticulin and epicuticle. In CFA-treated wing discs, in which chitin synthesis was inhibited, an amorphous, nonlamellar procuticle that contained nonpolymerized GlcNAc was observed (Mikólajczyk et al., 1994; Zimowska et al., 1994). It is clear that CFA blocked chitin synthesis in the cultured wing imaginal discs without any observable effects on other components of cellular activities or non-chitinous cuticle formation.

Retnakaran et al. (1989) reported that the inhibition of chitin synthesis by CFA was correlated with an accumulation of vesicles and loss of microvillae in the epithelium. This was thought to be connected with a failure to transport the chitin precursors to the site of synthesis. In related work with isolated integument from newly molted Periplaneta americana, BPUs inhibited chitin synthesis in isolated integument, but not in cell-free preparations. On the other hand, ion transport in intracellular vesicles was inhibited by DFB, and it was suggested that this action may be related to a blockage of exocytosis of vacuolar intracellular vesicles *in vivo*, which in turn results in a failure to synthesize chitin (Nakagawa et al. 1993; Nakagawa and Matsumura 1994). However, there was no apparent effect on exocytosis in *S. frugiperda* wing discs that were treated with CFA (Zimowska et al., 1994). In addition, although there was a temporary loss of microvillae, fully developed microvillae were observed in wing epithelium of CFA-treated wing discs which did not synthesize chitin. Thus, the *in vitro* experiments with imaginal wing discs indicate that CSIs can inhibit chitin synthesis

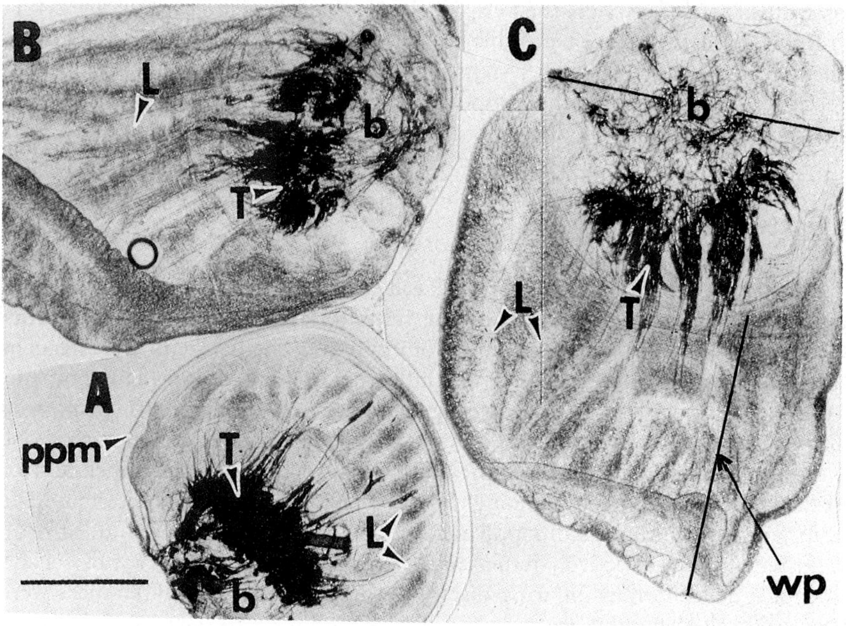

Fig. 3. General morphology of undeveloped imaginal wing discs from *Spodoptera frugiperda*. **A** Immediately after dissection. **B** After 48-h exposure *in vitro* to 20-OHE. **C** After 48-h simultaneous exposure to 20-OHE and CFA. 20-OHE treatment stimulated morphogenesis of the wing disc, which included tracheole (*T*) migration into lacunae (*L*), and evagination of the wing pouch (*wp*). During evagination, the peripodial membrane (*ppm*) slipped down and surrounded the basal portion (*b*) of the wing disc. Exposure of the wing discs to CFA did not affect their overall morphogenesis in response to 20-OHE. (Zimowska et al. 1994)

without any apparent effect on either vesicles or microvillae. On the other hand, earlier work showed that the integrity of the microtubular system was necessary for chitin synthesis in cultured wing discs of *P. interpunctella* (Oberlander et al., 1983).

5 Protein Synthesis

The necessity of protein synthesis for chitin synthesis has been documented in a number of studies. Oberlander et al. (1980) showed that inhibitors of RNA and protein synthesis blocked 20-OHE-induced chitin synthesis in cultured imaginal wing discs during the hormone-dependent period. Moreover, protein synthesis was required concurrently during the posthormone chitin synthetic period, although new RNA synthesis was no longer needed. Similar results with respect to the concurrent need of protein synthesis for the formation of chitin were demonstrated in crustaceans (Horst, 1990). The question then becomes whether CSIs might act by interfering with the synthesis of specific proteins required for chitin

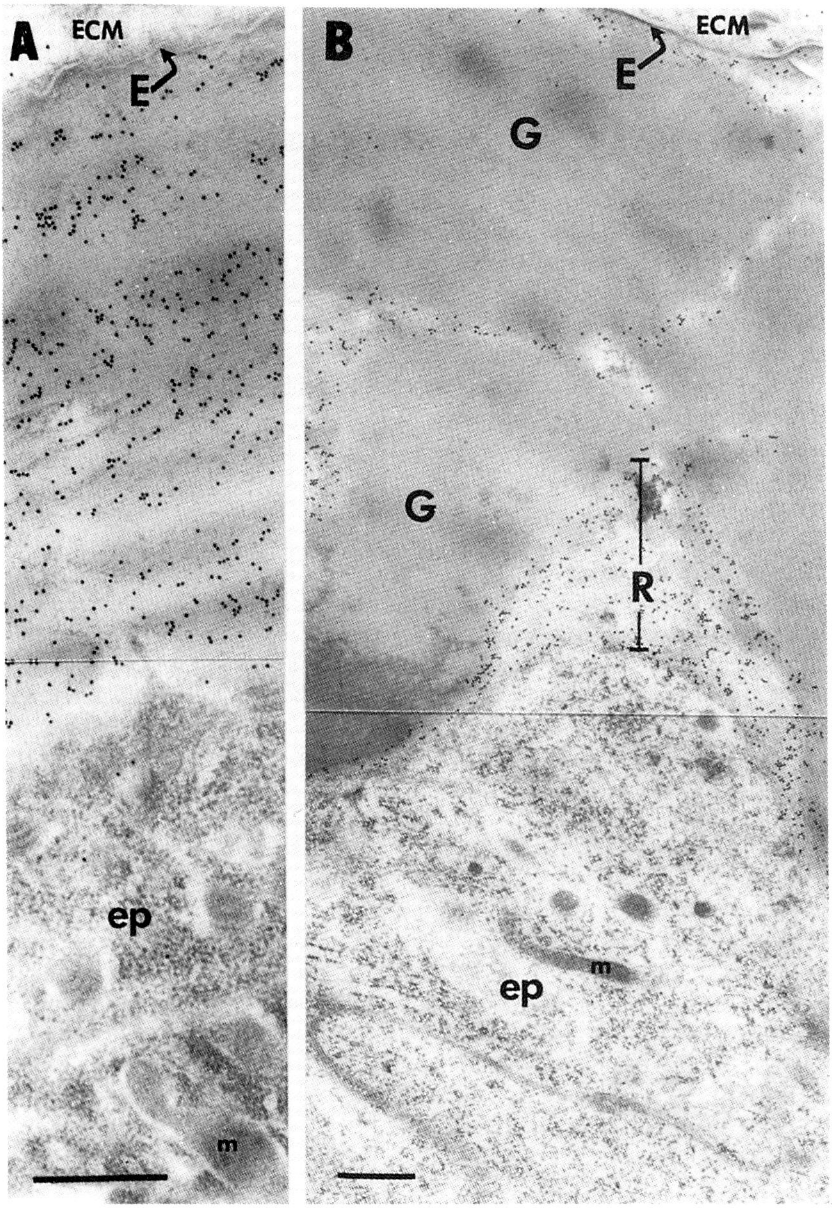

Fig. 4. Wheat germ agglutinin (WGA), a GlcNAc binding lectin, labeled transverse sections through the basal part of *Spodoptera frugiperda* wing discs cultured *in vitro*. **A** Cuticle formation was stimulated by a 24-h treatment with 20-OHE. **B** Treatment with CFA followed by a 24-h exposure to 20-OHE. 20-OHE caused deposition of a thick cuticle layer composed of lamellar procuticle (**P**), which contained chitin, based on labeling with WGA-gold. In the CFA-treated wing discs, instead of procuticle, large globular masses of proteinaceous material (**G**), which was not labeled with WGA-gold, were deposited under the epicuticle (**E**). Some WGA binding was observed in spaces among the globular masses, as well as in the regions (**R**) lying directly on the epithelium (*ep*). *ECM* Extracellular matrix; *m* mitochondrion. (Zimowska et al., 1994)

synthesis. In this connection, Porcheron et al. (1991) demonstrated that treatment of IAL-PID2 cells 20-OHE resulted in the secretion into the culture medium of a 5000 daltons N-acetyl-D-glucosamine-rich glycopeptide. A glycopeptide of similar molecular weight was observed also in intact imaginal discs treated with 20-OHE *in vitro* under conditions that induce chitin synthesis. Interestingly, treatment of the cells with TFB prevented the accumulation of this peptide. While the function of such glycopeptides *in vivo* is not known (Cassier et al., 1991), the reported observations raise the possibility of CSIs acting through interference with the production or secretion of specific proteins that are required for hormone-induced chitin synthesis. Based on our ultrastructural observations (Zimowska et al., 1994), we would expect that such a protein would most likely be involved in assembly of chitin microfibrils because imaginal discs treated with CFA and 20-OHE produced a proteinaceous, but nonlamellar, cuticle that contained nonpolymerized GlcNAc.

6
Conclusions

It is clear from the work with *in vitro* systems described above that the BPUs act specifically on chitin synthesis, and that they do not affect general cellular, or organismal activity in a way that can account for their inhibition of chitin synthesis. Based on our research on the effects of CSIs on imaginal discs, we believe that a proposed mode of action for the BPUs should account for the following: (1) CSIs inhibit chitin synthesis in cultured imaginal wing discs after only a very brief exposure whether it is applied before, during or after treatment with 20-OHE; (2) CSIs block chitin synthesis in wing imaginal discs, though cuticular protein deposition continues, and a nonlamellar cuticular layer is produced that contains nonpolymerized GlcNAc; (3) the availability of precursors for chitin synthesis. The evidence points to the likelihood of CSI interference with a lectin-like acceptor protein that promotes the formation of chitinous microfibrils. In this connection, Cohen (1993) postulated that translocation of chitin polymers across the cell membrane may be the site of action of BPUs. Thus, at this stage of our knowledge the most likely working hypothesis is that CSIs act by interrupting the synthesis and/or transport of specific proteins that are required for the assembly of GlcNAc monomers into polymeric chitin.

Acknowledgments

The authors wish to thank Drs. S. Ferkovich and, P. Shirk (United States Department of Agriculture) and G. Zimowska (University of Florida) for reviewing this manuscript. We also thank Wiley-Liss Inc. for permission to reprint Fig. 2-4.

References

Ascher KRS, Nemny NE (1984) The effect of CME 134 on *Spodoptera littoralis* eggs and larvae. Phytoparasitica 12: 13–27

Ascher KRS, Melamed-Madjar V, Nemny NE, Tam S (1987) The effect of benzoyl-phenyl urea molting inhibitors on larvae and eggs of the European corn borer, *Ostrini nubilalis* J Plant Dis Prot 94: 584–589

Binnington K, Retnakaran A(1991) Epidermis—a biologically active target for metabolic inhibitors. In: Binnington K, Retnakaran A (eds) Physiology of the insect epidermis. CSIRO, Melbourne, pp 307–326

Cassier P, Serrant P, Garcia R, Coudouel N, André M, Guillaumin D, Porcheron P, Oberlander H (1991) Morphological and cytochemical studies of the effects of ecdysteroids in a lepidopteran cell line (IAL-PID2). Cell Tissue Res 265: 361–369

Cohen E (1987) Chitin biochemistry: synthesis and inhibition. Annu Rev Entomol 322: 71–93

Cohen E (1993) Chitin synthesis and degradation as targets for pesticide action. Arch Insect Biochem Physiol 22: 245–261

Cohen E, Casida JE (1980) Properties of *Tribolium* gut chitin synthetase. Pestic Biochem Physiol 13: 121–128

DeLoach JR, Meola SM, Mayer RT, Thompson JM (1981) Inhibition of DNA synthesis by diflubenzuron in pupae of the stable fly *Stomoxys calcitrans* (L). Pestic Biochem Physiol 15: 172–180

Ferkovich SM, Oberlander H, Leach CE (1981) Chitin synthesis in larval and pupal epidermis of the Indian meal moth, *Plodia interpunctella* (Hübner), and the greater wax moth, *Galleria mellonella* (L.). J Insect Physiol 27: 509–514

Grosscurt AC (1978) Effect of diflubenzuron on mechanical penetrability, chitin formation, and structure of the elytra of *Leptinotarsa decemlineata*. J Insect Physiol 24: 827–831

Grosscurt AC, Anderson SO (1980) Effect of diflubenzuron on some chemical and mechanical properties of the elytra of *Leptinotarsa decemlineata*. Proc K Ned Akad Wet vol 83, pp 143-150

Grosscurt AC, Jongsma B (1987) Mode of action and insecticidal properties of diflubenzuron. In: Wright JE and Retnakaran A (eds) Chitin and benzoylphenyl ureas. Junk Dordrecht, pp 75–100

Horst MN (1981) The biosynthesis of crustacean chitin by a microsomal enzyme from larval brine shrimp. J Biol Chem 256: 1412–19

Horst MN (1990) Concurrent protein synthesis is required for *in vivo* chitin synthesis in postmolt blue crabs. J Exp Zool 256: 242–254

Ishaaya I, Casida JE (1974) Dietary TH 6040 alters composition and enzyme activity of housefly larval cuticle. Pestic Biochem Physiol 4: 484–490

Ishaaya I, Navon A, Gurevitz E (1986) Comparative toxicity of chlorfluazuron (IKI-7899) and cypermethrin to *Spodoptera littoralis, Lobesia botrana* and *Drosophila melanogaster*. Crop Prot 5: 385–388

Ishaaya I, Yablonski S, Ascher KRS (1987) Toxicological and biochemical aspects of novel acylureas on resistant and susceptible strains of *Tribolium castaneum*. In: Donahaye E, Navarro S (eds) Proc 4th Int Working Conf on Stored-product Protection Capsit, Jerusalem, pp 613–622

Klitschka GE, Mayer RT, Droleskey RE, Norman JO, Chen AC (1986) Effects of chitin synthesis inhibitors on incorporation of nucleosides into DNA and RNA in a cell line from *Manduca sexta* (L). Toxicology 39: 307–315

Leighton T, Marks E, Leighton F (1981) Pesticides: insecticides and fungicides as chitin synthesis inhibitors. Science 213: 905–907

Londershausen M, Kammann V, Spindler-Barth M, Spindler KD, Thomas H (1988) Chitin synthesis in insect cell lines. Insect Biochem 18: 631–636

Londershausen M, Turgerg A, Buss U, Spindler-Barth M, Spindler KD (1993) Comparison of chitin synthesis from an insect cell line and embryonic tick tissues. In: Mussarelli RAA (ed) Chitin enzymology. European Chitin Society, Ancona, pp 101–108

Lynn DE, Oberlander H (1983) The establishment of cell lines from imaginal wing discs of *Spodoptera frugiperda* and *Plodia interpunctella*. J Insect Physiol 29: 591–596

Marks EP, Balke J, Klosterman H (1984) Evidence for chitin synthesis in an insect cell line. Arch Insect Biochem Physiol 1: 225–230

Mayer RT, Chen AC, DeLoach JR (1981) Chitin synthesis inhibiting insect growth regulators do not inhibit chitin synthase. Experientia 37: 337–38

Mayer RT, Netter JJ, Leising HB, Schachtschabel DO (1984) Inhibition and uptake of nucleosides in cultured Harding-Passey melanoma cells by diflubenzuron. Toxicology 30: 1–6

Mayer RT, Witt W, Klitschka GE, Chen AC (1988) Evidence that chitin synthesis inhibitors affect cell membrane transport. In: Sehnal F, Zabza A, Denlinger DL (eds) Endocrinological frontiers in physiological insect ecology. Wroclaw Technical University Press, Wroclaw, pp 567–580

Mikólaczyk P, Oberlander H, Silhacek DL, Ishaaya I, Shaaya E (1994) Chitin synthesis in *Spodoptera frugiperda* wing imaginal discs. I. Chlorfluazuron, diflubenzuron, and teflubenzuron inhibit incorporation but not uptake of [^{14}C] N-acetyl-D-glucosamine. Arch Insect Biochem Physiol 25: 245–258

Mikólajczyk P, Zimowska G, Oberlander H, Silhacek DL (1995) Chitin synthesis in *Spodoptera frugiperda* wing imaginal discs. III. Role of the peripodial membrane. Arch Insect Biochem Physiol 28: 173–187

Mitlin N, Wiygul G, Haynes JW (1977) Inhibition of DNA synthesis in boll weevils (*Anthonomus grandis* B.) sterilized by dimilin. Pestic Biochem Physiol 7: 559–63

Mitsui T, Nobusawa C, Fukami J (1984) Mode of inhibition of chitin synthesis by diflubenzuron in the cabbage armyworm, *Mamestra brassicae* L. J Pestic Sci 9: 19–26

Mitsui T, Hitturu T, Nobusawa C, Yamaguchi I (1985) Inhibition of UDP-N-acetyl-glucosamine transport by diflubenzuron across biomembranes of the midgut epithelial cells in the cabbage armyworm, *Mamestra brassicae* L. J Pestic Sci 10: 55–60

Nakagawa Y, Matsumura F (1994) Diflubenzuron affects gamma-thioGTP stimulated Ca^{2+} transport *in vitro* in intracellular vesicles from the integument of the newly molted American cockroach, *Periplaneta americana* L. Insect Biochem Mol Biol 24: 1009–1015

Nakagawa Y, Matsumura F, Hashino Y (1993) Effect of diflubenzuron on incorporation of [^{3}H]-N-acetylglucosamine ([^{3}H]NAGA) into chitin in the intact integument from the newly molted American cockroach *Periplaneta americana*. Comp Biochem Physiol C106: 711–715

Oberlander H, Ferkovich SM, Van Essen F, Leach CE (1978) Chitin biosynthesis in imaginal discs cultured *in vitro*. W Roux's Arch 185: 95–98

Oberlander H, Ferkovich S, Leach CE, Van Essen F (1980) Inhibition of chitin biosynthesis in cultured imaginal discs; effects of alpha-amanitin, actinomycin-D, cycloheximide, and puromycin. W Roux's Arch 188: 81–86

Oberlander H, Lynn DE, Leach CE (1983) Inhibition of cuticle production in imaginal discs of *Plodia interpunctella* (cultured *in vitro*): effects of colcemid and vinblastine. J Insect Physiol 29: 47–53

Oberlander H, Silhacek DL, Leach E, Ishaaya I, Shaaya E (1991) Benzoylphenyl ureas inhibit chitin synthesis without interfering with amino sugar uptake in imaginal wing discs of *Plodia interpunctella*. Arch Insect Biochem Physiol 18: 219–227

Porcheron P, Oberlander H, Leach CE (1988) Ecdysteroid regulation of amino sugar uptake in a lepidopteran cell line derived from imaginal discs. Arch Insect Biochem Physiol 7: 145–155

Porcheron P, Moriniére M, Coudouel N, Oberlander H (1991) Ecdysteroid-stimulated synthesis and secretion of an N-acetyl-D-glucosamine-rich glycopeptide in a lepidopteran cell line derived from imaginal discs. Arch Insect Biochem Physiol 16: 257–271

Porcheron P, Morinière, Oberlander H (1993) Regulation of chitin synthesis in insects: ecdysteroid-regulated processing of N-acetyl-D-glucosamine by an insect epidermal cell line. In: Muzzarelli RAA (ed) Chitin enzymology. European Chitin Society, Ancona, pp 83–88

Post LC, de Jong BJ, Vincent WR (1974) 1-(2,6-disubstituted benzoyl)-3-phenyl urea insecticides: inhibitors of chitin synthesis. Pestic Biochem Physiol 4: 473–483

Quesada-Allue LA (1982) The inhibition of insect chitin synthesis by tunicamycin. Biochem Biophys Res Common 105: 312–319

Retnakaran A, Oberlander H (1993) Control of chitin synthesis in insects. In: Muzzarelli RAA (ed) Chitin enzymology. European Chitin Society, Ancona, pp 89–99

Retnakaran A, Wright JE (1987) Control of insect pests with benzoylphenyl ureas. In: Wright JE, Retnakaran A (eds) Chitin and benzoylphenyl ureas. Dr W Junk, Dordrecht, pp 205–282

Retnakaran A, MacDonald A, Nicholson D, Cuningham J (1989) Ultrastructural and autoradiographic investigations of the interference of chlorfluazuron with cuticle differentiation in the spruce budworm, *Choristoneura fumiferana*. Pestic Biochem Physiol 35: 172–184

Soltani N, Besson MT, Delachambre J (1984) Effects of diflubenzuron on the pupal–adult development of *Tenebrio molitor* (L): growth and development, cuticle secretion epidermal cell density, and DNA synthesis. Pestic Biochem Physiol 21: 256–26

Soltani N, Quennedey A, Delbecque JP, Delachambre J (1987) Diflubenzuron-induced alterations during *in vitro* development of *Tenebrio molitor* pupal integument. Arch Insect Biochem Physiol 5: 201–209

Spindler-Barth M (1993) Hormonal regulation of chitin metabolism in insect cell lines. In: Muzzarellli RAA (ed) Chitin enzymology. European Chitin Society, Ancona pp 75–82

Spindler-Barth M, Spindler KD, Londerhausen M, Thomas H (1989) Inhibition of chitin synthesis in an insect cell line. Pestic Sci 25: 115–121

Turnbull IF, Howells A J (1983) Integumental chitin synthase activity in cell-free extracts of larvae of the Australian sheep blowfly, *Lucilia cuprina*, and two other species of Diptera. Aust J Biol Sci 36: 251–62

Ward GB, Newman SM, Klosterman HJ, Marks EP (1988) Effect of 20-hydroxy ecdysone and diflubenzuron on chitin production by a cockroach cell line. *In Vitro* Cell Dev Biol 24: 326–332

Yu SJ, Terriere LC (1977) Ecdysone metabolism by soluble enzyme from three species of Diptera and its inhibition by the insect growth regulator TH-6040. Pestic Biochem Physiol 7: 48–55

Zimowska G, Mikólajczyk P, Silhacek DL, Oberlander H (1994) Chitin synthesis in *Spodoptera frugiperda* wing imaginal discs. II. Selective action of chlorfluazuron on wheat germ agglutinin binding and cuticle ultrastructure. Arch Insect Biochem Physiol 27: 89–108

CHAPTER 7

Bacillus thuringiensis: Use and Resistance Management

M.E. Whalon[1] and W.H. McGaughey[2]
[1]Department of Entomology and Pesticide Research Center, Michigan State University, East Lansing, Michigan, USA
[2]US Grain Marketing Research Laboratory, Agricultural Research Service, United States Department of Agriculture, Manhattan, Kansas, USA

1 Introduction

Bacillus thuringiensis (*B.t.*) is an aerobic, gram-positive, spore-forming bacterium found rather commonly in the environment. It produces a number of insect toxins, the most distinctive of which is a protein crystal formed during sporulation (Hannay and Fitz-James 1955; Bulla et al. 1977; Whiteley and Schnepf 1986). It is this crystalline protein inclusion, or δ-endotoxin, that is the principal active ingredient in formulations currently in use. Although *B.t.* was described by Berliner in 1911 and its potential as an insecticide was recognized relatively early, commercial development in the USA did not occur until the late 1950s and its success has been somewhat erratic (Burges 1986). During the early 1980s it was found that *B.t.* was amenable to genetic manipulation using recombinant DNA techniques because the genes that encode for δ-endotoxin production are borne on plasmids (Stahly et al. 1978; Gonzalez et al. 1981, 1982; Gonzalez and Carlton 1982, 1984; Held et al. 1982; Faust et al. 1983; Kronstad et al. 1983). Subsequently, techniques were developed for cloning these genes in other bacteria (Schnepf and Whiteley 1981; Held et al. 1982; Klier et al. 1982; Adang et al. 1985; McLinden et al. 1985; Obukowicz et al. 1986; Jahn et al. 1987; Sekar et al. 1987; Ahmad et al. 1989) and for transferring the genes into crop plants (Adang et al. 1987; Barton et al. 1987; Fischhoff et al. 1987; Vaeck et al. 1987, 1990; Perlak et al. 1990). This gene transfer technology has progressed rapidly and these biotechnological developments hold promise for facilitating genetic improvements in the potency and host spectrum of *B.t.* strains (Carlton 1988; Crickmore et al. 1990; Martens et al. 1990; Merryweather et al. 1990) and for developing crop varieties that are genetically engineered to produce *B.t.* toxins within their own tissues (Gasser and Fraley 1989; Boulter et al. 1990; Brunke and Meeusen 1991). Transgenic plants would overcome some of the stability problems associated with conventional application and improve control of pests that feed on plant parts or tissues that are difficult to treat using conventional methods.

The purpose of this chapter is to review the current use and status of insect resistance to *B.t.* δ-endotoxins and the related information on mode of action, specificity, and mechanisms of resistance to these toxins in order to provide a framework for discussing the development of resistance management strategies.

We include background information on the role of *B.t.* in pest management programs, its mode of action, and the classification and specificity of its δ-endotoxins. The cases of resistance reported to date are reviewed from the standpoints of the physiological mechanisms and the insect genetics involved. The theory and practice of resistance management are discussed as they may relate to preventing the evolution of insect resistance to *B.t.* toxins either in conventional pest management or transgenic plant applications. Although we recognize that *B.t* toxins have been variously formulated as single or mixtures of strains, encapsulation in other bacteria, chimeras, with or without spores, with or without CytA, with various UV protectants, with stickers, and with other insecticides like synthetic pyrethroids, we will not review this literature and refer the reader to commercial agrochemical company sources. The actual penetration of biological pesticides into global insecticide markets has been reviewed most recently by Powell and Jutsum (1993).

The desirability of *B.t.* toxins and the impetus for their use in pest control result primarily from their specificity for pest insects and safety to nontarget organisms. However, their specificity is not so limited as to make them unattractive for commercial development. Unlike the general case with insect pathogenic viruses, most *B.t.* toxins affect several economically important insect species, which improves their attractiveness for commercial development. The toxins have no known detrimental effects on humans or wild or domestic animals and they tend to be short-lived in the environment. Thus, they are attractive alternatives to chemical insecticides, which as a group are perceived to be more environmentally disruptive and hazardous to users and consumers. In addition, the limited range of activity of the toxins toward insects means that often they will kill pest species while having no effects on predatory or predaceous species. This feature makes them highly desirable for use as components of integrated pest management (IPM) programs.

These desirable attributes of *B.t.* toxins also carry over to transgenic plant applications where safety and activity spectrum are equally important considerations. However, the vast body of information available on the genetics, structure, and toxicity of *B.t.* toxins, and the fact that the δ-endotoxin genes are on plasmids which are relatively easily transferred into other organisms, have contributed to the rapid development of these toxins rather than other lesser known proteins to engineer insect-resistant plants.

2 Mode of Action

The mode of action of *B.t.* δ-endotoxins has been the subject of many studies over the last half-century and several theories have been postulated. Its site of action, in the insect midgut, and its disruptive effects on the insect midgut cell membrane were recognized relatively early (Heimpel and Angus 1959; Hoopingarner and Materu 1964; Ramakrishnan and Tiwari 1967; Sutter and Raun 1967; Atwa and Abdel-Rahman 1974; Narayanan and Jayaraj 1974). However, in spite of intense scrutiny, the specific molecular mechanisms involved in its toxicity to insects have only recently begun to be understood (see English and Slatin 1992 for an up-to-date review).

In the bacterium, δ-endotoxins are synthesized as large protein molecules and crystallized as parasporal inclusions. In susceptible insects, these inclusions dissolve in the midgut, releasing protoxins that range in size from 27 to 140 kDa and which are proteolytically converted into still smaller toxic polypeptides (Bulla et al. 1981; Andrews et al. 1985; Hofte and Whiteley 1989). There is extensive variation in the size and structure of the inclusion proteins, the intermediate protoxins, and the active toxins that are presumed to relate to the insect specificity of the toxins. There is some question as to the extent of proteolytic processing of Coleoptera active toxins (Carroll et al. 1989; Li et al. 1991).

Following activation, these toxins apparently bind with high specific affinity to receptors on the midgut epithelium (Hofmann and Luthy 1986; Hofmann et al. 1988a, b; Van Rie et al. 1989, 1990a, b; Ferre et al. 1991). These receptors are not very well understood, but are apparently glycoproteins (Knowles and Ellar 1986). Studies on CryIV (mosquito-specific) toxins suggest that a combination of phospholipid, glycoprotein, or perhaps other receptors may be involved in binding (Ward and Ellar 1986; Chilcott et al. 1990). Binding seems to be an essential step in toxicity and the studies cited above demonstrate close correlation between binding affinity and toxicity. However, binding does not assure toxicity. Studies by Garczynski et al. (1991) showed that in *Spodoptera frugiperda*, a somewhat insensitive species, toxin binding occurred but it did not result in insect mortality. Wolfersberger (1990) reported similar results with *Lymantria dispar* in which there was an inverse relationship between binding affinity and toxicity of two toxins.

Following binding to the midgut epithelial cells, the toxins generate pores in the cell membrane, disturbing cellular osmotic balance, and causing the cells to swell and lyse through a process that has been termed "colloid-osmotic lysis" (Knowles and Ellar 1987; Hofte and Whiteley 1989). The pore is not just a tear in the cell lipid bilayer, but an ion channel with different characteristics based upon the endotoxin and the time course of osmoregulation in the cell.

The three-dimensional structure of a CryIIIA (Coleoptera-active) toxin has been characterized (Li et al. 1991). It is a wedge-shaped molecule with three functional domains, each critical for binding and insertion or pore formation. Domain II probably binds the insect cell receptor. The apical portion of domain II is highly conserved across CryI–CryIV toxins and therefore is thought to be involved in binding. The hydrophobic and amphipathic helices of domain I penetrate the lipid membrane of the gut cell. Once this step occurs, the toxin is not easily removed from the cell surface, suggesting a rather rapid and permanent insertion process. Recent studies of CryIA(c) binding in the midgut brush border membrane of *Manduca sexta* indicate that the CryIA(c) binding protein is anchored in the membrane by a glycosyl-phosphatidylinositol (glycolipid) anchor (Garcyznski and Adang 1995). The insertion of domain III β-sheets causes a defect that progressively allows larger and larger diameter molecules (10 – 40 Å) to cross the membrane. These defects or lesions lead to gross ultrastructural changes and eventually to cell lysis. Death of the poisoned insect follows through starvation and septicemia (Endo and Nishitsutsuji-Uwo 1980).

The ultrastructural changes and time course of poisoning varies between toxins and insect species. The general characteristics are cessation of feeding within 1 h of ingestion, reduced activity within 2 h, and progressive sluggishness and paralysis within 6 h (Ebersold et al. 1977; Percy and Fast 1983; Lane et al. 1989; Bauer and Pankratz 1992). The epithelial cells swell with disrupted microvilli, cell lysis, and cell sloughing. With CryI, CryII, and CryIV toxins, ultrastructural changes may be observed within minutes of ingestion, but CryIIIA protein is apparently slower in action without microvilli or organelle disruption in *Chrysomela scripta*, the cottonwood leaf beetle (Bauer and Pankratz 1992).

3
Classification and Specificity of *B.t.* Toxins

The δ-endotoxins from different strains of *B.t.* have widely differing insect specificities, but they tend to be toxic primarily toward larvae of Lepidoptera. Inclusions from a few strains are toxic to certain species of Diptera or Coleoptera. Most of the formulations currently used in the USA are derived from one of three subspecies of *B.t.* Subspecies *kurstaki* is toxic toward many species of Lepidoptera and is used for controlling pests in vegetable crops, cotton, forests, home gardens, and stored grain. Subspecies *israelensis* is toxic toward several species of Diptera and is used primarily for controlling mosquitoes and blackflies. Subspecies *tenebrionis* is toxic to Colorado potato beetle (*Leptinotarsa decemlineata*) and has been introduced as a spray and transgenic plant for controlling that pest on potatoes. Many other strains of *B.t.* exist in nature and some that are known to have desirable activity spectra have been isolated, but they have not yet been commercialized.

Historically, *B.t.* strains have been classified on the basis of the flagellar or H-antigens of the vegetative cells. This system has given rise to the familiar nomenclature system of subspecies that is commonly used in the literature (de Barjac and Frachon 1990). Currently, 34 subspecies are recognized. However, this nomenclature system fails to consistently reflect the structure or insect specificity of the inclusion proteins. Screening programs have demonstrated extensive diversity in insecticidal spectra among isolates within several subspecies (for example, Dulmage and Cooperators 1981). An effort has been made and largely adopted in the scientific literature to classify the various crystal proteins and their genes based on their structure, antigenic properties, and activity spectrum (Hofte et al. 1988; Hofte and Whiteley 1989). Another classification system is being considered at this time by the Society of Invertebrate Pathologists.

In general, the Hofte and Whiteley classification system places the toxins produced by various strains of *B.t.* in four major groups. The proteins designated CryI are toxic toward Lepidoptera, the CryII proteins are toxic toward Lepidoptera and Diptera, the CryIII toward Coleoptera, the CryIV toward Diptera, the CryV toward Lepidoptera and Coleoptera, and CryVI has various activities. Each of these major groups has been divided into several toxin types. This classification system is far from static. It is evolving rapidly with the discovery of additional variations in structure and host spectrum within each of the groups. These toxins have been summarized by Adang (1991).

This system of classification according to toxin structure and spectrum corresponds only very generally with the conventional nomenclature system. The CryIV proteins are apparently limited to subsp. *israelensis* which is used commercially to control mosquitoes and black flies. The CryIII proteins are from subsp. *tenebrionis, tolworthi*, and strain EG2158, and are toxic toward *L. decemlineata*. However, according to de Barjac and Frachon (1990), *tenebrionis* is not a separate subspecies but is a pathovar of subsp. *morrisoni* that is distinguishable only by its toxicity to Coleoptera. The CryI and CryII proteins apparently are produced by many different subspecies. Additionally, a given isolate of these subspecies may produce several different CryI and/or CryII proteins. As examples, subsp. *kurstaki* isolate HD-1 produces CryIA(a), CryIA(b), CryIA(c), CryIIA, and CryIIB proteins, while isolate HD-73 produces only CryIA(c) (Hofte and Whiteley 1989). Subsp. *aizawai* isolate HD-133 produces CryIA(b), CryIC, and CryID proteins (Aronson et al. 1991). Such heterogeneity in toxin production is apparently responsible for some of the diversity in activity spectrum encountered in screening programs.

4 *B.t.* Specificity

The mechanisms involved in the specificity of these various toxins are only now being elucidated. Because it has been recognized for some time that the protein inclusions typically undergo extensive processing within the insect gut in order to activate them, it was logically assumed that different insect species might process the inclusions in a unique fashion giving rise to toxins affecting only that species. Indeed, studies by Haider et al. (1986) demonstrated that activation of *B.t.* subsp. *colmeri* inclusions by gut proteases from mosquitoes yielded proteins toxic primarily to mosquito cell lines, while gut proteases from *Pieris brassicae* yielded proteins toxic primarily to lepidopteran cells. *In vivo* studies to determine whether insect-specific differences in gut proteases might influence the specificity of activated toxins have been inconclusive (Jacquet et al. 1987; Lecadet and Martouret 1987; van Frankenhuyzen et al. 1991). However, the toxin composition of inclusions apparently does influence solubility (Aronson et al. 1991; van Frankenhuyzen et al. 1991). While there is no doubt that the ability to solubilize and activate the inclusion proteins must influence the susceptibility of insects to *B.t.*, the extent of that influence on host spectrum remains unresolved. More recent research has focused on the role of membrane binding as a determinant of specificity. In Lepidoptera, studies have shown that the binding affinity of individual toxins to receptor sites on the midgut membrane can account for the sensitivity of different insect species to various toxins (Knowles and Ellar 1986; Haider and Ellar 1987; Hofmann et al. 1988a,b; Van Rie et al. 1989, 1990a,b; Ferre et al. 1991).

Hofmann et al. (1988b) first established a relationship between binding affinity and differential toxicity in studies on the specificity of two δ-endotoxins toward *P. brassicae* and *M. sexta*. Subsequent studies by Van Rie et al. (1989, 1990a) confirmed the role of specific receptors in determining susceptibility to Lepidoptera-specific protoxins, and suggested that binding site specificity might not be a simple system in which each toxin binds to a unique receptor. In studies of the specificity of various toxins toward *Spodoptera littoralis*, *M. sexta*, and *Heliothis virescens*, they observed a high degree of heterogeneity among binding sites, suggesting that some sites may bind a single toxin whereas others may bind two or more toxins. Similarly, specific toxins may bind to more than one site in some insect species. Thus, in Lepidoptera, a gut membrane binding site recognition system apparently plays a key role in determining the specificity of various toxins, but probably not to the exclusion of other mechanisms such as proteolytic crystal activation or toxicity subsequent to binding.

5 Insect Resistance to *B.t.*

5.1 Resistance Reported to Date

The definition of insecticide resistance has varied with the circumstance of its origin. Field failures of conventionally applied insecticides have historically been called resistance irrespective of the level of resistance, operational factors, selection circumstance, resistance mechanisms, contributing biological and ecological factors, or insecticide/biocide classification. Field failure of conventionally selected host plant resistance has been termed pest biotype formation (see review by Diehl and Bush 1984). Laboratory and glasshouse selection of resistance with various biocides is usually defined in terms of a statistically significant increase in the resistance ratio of the selected to the sensitive population. In any case, resistance is the microevolutionary process of genetic adaptation through selection whether the biocides are deployed genetically or conventionally as a spray, fumigant, granular treatment, or presumably transgenic plant.

Insecticide resistance is one of the most formidable practical problems facing modern plant and health protection. The number of resistant insect and mite species continues to grow annually, and already more than 500 species have acquired resistance (Georghiou 1990b) with the actual number probably over 530 (Whalon and Hollingworth 1989–present). Microbial insecticides have not escaped this problem that many hoped would be limited to conventional chemicals. Within the last few years, at least 12 insect species have been selected for resistance to *B.t.* δ-endotoxins (Table 1).

Among these reported cases of resistance, only two involve resistance among wild populations. The Indianmeal moth, *Plodia interpunctella*, evolved low levels of resistance in grain bins as a result of treating the grain with *B.t.* (McGaughey 1985a,b; McGaughey and Beeman 1988). The diamondback moth, *Plutella xylostella*, is possibly the most notable because it evolved high levels of resistance in the field as a result of repeated use of *B.t.* in intense control programs. Resistance in that species has been reported essentially world-wide in the last 5 years (see for example Shelton and Wyman 1992; Shelton et al. 1993). Resistance in the other species has been reported only from laboratory selection experiments, but the capacity for resistance in these species is of concern because of the great economic and public health significance of these pests. Resistance levels in the

Table 1. Insect species selected for resistance to *B.t.*

Species	Reference
Plodia interpunctella	McGaughey (1985a)
Cadra cautella	McGaughey and Beeman (1988)
Plutella xylostella	Kirsch and Schmutterer (1988)
	Tabashnik et al. (1990)
Heliothis virescens	Stone et al. (1989)
Leptinotarsa decemlineata	Whalon et al. (1993)
Ostrinia nubialis	P.C. Bolin and W.D. Hutchison (1996), University Of Minnesota, pers. comm.
Culex quinquefasciatus	Georghiou and Vasquez (1982)
	Gill et al. (1992)
Aedes aegypti	Goldman et al. (1986)
Trichoplusia ni	Estada and Ferre (1994)
Spodoptera littoralis	Muller-Cohn et al. (1994)
Chrysomela scripta	Bauer et al. (1994)
Spodoptera exigua	Moar et al. (1995)

two mosquito species were relatively low and probably are no cause for concern at this time.

In addition to these reports of resistance being selected, either in the laboratory or field, studies have clearly documented differences in baseline susceptibility levels among insect populations that have not been previously exposed to *B.t.* treatment. Rossiter et al. (1990) reported small differences in susceptibility of *Lymantria dispar* populations. Stone and Sims (1993) reported differences in susceptibility ranging from 3.6- to 16-fold for field populations of *Heliothis virescens* and *Helicoverpa zea* to Dipel and purified endotoxin. Analysis of the response of *Choristoneura fumiferana* populations to strain HD-1-S-1980 and *L. decemlineata* to CryIIIA also suggests differences in susceptibility that could provide a basis for evolution of resistance (van Frankenhuyzen et al. 1995; Dively et al., unpubl. data, respectively).

It is not clear why the capacity for insect resistance to *B.t.* δ-endotoxins was not detected prior to the 1980s. *B.t.* has been used in control programs to some extent since the 1950s, and during that period many investigators had raised the question of the possibility of resistance. However, all efforts to select for resistance except those involving β-exotoxin against houseflies failed (Feigin 1963; Harvey and Howell 1965; Burges 1971; Briese 1981, 1986). Obviously, there are many biological, ecological, operational, and evolutionary considerations in the development of resistance. However, it is clear that resistance to *B.t.* δ-endotoxins can occur, and many factors may have contributed to the failed selection efforts that were reported prior to 1985. It would be a mistake to assume that the species involved do not have the genetic capacity to adapt to *B.t.* The selection attempts may have involved populations that had already gone through genetic bottlenecks of other insecticide resistance selection (Carlberg and Lindstrom 1987), long-term laboratory culturing (Salama and Matter 1991), starvation (Sneh and Schuster 1983), or adaptation to artificial diets which may have reduced variation and, therefore, the potential for resistance.

In field situations where resistance has failed to develop, we again cannot conclude that the species or population does not have the capacity to evolve resistance. On the contrary, this may be strong evidence that behavioral or genetic factors could be exploited for developing resistance management programs if these factors could be understood (Tabashnik and Croft 1985). In many of the instances where *B.t.* resistance has not developed, it would probably be found that selection pressure was low and infrequent.

5.2 Mechanisms of Resistance

Although acknowledged as a theoretical possibility, the likelihood of insect resistance to the *B.t.* δ-endotoxins was considered for many years to be remote (Burges 1971). This perception apparently had as part of its basis the existing understanding of what seemed to be a very complex mode of action of *B.t.* involving multiple toxins and multiple target sites. Thus, a single change or mutation would be expected to have little effect. Only in the highly unlikely event that several mutations occurred would resistance result (Boman 1981). However, it is now clearly recognized that rather than offering a mode of action so complex as to safeguard against the occurrence of resistance, the mode of action of *B.t.* provides many points at which behavioral or physiological changes might offer protection for the insects. Furthermore, the use of single-gene toxins in plant transformation would appear to bypass any advantages inherent in the natural mixtures of toxins produced in the parent bacteria.

Review of the mode of action of *B.t.* δ-endotoxins suggests three possible physiological or biochemical mechanism(s) of resistance. The first involves the pH- and protease-mediated dissolution and activation of the crystal. The resistant insects may have acquired mechanisms to detoxify the toxins, or lost the means to activate them, through enzymatic changes in the gut. Secondly, changes may have occurred in the gut cell membrane which interfere with binding of the toxic moiety. Thirdly, cellular changes may have occurred that influence the sensitivity of the cells to pore formation or their capacity to recover from the effects of the toxins.

One of the most obvious of the possible mechanisms of resistance is a change in gut enzymes that would cause differences in the dissolution and activation of the proteinaceous crystal. However, Johnson et al. (1990) made an exhaustive study of protease activity in the midguts of larvae of susceptible and resistant strains of *Plodia interpunctella* and the results indicated that resistance was not due to obvious changes in larval midgut protease activity. A more recent study by Oppert et al. (1994) of *P. interpunctella* resistant to *B.t.* subsp. *entomodicus* suggests that altered protoxin activation by midgut proteinases is indeed involved in some types of insect resistance to *B.t.* Different proteases can be produced in the insect gut depending on the plant material ingested (Broadway 1989). Such differences could influence susceptibility through slower activation or faster metabolism of the toxins.

Studies on the binding affinity of midgut receptors have been more fruitful. Studies by Van Rie et al. (1990b) and Ferre et al. (1991) have provided convincing evidence that in *P. interpunctella* and *P. xylostella*, resistance is due to a change in binding affinity of receptors or binding sites on the brush border membrane of the insect midgut. This appears to be the same mechanism that is involved in the host specificity of *B.t.* δ-endotoxins. Using ^{125}I-labeled δ-endotoxins and brush border membrane vesicles prepared from the midguts of the larvae, these studies demonstrated that in both insects there was a loss of binding affinity and a parallel decrease in susceptibility that was specific for a *B.t.* subsp. *kurstaki* type toxin, CryIA(b), that was a major constituent of the formulations used in selecting for resistance. In resistant *P. interpunctella*, binding sites for CryIC toxin remained functional and the insects were still susceptible to CryIC toxin. In resistant *P. xylostella*, binding sites for CryIC and CryIB toxins were shown to be still functional. In *H. virescens* resistant to subsp. *kurstaki* toxins, MacIntosh et al. (1991) also observed reduced toxicity and reduced binding affinity of CryIA(b) and CryIA(c) toxins. The mechanism of resistance in *L. decemlineata* is unknown at this time.

One of the most important points to emerge from these studies on strains of insects resistant to *B.t.* is that in each case the resistance appears to be somewhat specific for the toxin or toxins used in selection. The insects are not resistant to all δ-endotoxins. This is illustrated in studies on *P. interpunctella* that were selected for resistance to Dipel, a commercial formulation of the HD-1 isolate of subsp. *kurstaki* (McGaughey and Johnson 1987). The insects were resistant to δ-endotoxins of 32 isolates of subsp. *thuringiensis*, *kurstaki*, and *galleriae*, but they remained susceptible to some degree to at least 15 isolates of subsp. *kenyae*, *entomocidus*, *aizawai*, *tolworthi*, and *darmstadiensis*. Apparently, the strains that were still active against the resistant insects contained toxins other than the CryIA-type which had been used in their selection.

This high degree of resistance specificity has led to speculation that resistance can be managed by simultaneously using two or more toxins which do not cause cross-resistance to one another to reduce the probability of resistance evolving, or by simply incorporating another unrelated toxin once resistance has occurred (Georghiou 1990a; Stone et al. 1991; Van Rie 1991). The problem with this approach is that a certain amount of cross-resistance among toxins does occur in some insect species and general patterns of cross-resistance may not exist. In work already referred to on *H. virescens*, it was shown that certain toxins bind to multiple receptors and some receptors bind multiple toxins (Van Rie et al. 1989, 1990a). Thus, some toxins will select for cross-resistance to others. Indeed, work by Gould et al. (1991b) demonstrated that a strain of *H. virescens* selected for resistance to the CryIA(c) toxin typical of strain HD-73 (subsp. *kurstaki*) was cross-resistant to CryIA(a), CryIA(b), CryIB, CryIC, and CryIIA toxins. These findings provide little encouragement for the multiple toxin approach to resistance management in *H. virescens*. Many questions still remain regarding the degree of specificity or cross reactivity in this binding site recognition system. Selection of *L. decemlineata* with CryIIIA alone also resulted in broad cross-resistance to CryIIIB, CryIIIC, CryIIID, CryIIIE, and CryV, but mixtures with spores dramatically increased toxicity of CryIIIA protein (M. Whalon, D. Norris, and L. English, unpubl. data). These results suggest that selection with single or multiple toxins may result in rapid resistance development characterized by with broad cross-resistance, but formulations or strategies that incorporate spores may mitigate resistance development. If these results with single and multiple toxins extend to transgenic plants, broad spectrum resistance, particularly where "ultra" or "super" high expression levels are not achieved, may result. Further research is required to elucidate fully the cross-resistance relationships for various toxins in the major species of pest insects before this approach can be considered for managing resistance.

Historically, research has focused on membrane binding as the probable mechanism of resistance in Lepidoptera; it should be recognized that other mechanisms are possible. The series of steps involved in the mechanism of *B.t.* toxicity provides many opportunities for changes to occur. Indeed, two studies on resistance mechanisms in *H. virescens* suggest that other mechanisms may be involved in resistance of that species (MacIntosh et al. 1991; Gould et al. 1991b). Additionally, while the studies by Johnson et al. (1990) failed to demonstrate any differences in pH or proteolytic activity in the midguts of susceptible and resistant *P. interpunctella*, this dissolution and activation step as a possible mechanism of resistance to other toxins or in other insect species cannot be eliminated *a priori.*

Insects of all susceptible species have the capacity to recover from sublethal dosages of *B.t.* It is not known whether the mechanisms could be harbingers of adaptation. Vigor contributes to variability in susceptibility to *B.t.* products (Kinsinger and McGaughey 1979; Briese 1981; Salama and Matter 1991). Insect age is an important factor affecting susceptibility (McGaughey 1978; Ferro and Lyon 1991), and typical bioassay results show that *B.t.* gives a rather flat dose-mortality line with increasing age (Burges 1971). This phenomenon may be associated with increased gut cell numbers, gut environmental changes, or even a differential ability in the insect's capacity to slough off damaged gut cells, repair lesions, or resist septicemia (Chiang et al. 1986).

5.3 Behavioral Resistance

Historically, behavioral detection and avoidance of both formulated and purified *B.t.* toxins have been reported primarily in Lepidoptera. This may be a first line of potential resistance development. Although behavioral avoidance has not been associated with any of the *B.t.* resistance cases reported (Schwartz et al. 1991; Whalon et al. 1993), it could be a potential mechanism as *B.t.* is used more extensively in either conventional or transgenic delivery systems where feeding choices are available. The role of behavior in insect adaptation to insecticides and resistant host plants has been considered in more detail by Gould (1984) and Lockwood et al. (1984). In addition, the mechanism of feeding cessation that is characteristic of *B.t.* toxicology is not known. Its behavioral features could play an important role in resistance development especially if a hypersensitive reaction exists. Studies on behavioral factors that influence either the progression or stability of *B.t.* resistance are becoming common. Work by Gould and Anderson (1991) and Gould et al. (1991a) on *H. virescens* suggested that, where possible, susceptible larvae might avoid feeding on *B.t.*-treated diet. They interpreted this response to indicate slower adaptation in situations where the *B.t.* treatment was completely absent on parts of the plants. Initial studies on Colorado potato beetle were unable to demonstrate larval avoidance behavior of foliar applied *B.t.* (Hoy and Hall 1993; Whalon et al. 1993). A more recent study of Colorado potato beetle larvae feeding on transgenic *B.t.* potatoes indicated that the more resistant larvae tended to avoid feeding on high dose *B.t.* plants when given a range of doses from which to choose (Hoy and Head 1995). Further studies of adult beetle movement, using greenhouse simulated transgenic potato fieldborders and field plots with transgenic potato borders, demonstrated increased and sustained movement of both sexes after an initial feeding bout arrested movement (M. Bush, M. Whalon, and J. Weirenga, unpubl. data). Benedict et al. (1993b) reported changes in feeding behavior and plant abandonment on some transgenic cotton lines. Avoidance of *B.t.*-treated diet was not detected in *P. xylostella* (Schwartz et al. 1991). In forest ecosystems, spruce budworm, *Choristoneura fumiferana*, and Gypsy moth, *Lymantria dispar*, (Farrar and Ridgway 1995), avoided *B.t.* preparations while fall webworm, *Hyphantria cunea*, could not discriminate (Ramachandran et al. 1993).

From these studies it is clear that avoidance behavior to *B.t.* sprays and transgenic plants is present in many of the species studied to date, and these behaviors will have a significant impact both in the evolution of resistance in the field and resistance management programs. For instance, on corn earworm, *Ostrinia nubilalis*, Alstad and Andow (1995) presented a strategy for managing *B.t.* resistance that exploited natural movement and cultural management.

5.4 Resistance Genetics

The capacity for resistance is widespread in some species. In *P. interpunctella*, resistance has been selected in six colonies obtained from six different grain storage sites in the midwestern USA (McGaughey and Beeman 1988; McGaughey and Johnson 1992). Native populations of *P. interpunctella* in the central USA have an approximate seven-fold range in susceptibility (LC_{50}) to *B.t.*, which suggests considerable genetic variation (Kinsinger and McGaughey 1979; McGaughey 1985a). *Leptinotarsa decemlineata* also exhibits a wide range in susceptibility among field populations (Whalon et al. 1993; Dively et al., unpubl. data). Widespread capacity for resistance apparently occurs in *P. xylostella*, with reports of resistance from Hawaii (Tabashnik et al. 1990, 1991), the Philippines (Kirsch and Schmutterer 1988), and the continental USA (Shelton and Wyman 1992). Other species have been studied less intensively, but it will not be surprising if widespread capacity for resistance is eventually found in many species.

The inheritance of *B.t.* resistance in insects is not yet clearly understood. In *P. interpunctella*, resistance appears to be partially recessive and probably due to a single major factor (McGaughey 1985a, McGaughey and Beeman 1988). Sims and Stone (1991) characterized *B.t.* resistance in *H. virescens* as being partially recessive (incompletely dominant), but probably controlled by several genetic factors. Studies on *H. virescens* (Gould et al. 1991b) suggest that resistance in this species may be inherited as an additive trait. Studies on the genetics of *B.t.*

resistance in *P. xylostella* have shown that, like *P. interpunctella*, resistance is recessive and most likely controlled by a single major gene (Tabashnik et al. 1994). The genetics of resistance in *L. decemlineata* is autosomally inherited as an incompletely dominant gene with epistasis (Rahardja and Whalon 1995).

A significant obstacle in conducting definitive genetic studies thus far has been that most of the resistant lepidopteran colonies have been selected using Dipel, a commercial *B.t.* formulation that reportedly contains a mixture of toxins (Hofte and Whiteley 1989). Gould et al. (1991b) on *H. virescens* and Whalon et al. (1993) on *L. decemlineata* were the first to select pest populations using a single protein. When toxin mixtures are used, results may be complicated by different toxin frequencies and rates of resistance progression or modes of inheritance of resistance to the various components of the mixture. Indeed, the results with *P. interpunctella* provide evidence of this possibility (McGaughey and Beeman 1988). Resistance progressed at different rates and the degree of recessiveness differed among the colonies that were compared. In addition, the *P. interpunctella* data suggest a stepwise progression of resistance. The latter could be a result of multiple genes, resistance to multiple toxins, or simply fitness factors being sorted out. Tabashnik and McGaughey (1994) selected *P. interpunctella* with different *B.t.* strains containing an array of different endotoxins and analyzed the rate of resistance development using heritability estimates. They found that resistance development did not vary significantly between selection with the different strains containing up to six different toxins.

This work suggests that multiple toxin selections (mixtures and sequences) may not prevent or ameliorate resistance development, and that the initial resistance allele frequency was much higher in native populations than was once believed. However, selection of *Spodoptera exigua* with one of the same strains used by Tabashnik and McGaughey (1994), HD-1, but combined with spores did not result in resistance in 20 generations of selection (Moar et al. 1995). Yet when the same species was selected with CryIC toxin alone, high levels of resistance were achieved in 21 generations, and this strain exhibited cross-resistance to CryIA(b), CryIE-CryIC fusion protein, CryIH, and CrIIA.

5.5
B.t. Resistance Stability

In some cases, resistance is very stable when selection is discontinued. This seems to be true particularly when resistance has progressed to higher levels. At lower levels, there appears to be gradual reversion back to a more sensitive level. In *P. interpunctella*, there was a decline in field-selected resistance when colonies were first being reared in the laboratory, but in highly resistant laboratory colonies, resistance declined slowly or not at all when selection pressure was discontinued (McGaughey 1985a; McGaughey and Beeman 1988). Similar gradual declines in resistance of populations of *L. decemlineata*, *H. virescens*, and *P. xylostella* that were only moderately resistant have been reported (Sims and Stone 1991; Tabashnik et al. 1991; Whalon et al. 1993; Rahardja and Whalon 1995). Data are not available from field situations where immigration of susceptible insects or fitness of resistant individuals may be much more important in determining the stability of resistance. However, Tabashnik et al. (1991) suggested that in *P. xylostella*, gene flow was too weak to counter selection for resistance. Presumably, it also would be too weak to contribute significantly to restoration of susceptibility once resistance has been selected.

5.6 *B.t.* Resistance Costs

The evolutionary stability of pesticide resistance in the field is influenced by many factors including dominance, autosomal or maternal inheritance, linkage, developmental fitness costs, and, especially, reproductive fitness costs. In fact, resistance management, once resistance has occurred in the wild, is predicated on reversion to or near to the susceptible state when selection pressure is removed. Reversion occurs most readily where reproductive fitness costs are operating. Reproductive and developmental costs have been reported in *L. decemlineata* where egg numbers were significantly reduced and larval development time increased (Trisyono and Whalon 1997). Tabashnik et al. (1994) also found significant reproductive and developmental costs in diamondback moth. In general, developmental and reproductive costs are initially high under *B.t.* selection, and then decline presumably due to fitness assortment.

It should also be noted that several of the pests that are currently the object of control using *B.t.*, such as *Heliothis* spp., *L. decemlineata*, and *P. xylostella*, have successful histories of adapting to a wide range of insecticidal materials used against them. Thus, it would be logical to assume that there would be nothing about the genetics, population dynamics, or behavior of these pests that would prevent transgenic plant or conventional use of *B.t.* from selecting resistant populations just as synthetic organic insecticides have done.

6 Resistance Management

6.1 General Considerations

Since *B.t.* resistance is an evolutionary phenomena, it is controlled by four factors: (1) natural variation; (2) *B.t.* selection or differential mortality caused by *B.t.*, (3) survival of selected individuals; and (4) reproduction of *B.t.* selected survivors. Natural variation to *B.t.* within field populations of pests appears to be quite high, as previously discussed in Sections 5.1 and 5.4. In addition, the use of both high-dose *B.t.* transgenic plants and more effectively formulated and engineered sprays with greater toxicity and longer field-life will effectively increase *B.t.* selection on target populations. Since natural varitation in pest populations exists, many pests have already demonstrated the capacity for *B.t.* resistance, and, as selection intensity increases, resistance in key pest populations is virtually assured. This situation requires proactive measures to avert or delay resistance development (Kennedy and Whalon 1995). Governments and agricultural policy makers are beginning to understand this; therefore, many new policies are being developed that mandate some level of resistance management program as a condition of registration. Some of these policy options have been reviewed by Whalon and Norris (1996).

Resistance management is the delaying or preventing of resistance in pest species. This definition holds regardless of whether one is considering adaptation of insects to insecticides, conventionally devised varieties with plant defense mechanisms, or transgenic *B.t.* plants. Resistance management may also be viewed as a process of managing an important genetic resource, much the same as managing other renewable resources like forests, soil, water, minerals, and even genetic diversity, except that the object of conservation is susceptibility genes or alleles. From this viewpoint, a pest is a dynamic gene pool which can be managed in a positive (conservation) or negative (exploitation) manner. If exploited, significant socioeconomic and environmental consequences can result, i.e., resistance. If conserved, adverse effects are avoided or minimized. Comins (1977) viewed resistance as an inevitable consequence of insecticide use, and suggested that a major goal of any insect control policy should be to efficiently exploit the limited lifetime of insecticides.

The literature on resistance management strategies and tactics for synthetic organic insecticides is extensive. However, the hierarchical ecological, biological, or operational organization used by various authors can yield terminology differences. For example, Croft (1990) focused primarily on the ecosystem, community, and population levels while others have focused on the population to the genetic levels. Georghiou (1990b) took a strategic-operational focus and his "management by moderation" is probably synonymous with Croft's (1990) "reduced selection pressure" and Tabashnik's (1990) "reduced selection," which focus on the population to organism levels. Georghiou's "management by saturation" aligns with Roush's (1989) "destroy heritability" and Tabashnik's (1990) "reduce heritability," which focus on the genetic level. Georghiou's (1990b) "multiple attack" is a somewhat confusing term if used alone because it could imply a coupling of strategies or, as the text states, a more narrow combination and/or rotation of chemical tools.

Properly held, resistance management is a strategy within the philosophy of integrated pest management (IPM) because it contributes to IPM's ultimate goal of implementing the best set of management tactics to limit pest populations below economic injury levels while minimizing socioeconomic and environmental impact. IPM is usually understood at the ecosystem or community levels. Resistance management within the context of IPM usually involves four strategies which may or may not be combined: (1) diversification of mortality sources such that a pest is not selected by a single mortality mechanism; (2) reduction of selection pressure for each major mortality mechanism; (3) maintenance of susceptible individuals through refuges and/or immigration; and (4) development of resistance progress estimation and/or prediction through development of diagnostic tools, monitoring, and, perhaps, models. Several authors have emphasized developing resistance management programs that integrate into IPM systems to take advantage of reduced selection pressure and biological information development processes that such management programs can provide (Tabashnik and Croft 1982; Croft

1990; Tabashnik 1990; Denholm and Rowland 1992; McGaughey and Whalon 1992). Our experience also suggests that it is much easier educationally, economically, and tactically to modify an existing IPM sampling and delivery program to include resistance management tactics than to initiate a program predicated on resistance management alone. However, some of the specific tactics used in resistance management reflect primarily pesticide use terminology such as high dosage, moderate dosage, low dosage, mixtures, rotations, mosaics, or combinations of these tactics. If one moves directly to this pesticide tactical focus, the IPM framework for resistance management can be overlooked.

Because insect resistance to *B.t.* is a relatively recent phenomenon, the literature dealing with its management is mostly theoretical. For conventionally sprayed *B.t.*, approaches used for managing resistance to chemical insecticides should be applicable. Much of this chapter's discussion focuses on the potential for resistance and the need for resistance management strategies for transgenic crops. Transgenic plants expressing a *B.t.* toxin are already a reality and commercial release of transformed cotton, corn, rice and potato have already occurred. Gasser and Fraley (1989) reported significant progress in transforming 22 herbaceous dicots, three woody dicots, and five monocots, and the list is probably around 45 species today. Genetic engineering offers tremendous potential for producing plants resistant to pests, increasing yields, and enhancing natural starches, oils, and proteins (e.g., see the reviews by Gasser and Fraley 1989 and Meeusen and Warren 1989). In insect control, deploying toxin-producing genes in genetically engineered cultivars offers several advantages over conventional spraying, including (1) providing season-long insect control without costly, repeated application; (2) reducing impact on nontarget organisms, including predators and parasites, because only plant damaging species are exposed; and (3) protecting plant parts that are difficult or impossible to reach with conventional application of pesticides, such as roots, lower leaves, and new growth. While there have not been any reports of insect resistance to transgenic crops, it is generally assumed that pests will adapt to transgenic plants just as they have to conventionally deployed *B.t.* Several authors have discussed this risk and advised a cautious approach to using genetically engineered plants (e.g., Gould 1988a, b; Denholm and Rowland 1992; McGaughey and Whalon 1992).

Transgenic plant deployment of insect toxins has much in common with insect control through conventional host-plant resistance approaches, and many species of insects, viruses, bacteria, fungi, and nematodes have adapted to conventionally selected insect- or disease-resistant plant cultivars. Probably one of the best examples of this phenomenon is the success of the Hessian fly, *Mayetiola destructor*, in evolving resistant biotypes to at least 13 genes that confer resistance in wheat (Gallun 1977; Hatchett et al. 1981). Similar race formation has occurred against various pathogen-resistant crop cultivars (see the review by Kiyosawa 1982). Although host-plant resistance has been very successful in many instances, it is not without controversy, particularly concerning appropriate deployment strategies and tactics. Kennedy et al. (1987) reviewed various considerations in the deployment of insect-resistant plants.

6.2 Specific Strategies for Managing Resistance

Specific gene deployment strategies to improve durability of *B.t.* toxins are being widely discussed in industry, academia, and government forums, but only preliminary field research data are available on their comparative utility. Gould (1988a,b), Brunke and Meeusen (1991), Stone et al. (1991), and others have presented the rationale for many of these techniques. We have organized the transgenic strategies and tactics that could improve durability in Table 2, but since this is a rapidly developing field, the list of tactics is likely to change significantly over the next few years and economic and regulatory constraints will probably play key roles in determining the actual cultivars, strategies, and tactics eventually deployed.

The tactics being discussed are generally patterned after those used or proposed for use in managing chemical insecticide resistance and typically involve variations of (1) rotation or alternation of toxins, (2) mixtures or sequences of toxins; (3) high or low doses of toxin; (4) encouraging survival of susceptible insects by providing refuges or promoting immigration; and (5) alteration of the tissue, timing, or induction of *B.t.* toxin expression in transgenic plants.

6.2.1 *Rotation or Alternation*

Rotation or alternation of *B.t.* toxins with other toxins, insecticides, cultural, and/or biological control strategies is probably the simplest approach to resistance management and one that has been most widely used with chemical insecticides. Its effectiveness depends upon restoration of susceptibility in pest populations when selection pressure is discontinued or changed to another gene, toxin, or organism. This approach assumes that there is some fitness cost (pleiotropy)

Table 2. The potential strategies and tactics for deploying insecticidal genes in plants

	Gene tactics		
Gene strategies	Promoter	Expression	Operational tactics
Single gene	Constitutive	High dose	Uniform seed
Multigenic	Tissue specific	Low dose	Seed mixture
pyramid			multilines
stacked	Induction	Mixtures	
	wound		Refugia
Chimeric gene(s)	phenology		
	elicitor		Seed rotation
			sequential release
	Combinations above	Combinations above	Seed Mosaic
			spatial and temporal
			Combinations
			above
			within IPM

associated with resistance and that reversion to a more susceptible condition will occur if selection pressure is reduced. Studies on *B.t.*-resistant *L. decemlineata, P. interpunctella, P. xylostella*, and *H. virescens* have already shown that in the early stages of selection when resistance levels are still low, resistance is relatively unstable and susceptibility is quickly restored when selection is discontinued (McGaughey and Beeman 1988; Sims and Stone 1991; Tabashnik et al. 1991; Whalton et al. 1993; Rahardja and Whalon 1995). However, McGaughey and Beeman (1988) found that high levels of resistance in *P. interpunctella* were stable for long periods. Additionally, in highly resistant strains of *P. xylostella*, susceptibility is restored much more slowly in the absence of treatments, suggesting that rotations may not be an effective means for retarding resistance development (Tabashnik et al. 1992a).

6.2.2
Mixtures

Mixing toxins is another relatively simple approach that is possible in both conventional and transgenic plant deployment (Gould 1988a). In transgenic plants there are two ways to achieve a mixture. First, two or more different seed lines can be engineered with different toxins and the seeds mixed before planting to produce a multiline or seed mixture. This approach is applicable in crops grown from seed, but it poses practical problems for vegetatively propagated crops like potatoes because seed handling and physical mixing would require extensive investment by the industry. Secondly, two or more *B.t.* toxin genes or other insecticidal proteins could be engineered into the same plant cultivar to produce a stacked or pyramided transgenic plant (Bosch et al. 1994). A European patent application involving this approach has been filed (Van Mellaert et al. 1991). This strategy is also possible in conventional sprays with fused proteins (Puntambekar et al. 1995), and other classes of protein toxins could also be mixed with *B.t.* toxins like alpha anti-insect scorpion neurotoxin (Chejanovsky et al. 1995). Another possibility in mixtures is the pyramiding or stacking *B.t.* toxin genes with natural allelochemicals like terpenoids (Benedict et al. 1993a; Sachs et al. 1993), phenolic glycosides (Arteel and Lindroth 1992), or protease inhibitors (Tabashnik et al. 1992) in crops like cotton, aspen, and vegetables.

In general, experience with chemicals has shown that mixtures are likely to be more durable than a single toxin, but they are not always better than alterations (Curtis 1985; Mani 1985; Tabashnik 1989). Cox and Hatchett (1986) and Gould (1986a,b 1988a) used simulation models of conventionally developed host-plant resistance to compare some of the gene-deployment strategies and tactics listed in Table 2, including sequential release of two single-gene factors (seed rotation or alternation), random spatial mixtures of two single-gene factors (mosaics or multilines), and pyramiding (stacking) two genes in a single cultivar. They found that durability varied depending upon the number of resistant alleles that the pest had, the manner of inheritance (dominant or recessive), and epistasis. Pyramided cultivars were sometimes better than mixtures or sequences, particularly if susceptible plants were mixed in to provide refugia for susceptible insects (pyramided mixture). Each crop-pest system will require research to determine which mixing approach works best.

The durability of conventionally applied mixtures may depend upon equal persistence of the insecticides involved (Roush 1989), and this may not be a consideration with season-long, constitutively *B.t.*-expressing plants. If persistence differs, differential selection could occur, perhaps leading to accelerated resistance to both insecticides. Roush (1989) suggested that the most effective field approach to using mixtures might be to measure the persistence and efficacy of each independently before using them in combination. How many days or hours of differential attenuation, and how much difference in mortality constitute a real effect in resistance development is unknown. For transgenic plants, the problem is one of differences in level of expression or insect sensitivity rather than persistence. Furthermore, this problem may be compounded in situations where crops are attacked by two or more pest species that differ in susceptibility to the toxins used.

Multiple toxin approaches also require that toxins with different modes of action be available for mixing or replacement once resistance occurs. Some plant engineers assert that we will eventually have an almost inexhaustible supply of different *B.t.* toxins available and that other types of insecticidal proteins will be developed as well. Unfortunately, this is not generally true now. Extensive cross resistance among different *B.t.* toxins may reduce the likelihood that mixtures, alterations or rotations of *B.t.* toxins will effectively control resistance. Indeed, the evidence of binding site heterogeneity among different insect species (Van Rie et al. 1989, 1990a, b) and indications of extensive cross-resistance in *H. virescens* (Gould et al. 1991b) confirm that patterns of cross-resistance among *B.t.* toxins probably will differ among species of insects. Extensive research will be needed in order to elucidate these patterns before suitable mixtures or rotations can be recommended with any assurance of preventing or delaying resistance. Furthermore, studies on *P. xylostella* and *Plodia interpunctella* show that mixtures of *B.t.* toxins do not preclude the evolution of resistance and that they do not greatly improve durability (Tabashnik et al. 1991; McGaughey and Johnson 1992, Tabashnik et al. 1992a). *P. interpunctella* can evolve resistance to a variety of different *B.t.* toxins, both singly and in mixtures (Tabashnik and McGaughey 1994). Additional controversy has come through the proponents of spray *B.t.* versus transgenic *B.t.* (Roush 1994) and a conference sponsored by the United States Department of Agriculture (USDA was organized in 1996 around this question. The proceeding from this process will be available from the second author W.H. McGaughey sometime in 1997.

Reviewers tend to agree on the theoretical benefits of managing resistance within IPM frameworks that provide diverse sources of pest mortality (e.g., Tabashnik and Croft 1982; Denholm and Rowland 1992; Becker and Ludwig 1993; Tshernyshev 1995). However, actual studies on how pest adaptation to resistant cultivars is affected by combining host-plant resistance with chemical, biological, and cultural controls are limited and more research is needed in this area. Tanaka (1992) tested *B.t.* with various pheromone, light trap-out, and vacuum strategies in glasshouse studies. Others (Huang et al. 1994; Plapp 1993 and Trumble et al. 1994) all concluded that IPM together with spray or transgenic

B.t. would be the best strategy to avert resistance development. Gould et al. (1991c) and Johnson and Gould (1992) concluded from theoretical studies that natural enemies could either increase or decrease rates of adaptation to resistant cultivars, but, in general, selection for adaptation is lower when pest control is achieved by multiple mortality factors. Research is needed to address field situations where transgenic and conventionally applied *B.t.* may be used to control the same pest population. In some instances, the selection of pests on several alternate hosts treated with *B.t.* may exacerbate resistance development. This is especially important where interbreeding populations of pests may be exposed to the same or different *B.t.* toxins in a transgenic crop and on the same or other crops treated by conventional means (Kennedy and Whalon 1995). Ultimately, the best strategy is to stop using *B.t.*, as Tabashnik et al. (1994) advocated.

6.2.3 *Refugia*

Developing and maintaining refuges to insure the survival of susceptible insects may be the only certain resistance management tactic. Several studies using mathematical modeling have shown that providing refuges to encourage survival of susceptible genotypes or encouraging immigration or releasing susceptible insects into pest populations under selection can greatly slow the evolution of resistance (Comins 1977; Georghiou and Taylor 1977a, b, 1988a; Taylor and Georghiou 1979; Tabashnik and Croft 1982; Gould 1986a, b; Roush 1989; Stone et al. 1991; Denhom and Rowland 1992). In crops with limited distribution that are attacked by pests with wide host ranges and expansive dispersal behavior, natural refuges probably occur and additional refuges may not be necessary to enhance survival of susceptible genotypes. In pests with more restricted host ranges, benefits may accrue from providing untreated areas where susceptible insects can survive. Field research is needed to understand the optimum spatial or temporal scale of refuges because they will almost certainly differ for each pest-host situation. With *B.t.*-expressing transgenic plants and pest species that are not too mobile, mixtures of seeds of susceptible and resistant plants could be appropriate (Gould 1986a, 1988a; Tabashnik 1994). Tissue- or temporal-specific gene expression also might provide refuges in either space or time for survival of susceptible genotypes. For more mobile pest species, occasional rows, entire fields, or perhaps regions of untreated crops may be necessary to insure survival of unselected susceptible insects. This strategy would appear to be most useful within an IPM program where regular monitoring of pest population density and movement is already underway.

6.2.4 *Low Doses*

Reduction of rates, reduced frequency of application, reduced thoroughness of application, and transgenic plants with low expression of toxin are known collectively as "low dose" approaches. Low dose tactics which aim to reduce populations only slightly or slow larval development to the point that the number of generations per year is reduced have obvious advantages in terms of selection pressure. However, from a practical standpoint, pest managers often prefer highly efficacious products to prevent any damage. Therefore, grower acceptance of low dose tactics may depend upon effective incorporation into IPM programs. In instances where the pest population equilibrium position is far below the economic injury level or where naturally occurring biological control agents could suppress the pest population if it is partially controlled by *B.t.*, the low dosage tactic may be very durable. Here again, this tactic may require some *a priori* knowledge of resistance through previous research, the application of historical field experience, heuristics, and/or an appropriate sampling system.

6.2.5 *High Doses*

In resistance management, a high dose is usually defined as the rate which consistently kills resistant heterozygotes, the most abundant carriers of resistance (i.e., that renders resistance functionally recessive). It has long been recognized that the rate of resistance progression depends on its functional dominance (Taylor and Georghiou 1979; Georghiou 1980; Tabashnik and Croft 1982). The actual level or dose depends on the inheritance of resistance. It would be lowest in cases where resistance is recessive and highest where completely dominant. This approach is not generally successful with conventional application because of the cost of high dose formulations, lack of uniform coverage with current spray equipment, rapid attenuation of the toxin deposits, and because conventional applications cannot consistently target the most susceptible larval stages.

A high dose strategy in conjunction with untreated refuges has recently been advocated as a potential means of managing resistance development in transgenic plants (Denholm and Rowland 1992). Continuous expression of *B.t.* toxins in all tissues of transgenic plants may be sufficiently uniform and continuous to assure that heterozygotes, the most abundant carriers of resistance genes, are always killed. Then, by providing refuges to insure a continuous influx of susceptible genotypes to mate with the relatively rare homozygous resistant individuals, resistance could be effectively diluted or maintained at a very low level. Since homozygous resistant individuals are at a very low frequency early in the evolution of resistance and suitable refuges provide a continuous source of susceptible individuals, this tactic could be quite durable. However, some obvious problems are apparent where more than one *B.t.*-susceptible pest attacks a crop and where the susceptibility or mode of inheritance of resistance of these pests to *B.t.* differs.

An "ultra high dose" approach may be possible in instances of high insect sensitivity and optimized *B.t.* expression in transgenic plants. Such a dose is one sufficient to kill even homozygous resistant individuals and thus it would provide a condition where the pest does not have the evolutionary machinery to adapt. In other words, the transgenic plant essentially becomes a nonhost. The broad implications of the ultra high dose tactic have been historically unrealized in pest management. It has economically and socially staggering implications if successful. This approach may be plagued with the same philosophical, biological, and genetic conditions that led to our current situation with synthetic organic insecticides. That is, pest organisms usually have found a way to overcome single-tactic, unilateral mortality mechanisms, and it may be physically and biologically impossible to determine a *priori* whether the ultra high dose is high enough. Historically, many agricultural field biologists have observed that every conventionally applied chemical attenuates, and another pest or another resistance mechanism eventually can be marshalled to overcome man's tactics. The season-long stability of transgenic field expression remains uncertain (Murray et al. 1991), and any depletion of the toxin level even in occasional plants in a field could eventually lead to resistance development. We can neither deny the possibilities of the ultra high dose nor ignore the risks involved. Certainly, prerelease experimentation with existing *B.t.*-resistant strains of insects, cautious deployment strategies, and monitoring would be appropriate to avert resistance development. In general, cotton, corn, and potato *B.t.* transgenic registrations have followed this cautious approach.

6.2.6
Specific Gene Promoters

The timing, location, and induction of *B.t.* gene expression in plants are all exciting possibilities for future variations in transgenic crop deployment. Currently, most transgenic plants express genes constitutively, e.g., throughout development and in all plant tissues. This approach may cause great selection pressure on pest populations unless the effects can be ameliorated. Alternatively, it has been suggested that specific gene promoters (Table 2) may be available that would cause the toxins to be expressed only in certain plant tissues (tissue specific, e.g., Koziel et al. 1993), at certain growth stages (temporal specific), only in response to insect feeding (wound specific, e.g., Reynaerts and Jansens 1994), or only when induced by application of some elicitor (elicitor specific) (Gould 1988b; Stone et al. 1991; Van Rie 1991). Unfortunately, transgenic plants with these promoters may not be available for some time, and their utility in managing resistance will need to be evaluated. This approach seems particularly useful in crop plants like potatoes to avoid expressing the toxin in the edible tubers or in cotton to target the *B.t.* toxin in economically important plant parts like bolls. However, if tissue specific promoters cause *B.t.* toxins to be expressed in some tissues at a very high level, but are not completely effective in stopping expression in all other tissues, resistance could be accelerated rather than delayed because insects would be selected differentially based upon the tissue they utilize or their ability to detect the presence of the toxin (Hoy and Head 1995). Therefore, promoters could quickly increase the complexity of resistance selection based on host utilization behavior by different life stages of key insects, alternate host exposure for mobile insects, and differential processing of the *B.t.* gene through plant development. Considerable challenge also still exists in understanding translational and post translational processing and stability of mRNA and *B.t.* protein in transgenic plants (Murray et al. 1991). Any change in the stability, continued processing, or degradation of mRNA through plant development, maturity, and senescence could provide a change in selection pressure and result in resistance development.

To our knowledge, only limited studies have been done on behavioral responses of *B.t.*-resistant insects to transgenic plants. Gould and Anderson (1991), Gould et al. (1991a), and Whalon et al. (1993) have evaluated *B.t.*-susceptible and resistant strains of *H. virescens* and *L. decemlineata* for behavioral responses to *B.t.*-treated diets or plants conventionally treated with *B.t.* Sensitive and resistant strains of H. virescens avoided the *B.t.*-treated diets in choice/no choice studies, while *L. decemlineata* did not discriminate between treated and untreated plants in caged choice tests, but did discriminate in movement studies. Further work is needed to assess the role of behavior and other biological, ecological, and genetic factors in resistance development to *B.t.* and to *B.t.*-transgenic plants. Industry is understandably slow to release its engineered plants, particularly for resistance studies, but in our view these studies are essential for the wise deployment of potentially vulnerable cultivars.

6.3 Previous Resistance Management Programs

Pesticide resistance management has led to a unique cooperation between industry, academia and government. This is not to say that further cooperation is not needed, but that laudable progress has been made. Industry's insecticide (IRAC), fungicide (FRAC), and herbicide (HRAC) resistance action committees have established a record of responsiveness to critical resistance issues. More focused groups like the US Pyrethroid Efficacy Group (PEG) have significantly impacted resistance management implementation in cotton (Whalon and Hollingworth, 1989–present). In the USA a *B.t.* Management Working Group initiated through many industry scientists' efforts has fostered research on *B.t.* resistance and resistance management, but much is still necessary to significantly change current *B.t.* use practices.

Special committees within national societies throughout the world have fostered symposia, communication, and research. The US land grant university Experiment Station Committee on Policy (ESCOP) has sponsored a subcommittee on resistance and a multidisciplinary, multiregional communication committee (WRCC-60) on resistance and resistance management. This group, together with the IRAC Central Committee, initiated an international Resistant Pest Management (RPM) Newsletter which now has a circulation of over 2600 government, academic, and industry scientists and policy makers. The newsletter is funded through gifts and grants from government, industry, and academia in approximately equal proportions (Whalon and Hollingworth 1991) and subscriptions are free and/or available on the World Wide Web (http://www.msstate.edu/EntHome/Entomology.html). An International Organization for Resistant Pest Management (IOPRM in the USA or IRPM in Europe) with an administrative liaison in the US Environmental Protection Agency (EPA) convened a large forum in 1993 to develop working groups that could implement resistance management programs internationally. IOPRM has four working groups that have received public and private funding to implement programs in India, Poland, China, and Mexico.

7 Conclusions

Bacillus thuringiensis (*B.t.*) δ-endotoxins provide an environmentally safe alternative to chemical insecticides for controlling many species of pest insects. Recent biotechnological developments offer the promise of even greater use of these toxins in genetically transformed pest-resistant crops. However, the recent discovery that insects can adapt to these toxins raises concerns about the long-term durability of *B.t.* whether applied conventionally or transgenically.

B.t. δ-endotoxins are synthesized by the bacterium as crystalline protein inclusions that are toxic to various species of Lepidoptera, Coleoptera, and Diptera. Upon ingestion by susceptible insects, the crystals dissolve in the alkaline conditions of the gut and are proteolytically converted into smaller, toxic polypeptides. These polypeptides bind to glycoprotein receptors on the midgut cell membrane, generate pores in the membrane, and cause the cells to lyse. There is a great deal of diversity in host spectrum among *B.t.* toxins that is determined primarily (but perhaps not exclusively) through the specificity of midgut receptors for certain toxins. Changes in binding site specificity have also been implicated in resistance of lepidopteran pests to *B.t.*

Because of the great diversity of *B.t.* toxins occurring in nature, it has been suggested that these or other toxins might be used in mixtures, rotations, or sequences to delay pest adaptation. However, considerable controversy exists regarding how transgenic plants can or should be deployed to delay potential resistance development. Expression and dosage of gene products are functions of the various promoter, transcriptional, and translational factors associated with each resistance gene. The high dose, constituitively expressed gene tactic may be particularly vulnerable to resistance development since it could lead to selection of all plant feeding life stages of pests on all parts of the plants throughout the entire growing season. However, some researchers believe that the high dose or ultra high dose tactics are a compelling advantage of transgenic plants that could be used effectively in conjunction with refuges for susceptible insects without causing the pests to adapt. Although IPM combinations and mixtures of *B.t.* genes with other mortality mechanisms, together with refuges for susceptible insects, are probably better approaches to managing resistance, the economic and regulatory environment for transgenic plants may initially discourage all approaches other than plants constitutively expressing high doses of a single toxin.

The possible operational tactics for conventional and transgenic deployment are theoretically expansive and consitute many options for development and deployment (Table 2). None present clear advantages in all environments and with all pests except, perhaps, measures to encourage survival or immigration of susceptible genotypes. In addition, many of the operational tactics being considered will require industry investment or changed grower behavior to implement (Kennedy and Whalon 1995). These changes to current practice will face the same adoption challenges at the field level that other major innovations in pest management have required (Lambur et al. 1985). Given the slow pace of development of new insecticides, environmental and social concerns with synthetic organic insecticides, and the world wide competitive agricultural situation, the development of *B.t.* toxins and their management is at a crisis point. Scientifically responsible action cannot but advocate the resources, expanded research, and information delivery systems to sustainably manage these resources for the future of society.

References

Adang MJ (1991) *Bacillus thuringiensis* insecticidal crystal proteins: gene structure, action, and utilization. In: Maramorosch K (ed) Biotechnology for biological control of pests and vectors. CRC Press, Boca Raton, pp 3–24

Adang, MJ, Staver MJ, Rocheleau TA, Leighton J, Barker RF, Thompson DV (1985) Characterized full-length and truncated plasmid clones of the crystal protein of *Bacillus thuringiensis* subsp. *kurstaki* HD-73 and their toxicity to *Manduca sexta*. Gene 36: 289–300

Adang MJ, Firoozabady J, Klein J, DeBoer D, Sekar V, Kemp JD, Murray E, Rocheleau TA, Rashka K, Staffeld G, Stock C, Sutton D, Merlo DJ (1987) Expression of a *Bacillus thuringiensis* insecticidal crystal protein gene in tobacco plants. In Molecular strategies for crop protection. UCLA Symposia on molecular and cellular biology, vol 48. Alan R. Liss, New York, pp 345–353

Ahmad W, C Nicholls, DJ Ellar (1989) Cloning and expression of an entomocidal protein gene from *Bacillus thuringiensis galleriae* toxic to both Lepidoptera and Diptera. FEMS Microbiol Lett 59: 197–202

Alstad DN, Andow DA (1995) Managing the evolution of insect resistance to transgenic plants. Science 268: 1894–1896

Andrews, RE Jr., MM Bibilos, LA Bulla, Jr. (1985) Protease activation of the entomocidal protoxin of *Bacillus thuringiensis* subsp. *kurstaki*. Appl Environ Microbiol 50: 737–742

Aronson, AI, E-S Han, W McGaughey, D Johnson (1991) The solubility of inclusion proteins from *Bacillus thuringiensis* is dependent upon protoxin composition and is a factor in toxicity to insects. Appl Environ Microbiol 57: 981–986

Arteel, GE, RL Lindroth (1992) Effects of aspen phenolic glycosides on gypsy moth (Lepidoptera: Lymantriidae) susceptibility to *Bacillus thuringiensis*. Great Lakes Entomol. 25(4): 239–244

Atwa, WA, HA Abdel-Rahman (1974) Histopathological effects of *Bacillus thuringiensis* Berl. on larvae of *Pieris rapae* (L.) (Lep., Pieridae). Z Angew Entomol 76: 326–331

Barton, KA, HR Whiteley, N-S Yang (1987) *Bacillus thuringiensis* δ-endotoxin expressed in transgenic *Nicotiana tabacum* provides resistance to lepidopteran insects. Plant Physiol 85: 1103–1109

Bauer LS, Pankratz HS (1992) Ultrastructural effects of *Bacillus thuringiensis* var. *san diego* on midgut cells of the cottonwood leafbeetle. J Invertebr Pathol 60: 15–25

Bauer, LS, CN Koller, DL Miller, RM Hollingworth (1994) Laboratory selection of the cottonwood leaf beetles, *Chrysomela scripta*, for resistance to *Bacillus thuringiensis* var. *tenebrionis* delta-endotoxin. In: Abstracts 6th Int Collog Invertebr Pathol Montpellier, France, Aug 28 Sept 2, 1994, p 68

Becker, N, M Ludwig (1993) Investigations on possible resistance in *Aedes vexans* field populations after a 10-year application of *Bacillus thuringiensis israelensis*. Am Mosq Control Assoc 9(2): 221–224

Benedict, JH., ES Sachs, DW Altman, DR Ring, RR DeSpain, DJ Lawlor (1993a) Resistance of glandless transgenic *B.t.* cotton to injury from tobacco budworm. Proc 1993 Beltwide Cotton Conferences vol, 2, pp 814–816

Benedict, JH, ES Sachs, DW Altman, DR Ring, TB Stone, SR Sims (1993b) Impact of δ-endotoxin-producing transgenic cotton on insect–plant interactions with *Heliothis virescens* and *Helicoverpa zea* (Lepidoptera: Noctuidae). Environ Entomol 22(1): 1–9

Boman, HG (1981) Insect responses to microbial infection. In: Burges HD (ed) Microbial control of pests and plant diseases 1970-1980. Academic Press, New York, pp 769–784

Bosch, D, B Schipper, H van der Kleij, RA de Maagd, WJ Stiekema (1994) Recombinant *Bacillus thuringiensis* crystal proteins with new properties: possiblities for resistance management. BioTechnology 12: 915–918

Boulter, D, JA Gatehouse, AMR Gatehouse, VA Hilder (1990) Genetic engineering of plants for insect resistance. *Endeavour* 14: 185–190

Briese, DT (1981) Resistance of insect species to microbial pathogens. In Davidson, EW (ed) Pathogenesis of invertebrate microbial diseases. Allanheld Osmun, Totowa, New Jersey, pp 511–545

Briese, DT (1986) Host resistance to microbial control agents. In: Franz JM (ed) Fortschritte der Zoologie, vol 32 Biological plant and health protection. G Fischer, New York, pp 233–256

Broadway RM (1989) Characterization and ecological implications of midgut proteolytic activity in larval *Pieris rapae* and *Trichoplusia ni*. J Chem Ecol 15: 2101–2113

Brunke, KJ, RL Meeusen (1991) Insect control with genetically engineered crops. TIBTECH 9: 197–200

Bulla, LA, Jr., KJ Kramer, LI Davidson (1977) Characterization of the entomocidal parasporal crystal of *Bacillus thuringiensis*. J Bacteriol 130: 375–383

Bulla, LA, Jr., KJ Kramer, DJ Cox, BL Jones, LI Davidson, GL Lookhart (1981) Purification and characterization of the entomocidal protoxin of *Bacillus thuringiensis*. J Biol Chem 256: 3000–3004

Burges, HD (1971) Possibilities of pest resistance to microbial control agents. In: Burges HD, Hussey NW (eds) Microbial control of insects and mites. Academic Press, New York, pp 445–457

Burges, HD (1986) Impact of *Bacillus thuringiensis* on pest control with emphasis on genetic manipulation. MIRCEN J 2: 101–120

Carlberg, G, R Lindstrom (1987) Testing fly resistance to thuringiensin produced by *Bacillus thuringiensis*, serotype H-1. J Invertebr Pathol 49: 194–197

Carlton, BC (1988) Development of genetically improved strains of *Bacillus thuringiensis*. In: Biotechnology for crop protection. Am Chem Soc Symp Series. Am Chem Soc, Washington, DC, pp 260–279

Carroll, J, J Li, DJ Ellar (1989) Proteolytic processing of a coleopteran-specific delta-endotoxin produced by *Bacillus thuringiensis* var. *tenebrionis*. Biochem J 261: 99–105

Chejanovsky, N, N Zilberberg, H Rivkin, E Zlotkin, M Gurevitz (1995) Functional expression of an alpha anti-insect scorpion neurotoxin in insect cells and lepidopterous larvae. FEBS Lett 376: 181–184

Chiang, AS, DF Yen, WK Peng (1986) Defense reaction of midgut epithelial cells in the rice moth larva (*Corcyra cephalonica*) infected with *Bacillus thuringiensis*. J Invertebr Pathol 47: 333–339

Chilcott CN, Knowles BH, Ellar DJ, Drobniewski FA (1990) Mechanism of action of *Bacillus thuringiensis israelensis* parasporal body. In: deBarjac H, Southerland DJ (eds) Bacterial control of mosquitoes and black flies. Rutgers Univ Press, New Brunswick, New Jessey, pp 45–65

Comins, HN (1977) The development of insecticide resistance in the presence of migration. J Theor Biol 64: 177–197

Cox, TS, JH Hatchett (1986) Genetic model for wheat/Hessian fly (Diptera: Cecidomyiidae) interaction: strategies for deployment of resistance genes in wheat cultivars. Environ Entomol 15: 24–31

Crickmore, N, C Nicholls, DJ Earp, TC Hodgman, DJ Ellar (1990) The construction of *Bacillus thuringiensis* strains expressing novel entomocidal δ-endotoxin combinations. Biochem J 270: 133–136

Croft, BA (1990) Developing a philosophy and program of pesticide resistance management. In: Roush RT, Tabashnik BE (eds) Pesticide resistance in arthropods. Chapman and Hall, New York, pp 277–296

Curtis, CF (1985) Theoretical models of the use of insecticide mixtures for the management of resistance. Bull Entomol Res 75: 259–265

de Barjac, H, E Frachon (1990) Classification of *Bacillus thuringiensis* strains. Entomophaga 35: 233–240

Denholm I, MW Rowland (1992) Tactics for managing pesticide resistance in arthropods: theory and practice. Annu Rev Entomol 37: 91–112

Diehl, SR., GL Bush (1984) An evolutionary and applied perspective of insect biotypes. Annu Rev Entomol 29: 471–504

Dulmage, HT, and Cooperators (1981) Insecticidal activity of isolates of *Bacillus thuringiensis* and their potential for pest control. In: Burges HD (ed) Microbial control of pests and plant diseases 1970-1980. Academic Press, London, pp 193–223

Ebersold, HR, P Luthy, M Muller (1977) Changes in the fine structure of the gut epithelium of *Pieris brassicae* induced by the delta endotoxin of *Bacillus thuringiensis*. Bull Soc Entomol Suisse 50: 269–276

Endo, Y, J Nishiitsutsuji-Uwo (1980) Mode of action of *Bacillus thuringiensis* delta-endotoxin: histopathological changes in the silkworm midgut. J Invertebr Pathol 36: 90–103

English L, Slatin SL (1992) Mode of action of delta-endotoxins from *Bacillus thuringiensis*: a comparison with other bacterial toxins. Insect Biochem Mol Biol 22: 1–7

Estada, U, J Ferre (1994) Binding of insecticidal crystal proteins of *Bacillus thuringiensis* to the midgut brush border of the cabbage looper, *Trichoplusia ni* (Hübner) (Lepidoptera: Noctuidae), and selection for resistance to one of the crystal proteins. Appl Environ Microbiol 60: 3840–3846

Farrar, RR Jr., RL Ridgway (1995) Feeding behavior of gypsy moth (Lepidoptera: Lymantriidae) larvae on artificial diet containing *Bacillus thuringiensis*. Environ Entomol 24(3): 755–761

Faust, RM, K Abe, GA Held, T Iizuka, LA Bulla, CL Meyers (1983) Evidence for plasmid-associated crystal toxin production in *Bacillus thuringiensis* subsp. *israelensis*. Plasmid 9: 98–103

Feigin, JM (1963) Exposure of the house fly to selection by *Bacillus thuringiensis*. Ann Entomol Soc Am 56: 878-879

Ferre, J. MD Real, J Van Rie, S Jansens, M Peferoen (1991) Resistance to the *Bacillus thuringiensis* bioinsecticide in a field population of *Plutella xylostella* is due to a change in a midgut membrane receptor. Proc Natl Acad Sci USA 88: 5119–5123

Ferro, DN, SM Lyon (1991) Colorado potato beetle (Coleoptera: Chrysomelidae) larval mortality: operative effects of *Bacillus thuringiensis* subsp. *san diego*. J Econ Entomol 84: 806–809

Fischhoff, DA, KS Bowdish, FJ Perlak, PG Marrone, SM McCormick, JG Niedermeyer, DA Dean, K Kusano-Kretzmer, EJ Mayer, DE Rochester, SG Rogers, RT Fraley (1987) Insect tolerant transgenic tomato plants. BioTechnology 5: 807–813

Gallun, RL (1977) Genetic basis of Hessian fly epidemics. Ann NY Acad Sci 287: 223–229

Garcyznski, SF, MJ Adang (1995) *Bacillus thuringiensis* CryIA(c) δ-endotoxin binding aminopeptidase in the *Manduca sexta* midgut has a glycosyl-phosphatidylinositol anchor. Insect Biochem Mol Biol 25(4): 409–415

Garczynski, SF, JW Crim, MJ Adang (1991) Identification of putative insect brush border membrane-binding molecules specific to *Bacillus thuringiensis* δ-endotoxin by protein blot analysis. Appl Environ Microbiol 57: 2816–2820

Gasser, CS, RT Fraley (1989) Genetically engineering plants for crop improvement. Science 244: 1293–1299

Georghiou, GP (1980) Insecticide resistance and prospects for its management. Residue Rev 76: 131–145

Georghiou, GP (1990a) Resistance potential to biopesticides and consideration of countermeasures. In: JE Casida (ed) Pesticides and alternatives. Elsevier New York, pp 409–420

Georghiou, GP (1990b) Overview of insecticide resistance. In: MB Green, HM LaBaron WK Moberg (eds) Managing resistance to agrochemicals. Am Chem Soc Symp Ser 421, Am Chem Soc, Washington, DC , pp 18–41

Georghiou, GP, CE Taylor (1977a) Genetic and biological influences in the evolution of insecticide resistance. J Econ Entomol 70: 319–323

Georghiou, GP, CE Taylor (1977b) Operational influences in the evolution of insecticide resistance. J Econ Entomol 70: 653–658

Georghiou, GP, MG Vasquez (1982) Assessing the potential for development of resistance to *Bacillus thuringiensis* var. *israelensis* toxin (*B.t.i*) by mosquitoes. In: Mosquito control research: annual report. University of California, Davis, pp 80–81

Gill, SS, EA Cowles, PV Pietrantonio (1992) The mode of action of *Bacillus thuringiensis* endotoxins. Annu Rev Entomol 37: 615–636

Goldman, IF, J Arnold, BC Carlton (1986) Selection for resistance to *Bacillus thuringiensis* subspecies *israelensis* in field and laboratory populations of the mosquito *Aedes aegypti*. J Invertebr Pathol 47: 317–324

Gonzalez, JM, Jr., BC Carlton (1982) Plasmid transfer in *Bacillus thuringiensis*. Genet Cell Technol 1: 85–95

Gonzalez, JM, Jr., BC Carlton (1984) A large transmissible plasmid is required for crystal toxin production in *Bacillus thuringiensis* variety *israelensis*. Plasmid 11: 28–38

Gonzalez, JM, HT Dulmage, BC Carlton (1981) Correlation between specific plasmids and δ-endotoxin production in *Bacillus thuringiensis*. Plasmid 5: 351–365

Gonzalez JM Jr, Brown BJ, Carlton BC (1982) Transfer of *Bacillus thuringiensis* plasmids coding for δ-endotoxin among strains of *B. thuringiensis* and *B. cereus*. Proc Natl Acad Sci USA 79: 6951–6955

Gould, F (1984) Role of behavior in the evolution of insect adaptation to insecticides and resistant host plants. Bull Entomol Soc Am 30: 34–41

Gould, F (1986a) Simulation models for predicting durability of insect-resistant germ plasm: Hessian fly (Diptera: Cecidomyiidae) resistant winter wheat. Environ Entomol 15: 11–23

Gould, F (1986b) Simulation models for predicting durability of insect-resistant germ plasm: a deterministic diploid, two-locus model. Environ Entomol 15: 1–10

Gould, F (1988a) Evolutionary biology and genetically engineered crops: consideration of evolutionary theory can aid in crop design. BioScience 38: 26–33

Gould, F (1988b) Genetic engineering, integrated pest management and the evolution of pests. Trends in ecology and evolution (TREE). 3/TIBTECH 6: S15–S18

Gould, F, A Anderson (1991) Effects of *Bacillus thuringiensis* and HD-73 delta-endotoxin on growth, behavior, and fitness of susceptible and toxin-adapted strains of *Heliothis virescens* (Lepidoptera: Noctuidae). Environ Entomol 20: 30–38

Gould, F, A Anderson, D Landis, H Van Mellaert (1991a) Feeding behavior and growth of *Heliothis virescens* larvae on diets containing *Bacillus thuringiensis* formulations or endotoxins. Entomol Exp Appl 58: 199–210

Gould, F, A Martinez-Ramirez, A Anderson, J Ferre, F Silva, W Moar (1991b) Resistance to *B.t.* toxins in *Heliothis virescens*. Annu Meet of Entomol Soc Am, Reno, Nevada, Pap No 606

Gould, F, GG Kennedy, MT Johnson (1991c) Effects of natural enemies on the rate of herbivore adaptation to resistant host plants. Entomol Exp Appl 58: 1–14

Haider, MZ, DJ Ellar (1987) Analysis of the molecular basis of insecticidal specificity of *Bacillus thuringiensis* crystal δ-endotoxin. Biochem J 248: 197–201

Haider, MZ, BH Knowles, DJ Ellar (1986) Specificity of *Bacillus thuringiensis* var. *colmeri* insecticidal δ-endotoxin is determined by differential proteolytic processing of the protoxin by larval gut proteases. Eur J Biochem 156: 531–540

Hannay, CL, P Fitz-James (1955) The protein crystals of *Bacillus thuringiensis* Berliner. Can J Microbiol 1: 694–710

Harvey, TL, DE Howell (1965) Resistance of the house fly to *Bacillus thuringiensis* Berliner. J Invertebr Pathol 7: 92-100

Hatchett, JH, TJ Marin RW Livers (1981) Expression and inheritance of resistance to Hessian fly in synthetic hexaploid wheat derived from *Triticum tauschii* (Coss) Schmal. Crop Sci 21: 731–734

Heimpel, AM, TA Angus (1959) The site of action of crystalliferous bacteria in Lepidoptera larvae. J Insect Pathol 1: 152–170

Held, GA, LA Bulla, Jr., E Ferrari, J Hoch, AI Aronson, SA. Minnich (1982) Cloning and localization of the lepidopteran protoxin gene of *Bacillus thuringiensis* subsp. *kurstaki*. Proc Natl Acad Sci USA 79: 6065–6069

Hofmann, C, P Luthy (1986) Binding and activity of *Bacillus thuringiensis* delta-endotoxin to invertebrate cells. Arch Microbiol 146: 7–11

Hofmann, C, P Luthy, R Hutter, V Pliska. (1988a) Binding of the delta-endotoxin from *Bacillus thuringiensis* to brush-border membrane vesicles of the cabbage butterfly (*Pieris brassicae*). Eur J Biochem 173: 85–91

Hofmann, C, H Vanderbruggen, H Hofte, J Van Rie, S Jansens, H Van Mellaert (1988b) Specificity of *Bacillus thuringiensis* δ-endotoxins is correlated with the presence of high-affinity binding sites in the brush border membrane of target insect midguts. Proc Natl Acad Sci USA 85: 7844–7848

Hofte, H, HR Whiteley (1989) Insecticidal crystal proteins of *Bacillus thuringiensis*. Microbiol Rev 53: 242–255

Hofte, H, J Van Rie, S Jansens, A Van Houtven, H Vanderbruggen, M Vaeck (1988) Monoclonal antibody analysis and insecticidal spectrum of three types of lepidopteran-specific insecticidal crystal proteins of *Bacillus thuringiensis*. Appl Environ Microbiol 54: 2010–2017

Hoopingarner, R, MEA Materu (1964) The toxicology and histopathology of *Bacillus thuringiensis* Berliner in *Galleria mellonella* (Linnaeus). J Insect Pathol 6: 26–30

Hoy, CW, FR Hall (1993) Feeding behavior of *Plutella xylostella* and *Leptinotarsa decemlineata* on leaves treated with *Bacillus thuringiensis* and esfenvalerate. Pestic Sci 38: 335–340

Hoy, CW, G Head (1995) Correlation between behavioral and physiological responses to transgenic potatoes containing *Bacillus thuringiensis* δ-endotoxin in *Leptinotarsa decemlineata* (Coleoptera: Chrysomelidae). J Econ Entomol 88(3): 480–486

Huang, H, Z Smilowitz, MC Saunders, R Weisz (1994) Field evaluation of insecticide application strategies on development of insecticide resistance by Colorado potato beetle (Coleoptera: Chrysomelidae). J Econ Entomol 87(4): 847–857

Jahn, N, W Schnetter, K Geider (1987) Cloning of an insecticidal toxin gene of *Bacillus thuringiensis* subsp. *tenebrionis* and its expression in *Escherichia coli* cells. FEMS Microbiol Lett 48: 311–315

Jaquet, F, R Hutter, P Luthy (1987) Specificity of *Bacillus thuringiensis* delta-endotoxin. Appl Environ Microbiol 53: 500–504

Johnson, DE, GL Brookhart, KJ Kramer, BD Barnett, WH McGaughey (1990) Resistance to *Bacillus thuringiensis* by the Indianmeal moth, *Plodia interpunctella*: comparison of midgut proteinases from susceptible and resistant larvae. J Invertebr Pathol 55: 235–243

Johnson, M, F Gould (1992) Interaction of genetically engineered host plant resistance and natural enemies of *Heliothis virescens* (Lepidoptera: Noctuidae) in tobacco. Environ Entomol 21(3): 586–597

Kennedy, GG, ME Whalon (1995) Managing pest resistance to *Bacillus thuringiensis* endotoxins: constraints and incentives to implementation. J Econ Entomol 88(3): 454–460

Kennedy, GG, F Gould, OMB Deponti, RE Stinner (1987) Ecological, agricultural, genetic, and commercial considerations in the deployment of insect-resistant germplasm. Environ Entomol 16: 327–338

Kinsinger, RA, WH McGaughey (1979) Susceptibility of populations of Indianmeal moth and almond moth to *Bacillus thuringiensis*. J Econ Entomol 72: 346–349

Kirsch, K, H Schmutterer (1988) Low efficacy of a *Bacillus thuringiensis* (Berl.) formulation in controlling the diamondback moth, *Plutella xylostella* (L.), in the Philippines. J Appl Entomol 105: 249-255

Kiyosawa, S (1982) Genetics and epidemiological modeling of breakdown of plant disease resistance. Annu Rev Phytopathol 20: 93–117

Klier, A, F Fargette, J Ribier, G Rapoport (1982) Cloning and expression of the crystal protein genes from *B. thuringiensis* strain Berliner 1715. EMBO J 1: 791–799

Knowles, BH, DJ Ellar (1986) Characterization and partial purification of a plasma membrane receptor for *Bacillus thuringiensis* var. *kurstaki* lepidopteran-specific δ-endotoxin. J Cell Sci 83: 89–101

Knowles, BH., DJ Ellar (1987) Colloid-osmotic lysis is a general feature of the mechanism of action of *Bacillus thuringiensis* delta endotoxin with different insect specificity. Biochem Biophys Acta 924: 509–518

Koziel, MG, GL Beland, C Bowman, NB Carozzi, R Crenshaw, L Crossland, J Dawson, N Desai, M Hill, S Kadwell, K Launis, K Lewis, D Maddox, K McPherson, MR Meghji, E Merlin, R Rhodes, GW Warren, M Wright, SV Evola (1993) Field performance of elite transgenic maize plants expressing an insecticidal protein derived from *Bacillus thuringiensis*. BioTechnology 11: 194–200

Kronstad, JW, HE Schnepf, HR Whiteley (1983) Diversity of locations for *Bacillus thuringiensis* crystal protein genes. J Bacteriol 154: 419–428

Lambur, MT, ME Whalon, FA Fear (1985) Diffusion theory and integrated pest management: illustrations from the Michigan fruit IPM program. Bull Entomol Soc Am 31: 40–45

Lane, NJ, JB Harrison, WM Lee (1989) Changes in microvilli and Golgi-associated membranes of lepidopteran cells induced by an insecticidally active bacterial delta-endotoxin. J Cell Sci 93: 337–347

Lecadet, MM, D Martouret (1987) Host specificity of the *Bacillus thuringiensis* δ-endotoxin toward lepidopteran species: *Spodoptera littoralis* (Bdv.) and *Pieris brassicae* (L.). J Invertebr Pathol 49: 37–48

Li, J, J Carroll, DJ Ellar (1991) Crystal structure of insecticidal delta endotoxin from *Bacillus thuringiensis* at 2.5 Å resolution. Nature 353: 815–821

Lockwood, JA, TC Sparks, RN Story (1984) Evolution of insect resistance to insecticides: a reevaluation of the roles of physiology and behavior. Bull Entomol Soc Am 30: 41–51

MacIntosh, SC, TB Stone, RS Jokerst, RL Fuchs (1991) Binding of *Bacillus thuringiensis* proteins to a laboratory-selected line of *Heliothis virescens*. Proc Natl Acad Sci USA 88: 8930–8933

Mani, GS (1985) Evolution of resistance in the presence of two insecticides. Genetics 109: 761–783

Martens, JWM., G Honee, D Zuidema, JWM van Lent, B Visser, JM Vlak (1990) Insecticidal activity of a bacterial crystal protein expressed by a recombinant baculovirus in insect cells. Appl Environ Microbiol 56: 2764–2770

McGaughey, WH (1978) Effects of larval age on the susceptibility of almond moths and Indianmeal moths to *Bacillus thuringiensis*. J Econ Entomol 71: 923–925

McGaughey, WH (1985a) Insect resistance to the biological insecticide *Bacillus thuringiensis*. Science 229: 193–195

McGaughey, WH (1985b) Evaluation of *Bacillus thuringiensis* for controlling Indianmeal moths (Lepidoptera: Pyralidae) in farm grain bins and elevator silos. J Econ Entomol 78: 1089–1094

McGaughey, WH, RW Beeman (1988) Resistance to *Bacillus thuringiensis* in colonies of Indianmeal moth and almond moth (Lepidoptera: Pyralidae). J Econ Entomol 81: 28–33

McGaughey, WH, DE Johnson (1987) Toxicity of different serotypes and toxins of *Bacillus thuringiensis* to resistant and susceptible Indianmeal moths (Lepidoptera: Pyralidae). J Econ Entomol 80: 1122–1126

McGaughey, WH, DE Johnson (1992) Indianmeal moth (Lepidoptera: Pyralidae) resistance to different strains and mixtures of *Bacillus thuringiensis*. J Econ Entomol 85(5): 1594–1600

McGaughey, WH, ME Whalon (1992) Managing insect resistance to *Bacillus thuringiensis* toxins. Science 258: 1451–1455

McLinden, JH, JR Sabourin, BD Clark, DR Gensler, WE Workman, DH Dean (1985) Cloning and expression of an insecticidal k-73 type crystal protein gene from *Bacillus thuringiensis* var. *kurstaki* into *Escherichia coli*. Appl Environ Microbiol 50: 623–628

Meeusen, RL, G Warren (1989) Insect control with genetically engineered crops. Annu Rev Entomol 34: 373–381

Merryweather, AT, U Weyer, MPG Harris, M Hirst, T Booth, RD Possee (1990) Construction of genetically engineered baculovirus insecticides containing the *Bacillus thuringiensis* subsp. *kurstaki* HD 73 delta-endotoxin. J Gen Virol 71: 1535–1544

Moar, W, M Putsztai-Carey, H Van Faassen, D Bosch, R Frutos, C Rang, K Luo, MJ Adang (1995) Development of *Bacillus thuringiensis* CryIC resistance by *Spodoptera exigua* (Hübner) (Lepidoptera: Noctuidae). Appl Environ Microbiol 61(6): 2086–2092

Muller-Cohn, J, J Chaufaux, C Buisson, N Gilois, V Sanchis, D Lereclus (1994) *Spodoptera littoralis* resistance to the *Bacillus thuringiensis* CryIC toxin and cross-resistance to other toxins. In: Abstr 6th Int Coll Invertebr Pathol, Montpellier, France, Aug 28-Sept. 2, 1994, p 66

Murray, EE, T Rocheleau, M Eberle, C Stock, V Sekar, M Adang (1991) Analysis of unstable RNA transcripts of insecticidal crystal protein gene of *B.t.* in transgenic plants and electroporated protoplasts. Plant Mol Biol 16: 1035–1050

Narayanan, K, S Jayaraj (1974) Mode of action of *Bacillus thuringiensis* Berliner in citrus leaf caterpillar, *Papilio demoleus* L. (Papilionidae: Lepidoptera). Indian J Exp Biol 12: 89–91

Obukowicz, MG, FJ Perlak, K Kusano-Kretzmer, EJ Mayer, SL Bolten, LS Watrud (1986) Tn5-mediated integration of the delta-endotoxin gene from *Bacillus thuringiensis* into the chromosome of root-colonizing pseudomonads. J Bacteriol 168: 982–989

Oppert, BS, KJ Kramer, DE Johnson, SC MacIntosh, WH McGaughey (1994) Altered protoxin activation by midgut enzymes from a *Bacillus thuringiensis* resistant strain of *Plodia interpunctella*. Biochem Biophys Res Commun 198(3): 940–947

Percy, J, PG Fast (1983) *Bacillus thuringiensis* crystal toxin: ultrastructural studies of its effect on silkworm midgut cells. J Invertebr Pathol 41: 86–98

Perlak, FJ, RW Deaton, TA Armstrong, RL Fuchs, SR Sims, JT Greenplate, DA Fischhoff (1990) Insect resistant cotton plants. BioTechnology 8: 939–942

Plapp, FW (1993) Alternate strategies for insect control and resistance management: possibilities and future prospects. Proc 1993 Beltwide Cotton Conferences, vol 2, pp 698-701

Powell, KA, AR Jutsum (1993) Technical and commercial aspects of biocontrol products. Pestic Sci 37: 315–321

Puntambekar, US, SN Mukherjee, PK Ranjekar (1995) Toxicity of *Bacillus thuringiensis* and protoplast fusant against *Spodoptera litura* (F.). Lett Appl Microbiol 21: 348–350

Rahardja, U, ME Whalon (1995) Inheritance of resistance to CryIIIA δ-endotoxin in Colorado potato beetle (Coleoptera: Chrysomelidae). J Econ Entomol 88(1): 21–26

Ramachandran, R, KF Raffa, MJ Miller, DD Ellis, BH McCown (1993) Behavioral responses and sublethal effects of spruce budworm (Lepidoptera: Tortricidae) and fall webworm (Lepidoptera: Arctiidae) larvae to *Bacillus thuringiensis* CryIA(a) toxin in diet. Environ Entomol 22(1): 197–211

Ramakrishnan, N, LD Tiwari (1967) Histological changes in *Plusia orichalcea* caused by *Bacillus thuringiensis*. J Invertebr Pathol 9: 579–580

Reynaerts, A, S Jansens (1994) Engineered insect resistance in tomato. Acta Hortic 376: 347–352

Rossiter, M, WG Yendol, MR Dubois (1990) Resistance to *Bacillus thuringiensis* in gypsy moth (Lepidoptera: Lymantriidae): genetic and environmental causes. J Econ Entomol 83(6): 2211–2218

Roush, RT (1989) Designing resistance management programs: how can you choose? Pestic Sci 26: 423–441

Roush, RT (1994) Managing pests and their resistance to *Bacillus thuringiensis*: can transgenic crops be better than sprays? Biocontrol Sci Technol 4: 501–516

Sachs, ES, JH Benedict, DW Altman (1993) Gene pyramiding to improve insect resistance of *B.t.* cotton. Proc 1993 Beltwide Cotton Conferences, vol 2, pp 808–812

Salama, HS, MM Matter (1991) Tolerance level to *Bacillus thuringiensis* Berliner in the cotton leafworm *Spodoptera littoralis* Boisduval (Lepidoptera: Noctuidae). J Appl Entomol 111: 225–230

Schnepf, HE, HR Whiteley (1981) Cloning and expression of the *Bacillus thuringiensis* crystal protein gene in *Escherichia coli*. Proc Natl Acad Sci USA 78: 2893–2897

Schwartz, JM, BE Tabashnik, MW Johnson (1991) Behavioral and physiological responses of susceptible and resistant diamondback moth larvae to *Bacillus thuringiensis*. Entomol Exp Appl 61: 179–187

Sekar, V, DV Thompson, MJ Maroney, RG Bookland, MJ Adang (1987) Molecular cloning and characterization of the insecticidal crystal protein gene of *Bacillus thuringiensis* var. *tenebrionis*. Proc Natl Acad Sci USA 84: 7036–7040

Shelton, AM, JA Wyman (1992) Insecticide resistance of diamondback moth in North America. In: Talekar, NS (ed) Management of diamondback moth and other crucifer pests. Proc 2nd Int Works, Asian Vegetable Research and Development Center, Shanhua, Taiwan

Shelton, AM, JL Robertson, JD Tang, C Perez, SD Eigenbrode, HK Preisler, WT Wilsey, RJ Cooley (1993) Resistance of diamondback moth (Lepidoptera: Plutellidae) to *Bacillus thuringiensis* subspecies in the field. J Econ Entomol 86(3): 697–705

Sims, SR, TB Stone (1991) Genetic basis of tobacco budworm resistance to an engineered *Pseudomonas fluorescens* expressing the δ-endotoxin of *Bacillus thuringiensis kurstaki.* J Invertebr Pathol 57: 206–210

Sneh, B, S Schuster (1983) Effect of exposure to sublethal concentration of *Bacillus thuringiensis* Berliner ssp. *entomocidus* on the susceptibility to the endotoxin of subsequent generations of the Egyptian cotton leafworm *Spodoptera littoralis* Boisd. (Lep.: Noctuidae). Z Angew Entomol 96: 425–428

Stahly, DP, DW Dingman, LA Bulla, Jr., AI Aronson (1978) Possible origin and function of the parasporal crystals in *Bacillus thuringiensis.* Biochem Biophys Res Commun 84: 581–588

Stone, TB, SR Sims (1993) Geographic susceptibility of *Heliothis virescens* and *Helicoverpa zea* (Lepidoptera: Noctuidae) to *Bacillus thuringiensis.* J Econ Entomol 86(4): 989–994

Stone, TB, SR Sims, PG Marrone (1989) Selection of tobacco budworm for resistance to a genetically engineered Pseudomonas fluorescens containing the δ-endotoxin of *Bacillus thuringiensis* subsp. *kurstaki.* J Invertebr Pathol 53: 228–234

Stone, TB, SR Sims, SC MacIntosh, RL Fuchs, PG Marrone (1991) Insect resistance to *Bacillus thuringiensis.* In: Maramorosch, K (ed) Biotechnology for biological control of pests and vectors. CRC Press, Boca Raton, Florida, pp 53–66

Sutter, GR, ES Raun (1967) Histopathology of European-corn-borer larvae treated with *Bacillus thuringiensis.* J Invertebr Pathol 9: 90–103

Tabashnik, BE (1989) Managing resistance with multiple pesticide tactics: theory, evidence, and recommendations. J Econ Entomol 82: 1263–1269

Tabashnik, BE (1990) Modeling and evaluation of resistance management tactics. In: Roush RT, Tabashnik BE (eds) Pesticide resistance in arthropods. Chapman and Hall, New York, pp 153–182

Tabashnik, BE (1994) Delaying insect adaptation to transgenic plants: seed mixtures and refugia reconsidered. Proc R Soc Lond B 255: 7–12

Tabashnik, BE, BA Croft (1982) Managing pesticide resistance in crop–arthropod complexes: interactions between biological and operational factors. Environ Entomol 11: 1137-1144

Tabashnik, BE, BA Croft (1985) Evolution of pesticide resistance in apple pests and their natural enemies. Entomophaga 30: 37–49

Tabashnik, BE, WH McGaughey (1994) Resistance risk assessment for single and multiple insecticides: responses of Indianmeal moth (Lepidoptera: Pyralidae) to *Bacillus thuringiensis.* J Econ Entomol 87(4): 834–841

Tabashnik, BE, NL Cushing, N Finson, MW Johnson (1990) Field development of resistance to *Bacillus thuringiensis* in diamondback moth (Lepidoptera: Plutellidae). J Econ Entomol 83: 1671–1676

Tabashnik, BE, N Finson, MW Johnson (1991) Managing resistance to *Bacillus thuringiensis*: lessons from the diamondback moth (Lepidoptera: Plutellidae). J Econ Entomol 84: 49–55

Tabashnik, BE, N Finson, MW Johnson (1992) Two protease inhibitors fail to synergize *Bacillus thuringiensis* in diamondback moth (Lepidoptera: Plutellidae). J Econ Entomol 85(6): 2082–2087

Tabashnik, BE, N Finson, J M Schwartz, MA Caprio, MW Johnson (1992a) Diamondback moth resistance to *Bacillus thuringiensis* in Hawaii. In Talekar NS (ed) Proc 2nd Int Works Tainan, Taiwan, December 10-14, 1990. Taipei: Asian Vegetable Research and Development Center

Tabashnik, BE, N Finson, FR Groeters, WJ Moar, MW Johnson, K Luo, MJ Adang (1994) Reversal of resistance to *Bacillus thuringiensis* in *Plutella xylostella.* Proc Natl Acad Sci USA 91: 4120–4124

Tanaka, H (1992) Occurrence of resistance to *Bacillus thuringiensis* in diamondback moth, and results of trials for integrated control in a watercress greenhouse. In Talekar, NS (ed) Proc 2nd Int Works Tainan, Taiwan, December 10-14, 1990. Taipei: Asian Vegetable Research and Development Center

Taylor, CE, GP Georghiou (1979) Suppression of insecticide resistance by alteration of gene dominance and migration. J Econ Entomol 72: 105–109

Trisyono, A, ME Whalon (1997) Fitness costs of resistance to *Bacillus thuringiensis* in Colorado potato beetle (Coleoptera: Chrysomelidae). J Econ Entomol (in press)

Trumble, JT, WG Carson, KK White (1994) Economic analysis of a *Bacillus thuringiensis*-based integrated pest-management program in fresh-market tomatoes. J Econ Entomol 87(6): 1463–1469

Tshernyshev, WB (1995) Ecological pest management (EPM): general approaches. J Appl Entomol 119: 379–381

Vaeck, M, A Reynaerts, H Hofte, S Jansens, M De Beukeleer, C Dean, M Zabeau, M Van Montagu, J Leemans (1987) Transgenic plants protected from insect attack. Nature 328: 33–37

Vaeck, M, A Reynaerts, H Hofte (1990) Protein engineering in plants: expression of *Bacillus thuringiensis* insecticidal protein genes. Cell Cult Somatic Cell Genet Plants 6: 425–439

Van Frankenhuyzen, K, JL Gringorten, RE Milne, D Gauthier, M Pusztai, R Brousseau, L Masson (1991) Specificity of activated CryIA proteins from *Bacillus thuringiensis* subsp. *kurstaki* HD-1 for defoliating forest Lepidoptera. Appl Environ Microbiol 57: 1650–1655

Van Frankenhuyzen, K, CW Nystrom, BE Tabashnik (1995) Variation in tolerance to *Bacillus thuringiensis* among and within populations of the spruce budworm (Lepidoptera: Tortricidae) in Ontario. J Econ Entomol 88: 97–105

Van Mellaert, H, J Botterman, J Van Rie, H Joos (1991) Prevention of *B.t.* resistance development. European Patent Application 90401427.1

Van Rie, J (1991) Insect control with transgenic plants: resistance proof? TIBTECH 9: 177–179

Van Rie, J, S Jansens, H Hofte, D Degheele, H Van Mellaert (1989) Specificity of *Bacillus thuringiensis* δ-endotoxins: importance of specific receptors on the brush border membrane of the mid-gut of target insects. Eur J Biochem 186: 239–247

Van Rie, J, S Jansens, H Hofte, D Degheele, H Van Mellaert (1990a) Receptors on the brush border membrane of the insect midgut as determinants of the specificity of *Bacillus thuringiensis* delta-endotoxins. Appl Environ Microbiol 56: 1378–1385

Van Rie, J, WH McGaughey, DE Johnson, BD Barnett, H Van Mellaert (1990b) Mechanism of insect resistance to the microbial insecticide *Bacillus thuringiensis*. Science 247: 72–74

Ward ES, DJ Ellar (1986) *Bacillus thuringiensis* var. *israelensis* delta endotoxin: nucleotide sequence and characterization of the transcripts in *Bacillus thuringiensis* and *Escherichia coli*. J Mol Biol 191: 1–11

Whalon ME, Hollingworth RL (eds) Resistant Pest Management Newsletter vol 1–3. (1989 present) Pesticide Research Center, Michigan State University, E Lansing, Michigan

Whalon ME, DL Norris (1996) Resistance management for transgenic *Bacillus thusingiensis* plants. *Biotechnol* Monitor 29: 8–12

Whalon, ME, DL Miller, RM Hollingworth, EJ Grafius, JR Miller (1993) Selection of a Colorado potato beetle (Coleoptera: Chrysomelidae) strain resistant to *Bacillus thuringiensis*. J Econ Entomol 86(2): 226–233

Whiteley, HR, HE Schnepf (1986) The molecular biology of parasporal crystal body formation in *Bacillus thuringiensis*. Annu Rev Microbiol 40: 549–576

Wolfersberger, MG (1990) The toxicity of two *Bacillus thuringiensis* δ-endotoxins to gypsy moth larvae is inversely related to the affinity of binding sites on midgut brush border membranes for the toxins. Experientia 46: 475–477

CHAPTER 8

Pyrrole Insecticides: A New Class of Agriculturally Important Insecticides Functioning as Uncouplers of Oxidative Phosphorylation

D.A. Hunt and M.F. Treacy
American Cyanamid Company, Agricultural Products Research Division,
P.O. Box 400, Princeton, New Jersey 08543-0400, USA

1 Introduction

In the never-ending search for new insecticides and insecticidal leads, researchers have consistently relied upon nature for guidance (Addor 1995; Henrick 1995). This strategy in general has provided the research community with a plethora of compound classes possessing biological activity, with fermentation chemistry serving as one of the foremost sources for compounds derived from nature. Fermentation technology has proven invaluable for the generation of many classes of compounds encompassing broad ranges of biological activities amendable for exploitation by both the pharmaceutical and agrochemical industries.

As part of a multifaceted approach to discover new insecticides and insecticide lead structures, the fermentation research group formerly associated with Cyanamid's Medical Research Division isolated dioxapyrrolomycin in 1985 from a *Streptomyces fumanus* (Sveshnikova) culture derived from a soil sample taken in Oklahoma (Fig. 1; Carter et al. 1987).

This compound exhibits moderate levels of insecticidal activity against a broad spectrum of insects and acarina. During the formative stages of investigations with this compound, it was found to possess an unacceptable toxicity profile (mouse oral LD_{50} = 14 mg kg^{-1}). This finding, coupled with the moderate levels of insecticidal activity, precluded further development of dioxapyrrolomycin *per se*. While disappointing from a development perspective, the structural novelty and simplicity, along with the discovery that the mode of action of dioxapyrrolomycin is uncoupling of oxidative phosphorylation, provided chemists and biologists with the impetus for a broad-scale synthesis and screening program (Addor et al. 1992; Kuhn et al. 1993; Black et al. 1994; Hunt 1994).

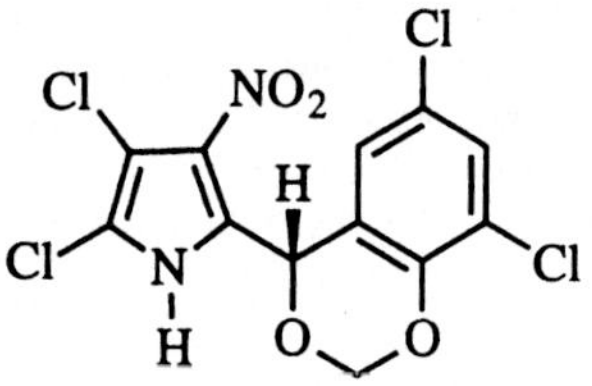

Dioxapyrrolomycin

Fig. 1. Structure of dioxapyrrolomycin

2 Relationship of Physicochemical Parametes to Uncoupling

After the physiological outcome of uncoupling was demonstrated and defined (Loomis and Lipmann 1948), the definition was refined to include an explanation of evens on a cellular level (Mitchell 1961). As a further refinement, studies with a series of phenol derivatives showed that inhibition of phosphorylation in yeast, and hence uncoupling, was intimately related to the two physicochemical arameters of acidity (pKa) and lipophilicity (log P) and that the observed efficiency of uncoupling is improved by optimizing both parameters (Hansch et al. 1965). Subsequent studies in other laboratories on a variety of weak organic acids encompassing a variety of structural classes have served to confirm Hansch's findings (Tollenaere 1973; Labbe-Bois et al. 1975; Corbett et al. 1984).

With an understanding of the mode of action as defined by Hansch and confirmed by rat liver mitochondrial assays, pyrroles substituted with a variety of functional groups were prepared in order to develop a comprehensive structure–activity relationship (Gange et al. 1995) focused on the optimization of lipophilicity and acidity, the two crucial physicochemical parameters.

3 Structure–Activity Relationships

3.1 Variation of Substitution at the Two-Position of the Pyrrole Ring

Initial studies focused on addressing the requirement of the benzofused 1,3-dioxane ring system with regard to insecticidal activity, resulting in the preparation of a series of 2-benzoylpyrroles, examples of which are illustrated in Table 1. As seen from the screening data, this series did not result in especially broad spectrum compounds. Concurrently, the three trichloropyrroles shown in Table 2 were prepared in order to address the dependency of activity on the presence and nature of the electron withdrawing group (EWG) at the 3-position on the pyrrole ring. Thus, while the 3-nitro- and 3-cyanotrichloropyrroles proved to be moderately active, the 3-carbomethoxy derivative was completely devoid of activity.

Table 1. 2-Benzoylpyrroles, per cent control at 100 ppm

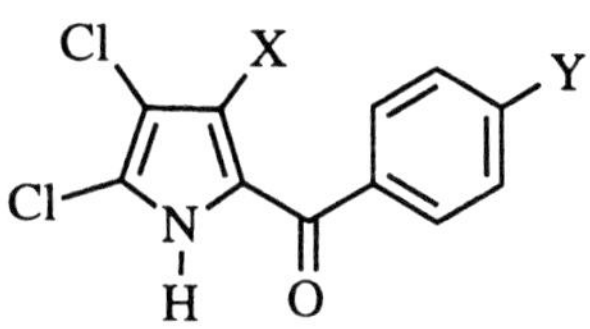

X	Y	SAW[a]	TBW[b]
CN	CN	100	0
/Cl	Cl	100	0

[a]Southern armyworm (Spodoptera eridania) third instar.
[b]Tobacco budworm (Heliothis virescens), third instar.
[c]Arthropods exposed to treated foliage.

Table 2. Trichloropyrroles with electron withdrawing groups at the three position; per cent control at 100 ppm

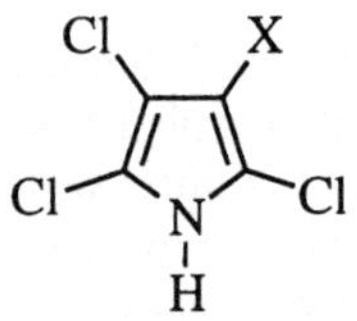

X	SAW[a]	TBW[b]	Mite[c]	Leafhopper[d]
CN	100	0	0	100
NO_2	0	0	60	0
CO_2CH_3	0	0	0	0

[a]Southern armyworm (Spodoptera eridania) third instar.
[b]Tobacco budworm (Heliothis virescens), third instar.
[c]Twospotted mite (Tetranychus urticae), motiles.
[d]Western potato leafhopper (Empoasca abrupta), adults and nymphs.
[e]Arthropods exposed to treated foliage,

3.2 Variation of Aryl Substitution for 2-Arylpyrroles

Based on the chemistry used to prepare the benzoylpyrroles (Benages and Albonico 1978) and as a further structural simplification, a series of 2-aryl-3-cyano- and 2-aryl-3-nitro-4,5-dihalopyrroles were prepared with examples shown in Table 3. Within this series, compounds with halogen substitution on the 2-phenyl ring typically exhibited good levels of activity against Lepidoptera with little Acarina activity, while alkyl group substitution on the phenyl ring was detrimental to the overall insecticidal activity. The lack of efficacy associated with this latter functional group substitution may be attributed to deactiviation of the substrate by metabolic oxidation of the alkyl group and/or a decrease in the pKa of the pyrrole–NH due to the electron donating propensity of alkyl-substituted phenyl ring systems *via* inductive effects. The observation that pyrrole acidity, and therefore insecticidal efficacy, is markedly affected by functional group substitution on phenyl ring is supported by a comparison of the pKa's of 4-chlorophenyl-(6.3) versus 4-tolyl-(7.4)-substituted pyrroles. These pKa's were determined in a mixed solvent system and values extrapolated to measurement in a 100% water system.

To further support the relationship of activity to phenyl ring substitution, the introduction of either a trifluoromethyl or trifluoromethoxy group on the 2-phenyl ring resulted in altering the spectrum of control. While these derivatives afforded Lepidoptera control comparable with that observed for the halo- and dihalophenyl analogs, they also provided moderate levels of mite control. However, accompanying this increase in the spectrum of control was an observation of phytotoxicity, an issue discussed in Section 3.5. Both the nature and the position of the phenyl ring substitution were found to be crucial to the attainment of insecticidal activity. In general, compounds containing strong electron donating groups on either the phenyl or pyrrole ring in any position are weakly active at best. Monohalo-, dihalo- or trifluoromethyl substitution on the phenyl ring afford compounds which are highly active. Compounds with substitution at the four-position of the phenyl ring tend to possess better broad spectrum activity compared with analogs with an open four-position.

Table 3. 2-Aryl-3-cyano/nitro-4, 5-dihalopyrroles, per cent control at 100 ppm[a]

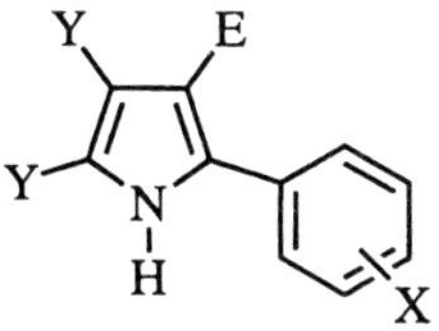

Y	X	E	SAW[b]	TBW[c]	Mite[d]	Leafhopper[e]
Cl	H	CN	100	0	0	0
Cl	H	NO_2	100	70	0	0
Cl[f]	4-Cl	CN	100	90	0	0
			(5.4)	(24)		
Cl	4-Cl	NO_2	100	90	60	90
			(8.0)	(21)		
Cl[g]	3,4-diCl	CN	100	100	0	0
			(4.7)	(15)		
Cl	3,4-ciCl	NO_2	100	100	0	100
			(8.2)	(63)		
Cl	4-CF_3	CN	100	100	100	100
Cl	4-CF_3	NO_2	100	100	100	100
			(10.7)	(41)	(19)	
Cl	4-OCF_3	CN	100	100	100	100
Cl[h]	4-CH_3	CN	0 0	0	0	
Cl	4-$COCH_3$	CN	0 0	0	0	
Cl	4-CN	CN	100	90	0	0
Cl	4-NO_2	CN	100	80	0	0
Br	4-Cl	CN	100	100	0	0
Br	4-Cl	NO_2	100	100	0	0

Arthropods exposed to treated foliage.
[a]Selected LD_{50} values (ppm) are in parentheses.
[b]Southern armyworm (Spodoptera eridania), third instar.
[c]Tobacco budworm (Heliothis virescens), third instar.
[d]Two-spotted mite (Tetranychus urticae), motiles.
[e]Western potato leafhopper (Empoasca abrupta), adults and nymphs.
[f]pKa = 6.3.
[g]pKa = 6.1.
[h]pKa = 7.4.

3.3 Variation of the Electron Withdrawing Group at the Three-Position of the Pyrrole Ring

Yet another structural requirement to attain good levels of insect control in this series, as the earlier work on three-substituted trichloropyrroles suggested, is the necessity of a potent electron withdrawing group at the three-position of the pyrrole ring. Again, the observed levels of activity are highly dependent on the nature of the electron withdrawing group. For example, compounds with 3-cyano or 3-nitro substitution impart very good control of *Heliothis* and the 3-trifluoromethyl-sulfonyl derivative has broad spectrum activity, while the 3-trifluoromethylthio or 3-acetyl cogeners provide little to no insecticidal activity.

3.4 Variation of Substituents at the Four- and Five-Positions of the Pyrrole Ring

The effects of varying the functional groups at the 4- and 5-positions on 2-(4-

chlorophenyl) pyrroles containing electron withdrawing groups in the 3-position are shown in Table 4. As previously illustrated in Table 3, the 4,5-dichloro substitutions afford compounds possessing good insecticidal and acaricidal activity. Virtually no difference in activity is observed for the corresponding 4,5-dibromo derivatives. The cyano electron withdrawing group can be juxtaposed with the 4-halogen substituent to provide compounds with activities largely limited to lepidopterous larvae. Likewise, 2-aryl-5-cyano compounds possessed good lepidopteran activity. Of interest in this series is the observation that 2-aryl-4-nitro analogs are less active compared with their cyano counterparts. Additionally, moving a variety of substituted phenyl groups to the three-position of the pyrrole ring for either nitro- or cyano-substituted pyrroles typically results in an order of magnitude decrease in activity.

3.5 Activity of 2-Arylpyrroles with Trifluoromethyl Substitution

With biological efficacy of various halogenated 2-arylpyrroles now established, 2-arylpyrroles containing a trifluoromethyl group attached to the pyrrole ring were prepared based on previous observations of enhanced biological efficacy when a trifluoromethyl group is substituted for a halogen atom (Filler and Kobayashi 1986). The basis for this observation may be attributed to enhancement of both the electrophilic and lipophilic nature of the CF_3 group compared with halogen. In order to illustrate this effect, the activity profile of a series of trifluoromethyl-substituted pyrroles bearing a 2-(4-chlorophenyl) substituent is shown in Table 5. As might be predicted based on the previously established structure–activity observations, the relative positions of these substituents on the pyrrole ring have a dramatic effect on the acidity of the pyrrole –NH hydrogen and therefore an effect on the insecticidal activity. The nature and position of the phenyl ring substitution are critical for this series of compounds as well.

As indicated in Section 3.2, problems were encountered during the development of the pyrrole chemistry with phytotoxicity associated with some of the

Table 4. 4,5-disubstituted-2-(4-chlorophenyl)pyrroles; per cent control at 100 ppm

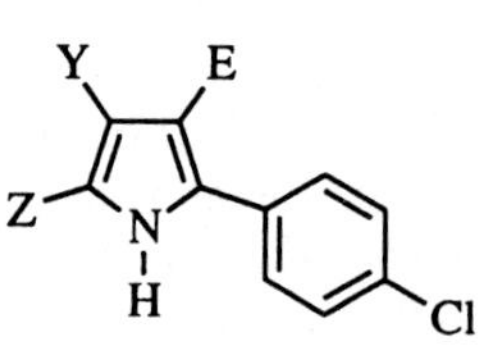

Y	Z	E	SAW[a]	TBW[b]	Mite[c]	Leafhopper[d]
Br	Br	CN	100	100	0	0
Cl	Cl	CF3SO2	100	100	100	100
Br	Br	CF3S	100	0	60	0
Cl	H	CN	0	0	0	0
Br	CN	CN	100	60	90	70
Br	NO2	CN	100	0	0	80

Arthropods exposed to treated foliage.
[a]Southern armyworm (Spodoptera eridania), third instar.
[b]Tobacco budworm (Heliothis virescens), third instar.
[c]Two-spotted mite (Tetranychus urticae), motiles.
[d]Western potato leafhopper (Empoasca abrupta), adults and nymphs.

Table 5. Regioisomeric pyrroles; per cent mortality at 100 ppm

	SAW[a]	TBW[b]
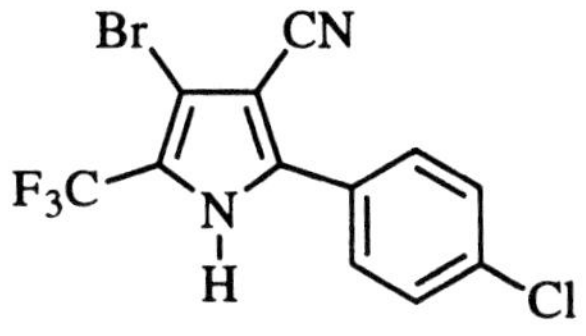	100	100
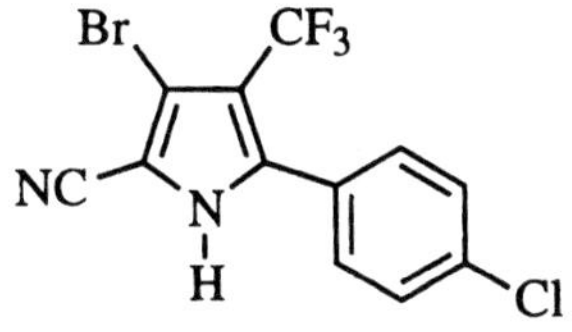	0	0
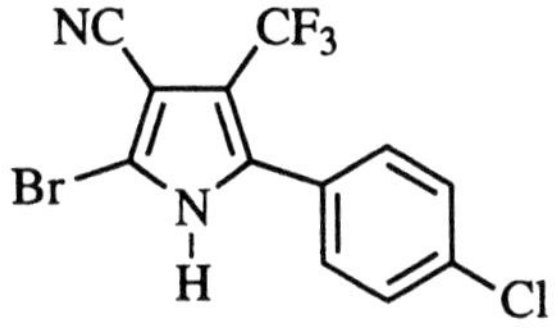	0	0
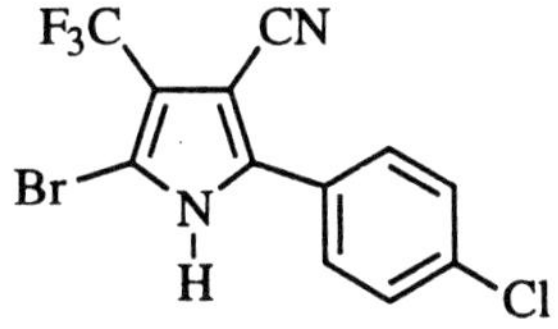	100	0
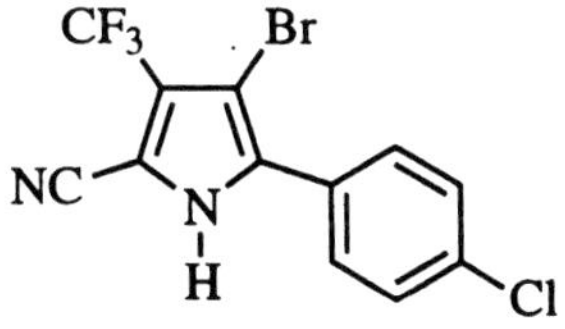	100	0

Arthropods exposed to treated foliage.
[a]Southern armyworm (Spodoptera eridania), third instar.
[b]tobacco budworm (Heliothis virescens), third instar.

N-unsubstituted pyrroles. There are indications that this occurs with certain classes of uncouplers and has been remedied by modification of pesticidally effective derivatives with functional groups labile to metabolic or hydrolytic cleavage *in vivo* (Hunt 1994). The use of the proinsecticide concept to ameliorate

Table 6. N-Derivatives of some pyrroles; per cent control at 100 ppm

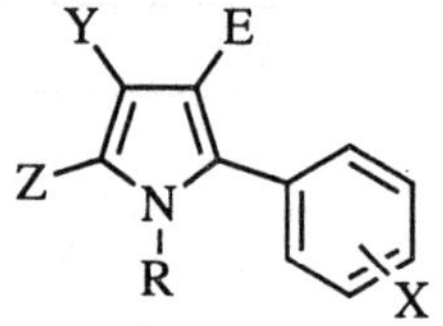

R	X	Y	Z	E	SAW[a]	TBW[b]	Mite[c]	Leafhopper[d]
H	4-Cl	Br	CF_3	CN	100	100	100	100
CH_3	4-Cl	Br	CF_3	CN	100	100	0	100
$C_2H_5OCH_2$	4-Cl	Br	CF_3	CN	100	100	100	100
CH_2Ph	3,4-diCl	Cl	ClCN		100	0	0	0
CH_2CO_2Et	3,4-diCl	Cl	ClCN		100	50	0	0
$CONEt_2$	3,4-diCl	Cl	ClCN		100	100	100	100
H	4-CF_3	Cl	ClCN		100	100	100	100
$C_2H_5OCH_2$	4-CF_3	Cl	ClCN		100	100	0	100
H	4-Cl	Br	CF_3	CF_3SO_2	100	100	70	0
$C_2H_5OCH_2$	4-Cl	Br	CF_3	CF_3SO_2	100	100	100	50

Arthropods exposed to treated foliage.
[a]Southern armyworm (Spodoptera eridania), third instar.
[b]Tobacco budworm (Heliothis virescens), third instar.
[c]Twospotted mite (Tetranychus urticae), motiles.
[d]Western potato leafhopper (Empoasca abrupta), adults and nymphs.

problems such as phytotoxicity and acute mammalian toxicity associated with uncouplers used directly as pesticides has proven to be a valuable strategy. As applied to the pyrroles, this strategy affords two additional advantages: (1) an increase in the lipophilicity of the compound compared with the parent enhances cuticular and cellular penetration, and (2) the release of the active toxicant moiety (the –NH pyrrole) is selectively carried out by the insect. Table 6 provides results of N-derivatization of some pyrroles and, as clearly seen for the 2-(3,4-dichlorophenyl)pyrroles, indicates that activity is associated with the insect's ability to remove the N-protecting group which, in turn, is dependent on the nature of the N-substituent.

4
Overview of Structure-Activity Trends

Thus, using dioxapyrrolomycin as a template, several pyrroles of the general structure 1 have been prepared (Fig. 2). Based on extensive laboratory and field studies, the following structure–activity trends are observed:

1. X=Y=Cl or Br afford compounds of nearly equal activity. The most activity is associated with compounds where X=Br or Cl and Y=CF_3.
2. When EWG=NO_2, CN, or $S(O)_xCF_3$, compounds demonstrate approximately equivalent activity.

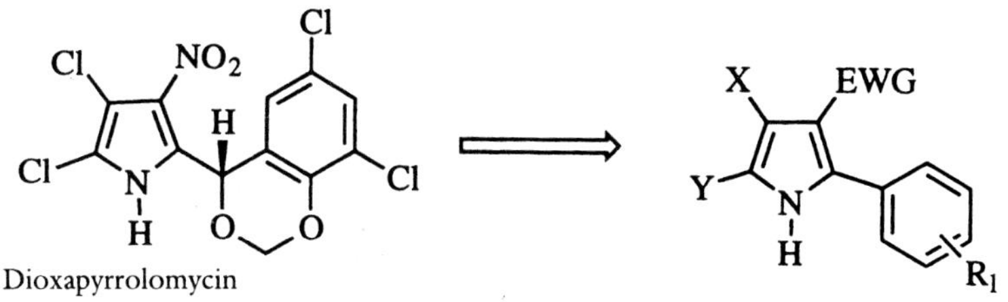

Fig. 2. General Structure 1 of Several pyrroles, evolving from dioxapyrrolomycin

3. Poor activity is typically observed for compounds where R_1 is an electron donating group (e.g., H, CH_3), while groups affording some electron withdrawal and appropriate lipophilicity parameters (e.g., Cl, Br, CH_3) provide compounds with the optimum activity. The 4- and 3,4-disubstituted analogs typically provide good activity.
4. The pyrroles in this series have weakly acidic –NHs which are required for uncoupling. There have been strong correlations among pKa, logP, and insecticidal activity, with the highest levels of activity observed in a relatively well-defined range.

5 Bioactivity and Pharmacodynamics of CL 303,630

5.1 Pesticidal Properties and Target Spectrum

A commercial success resulting from Cyanamid's extensive pyrrole synthesis program is that of the aryl-substituted cyanopyrrole CL 303,630 [4-bromo-2-(4-chlorophenyl)-1-(ethoxymethyl)-5-(trifluoromethyl)pyrrole-3-carbonitrile]. Laboratory assays have shown CL 303,630 to possess potent, broad-spectrum activity against important pestiferous insects and mites. For example, this pyrrole analog exhibits inherent bioactivity against many species of Coleoptera, Lepidoptera, Acarina, and Thysanoptera at levels equal to, or better than, several commercial pyrethroid, organophosphate, carbamate, pyrazole, and organotin pesticides (Table 7).

Field trials have demonstrated foliar applications of CL 303,630, at approximately 60-400 g/ha, to be very effective in controlling more than 70 species of phytophagous insects and mites in crops such as cotton, vegetables, tree fruit, citrus, cereals, and ornamentals (Miller et al. 1990; Farlow et al. 1991; Anonymous 1992). CL 303,630 is currently in commerical development and market-entry phases on a global basis.

5.2 Mechanism of Action

Studies *in vitro* have shown that CL 303,630 (i.e., active metabolite thereof) acts at the mitochondial level and expresses its biochemical lesion by uncoupling

Table 7. Examples of CL 303,630 pest spectrum

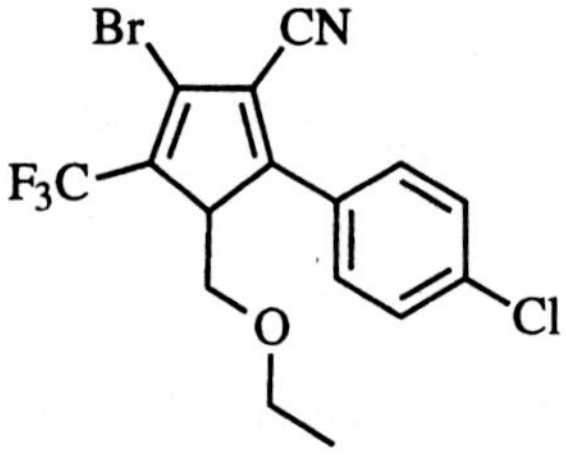

CL 303,630

	LC_{50}(95% CL) (ppm)[a]
Leptinotarsa decemlineata (third instars)[b]	
CL 303,630	3.1(2.6–3.6)
Aldicarb	104.2(86.6–168.1
Endosulfan	127.7(105.6–172.6)
Heliothis virescens(third instars)	
CL 303,630	7.5(6.7–8.5)
Cypermethrin	7.5(4.5–11.7)
Profenofos	32.5(27.4–37.3)
Methomyl	31.8(26.1–39.5)
Tetranychus urticae (motiles)	
CL 303,630	3.2(2.5–3.9)
Dicofol	13.7(11.7–16.0)
Fenbutatin-oxide	112.8(86.5–146.3)
Tebufenpyrad	6.3(5.2–7.5)
Thrips palmi (nymphs and adults)	
CL 303,630	1.8(–)
Sulprofos	34.1(–)

[a]Insects exposed to leaf tissue previously dipped in treatments.
[b]*Leptinotarsa decemlineata larvae collected from potato field; all other insects lab-reared.*

oxidative phosphorylation. The activated pyrrole disrupts the proton gradient across mitrochondrial membranes and impairs the ability of the mitochondia to produce ATP, which leads to cell destruction and ultimately death of the affected pest organism.

As noted in Section 2, a molecule must be lipophilic and acidic in order to be an uncoupler of oxidative phosphorylation. With the proper balance in these two physical properties, such a molecule should be able to traverse mitochondrial membranes and dissociate protons. Although CL 303,630 is indeed lipophilic, with its log P measured to be approximately 4.6 over a pH range of 2.4–11.0 (Treacy et al. 1994), it is not acidic. Conversely, a nonethoxymethyl derivative of CL 303,630 (or –NH pyrrole form of CL 303,630, designated CL 303,268) is both lipophilic and a weak acid. LogP value of protonated (unionized) form of CL 303,268 is about 5 and its pKa is 7.6 (Black et al. 1994; Treacy et al. 1994).

CL 303,630 is a proinsecticide which must be activated by the herbivore via oxidative removal of the N-ethoxymethyl group. Such oxidative metabolism produces the lipophilic and weakly acidic –NH pyrrole, CL 303,268, which in turn produces its lethal impact on the affected herbivore *via* mitochondial uncoupling. Indeed, foliage-chewing insects are known to be very capable oxidizers of xenobiotics (Wilkinson 1983). Several assays appear to confirm existence of metabolic activation of CL 303,630 in phytophagous insects. For example, when larval tobacco budworms, *Heliothis virescens*, were fed [^{14}C] CL 303,630, the total amount of radiolabel recovered from these insects 2 days later consisted of 71% CL 303,268 and 13% CL 303,630, with the remainder unidentified metabolites (Treacy et al. 1994). Studies reported by Black et al. (1994) also showed that CL 303,268 is a major metabolite produced from CL 303,630 in insects. The microsomal mono-oxygenase inhibitor, piperonyl butoxide, was found to dramatically reduce potency of CL 303,630 (but not CL 303,268) against adult Colorado potato beatle, *Leptinotarsa decemlineata*; thus further suggesting that oxidative metabolism of CL 303,630 by an insect is responsible for liberation of the active –NH pyrrole. Finally, based on LC_{50} values resulting from comparative direct-exposure bioassays, CL 303,268 has been shown to exhibit pesticidal potency at levels equal to that of CL 303,630 against several insect and mite species.

Since respiration stimulation is a key phenomenon associated with physiological effects induced by uncouplers, Black et al. (1994) studied effects of CL 303,630 and CL 303,268 on respiration in isolated rat liver mitochondria. Quantifying effects with a Clarke oxygen electrode, these workers found that CL 303,268 caused a doubling of state 4 respiration rate (i.e., 100% increase) at approximately 20 nM (Table 8). At higher concentrations, CL 303,268 began to inhibit respiration. Conversely, results from these rat liver assays *in vivo* showed CL 303,630 to be virtually inactive as an uncoupler of oxidative phosphorylation.

5.3 Insect Age-Dependent Sensitivity and Insecticide Resistance

As shown in Table 9, CL 303,630 exhibits relatively good bioactivity against lepidopteran larvae such as *H. virescens* (Treacy et al. 1991). Interestingly, these data show that third-instar *H. virescens* are only 2.6-fold less sensitive than first-

Table 8. Uncoupling (UC) of rat liver mitochondria

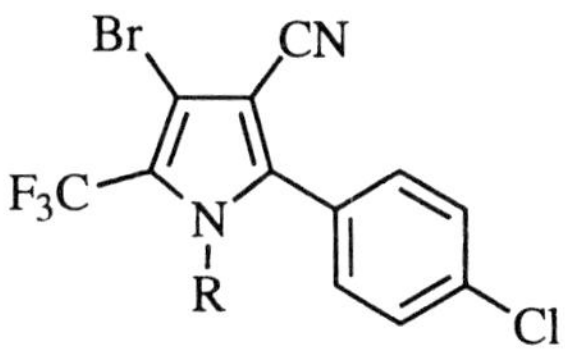

R	CL
$CH_2OC_2H_5$	303,630
H	303,268

	UC_{100}(nM)[a]
CL 303,268	20
CL 303,630	45 000

[a]Cancentration to increase state 4 respiration by 100%.

Table 9. Bioactivity against different age *Heliothis virescens*

	LC_{50}(95% CL) (ppm)[a]
First instars	
CL 303,630	2.8(2.4–3.1)
Profenofos	2.9(2.6–3.4)
Methomyl	2.4(1.8–4.0)
Third instars	
CL 303,630	7.5(6.7–8.5)
Profenofos	32.5(27.4–37.3)
Methomyl	31.8(26.1–39.5)

[a]Insects exposed to leaf tissue previously dipped in treatments.

instars to CL 303,630. Conversely, increases in LC_{50} values between first- and third-instar *H. virescens* averaged 12.2-fold for the organophosphate, profenofos, and the carbamate, methomyl.

Age-dependent tolerance by *H. virescens* to CL 303,630 occurred at a slower rate than age-dependent tolerance to profenofos or methomyl. It has been demonstrated that lepidopteran larvae are capable of metabolizing xenobiotics primarily by microsomal oxidases and glutathion S-transferase, and that larger/older larvae possess greater detoxification capabilities than smaller/younger larvae (Yu 1983; Muehleisen 1987; Hung et al. 1990). Results shown in Table 9 suggest that CL 303,630 may be less susceptible to increased detoxification capabilities of older larvae (vs. younger larvae) than certain organophosphate or carbamate insecticides. Additionally, increased oxidative metabolic abilities of older larvae may hasten activiation of CL 303,630 to CL 303,268 (vs. more rapid detoxification of organophosphates and carbamates).

Laboratory assays have shown that CL 303,630 is very potent against strains of insects and mites known to exhibit high levels of resistance to pyrethroids, organophosphates, and carbamates. For example, CL 303,630 was equally active against pyrethroid-susceptible and -resistant strains of *H. virescens* (Table 10). The particular strain of pyrethroid-resistant *H. virescens* (PEG colony, Zeneca) used in this study was identified as possessing both altered target site and

Table 10. Bioactivity of CL 303,630 against pyrethroid-susceptible and-resistant stains of *Heliothis virescens*

	LC_{50}(95% CL) (ppm)[a]	LC_{90}(95% CL) (ppm)[a]
Pyrethroid-susceptible strain[b]		
CL 303,630	3.5(3.2–3.9)	9.1(7.7–11.3)
Cypermethrin	5.0(4.5–5.6)	13.9(11.3–18.5)
Pyrethroid-resistant strain[b]		
CL 303,630	4.4(3.9–4.9)	10.5(9.0–13.2)
Cypermethrin	271.2(210.3–425.1)	1408.0(723.1–885.0)

[a]Insects exposed to leaf tissue previously dipped in treatments.
[b]Third instars were evaluated in this assay.

metabolic mechanisms of resistance. Campanhola and Plapp (1989) indicated that target site resistance is a common mechanism of pyrethroid resistance for all life stages of *H. virescens*, while metabolic resistance is manifested at greater levels in later larval stages (e.g., third-instars or older). Results shown in Table 10 demonstrate that third-instar larvae from the PEG strain are very susceptible to CL 303,630 even though they are highly resistant to the pyrethroid cypermethrin. Thus, it appears that biological activity of CL 303,630 is not affected by either target site or metabolic resistance mechanisms possessed by this strain of pyrethroid-resistant *H. virescens.*

Recently, it has been shown (Pimprale et al. 1997) that adults of another pyrethroid-resistant strain of *H. virescens* were 35- to 70-fold (depending on exposure time) more sensitive to CL 303,630 than adults from a series of pyrethroid-susceptible strains. These researchers speculated that the negative correlation between pyrethroid resistance and CL 303,630 sensitivity may be due to the greater pyrrole activation capability (i.e., increased oxidative metabolism) of this particular pyrethroid-resistant strain vs. pyrethroid-susceptible strain.

Studies have also been conducted with strains of phytophagous mites known to possess resistance to selected acaricides. In a series of leaf-dip assays, CL 303,630 was found to be equally potent against acaricide-susceptible, organophosphate-resistant, and carbamate-resistant strains of twospotted spider mite, *Tetranychus urticae* (Table 11). Conversely, phosphate- and carbamate-resistant *T. urticae* exhibited 70- and 179-fold levels of resistance to phosalone and formetanate, respectively.

6 Conclusions

Pyrrole derivatives, particularly 2-arylpyrroles, constitute a novel class of insecticides and acaricides based on the lead natural product dioxapyrrolomycin. The mode of action of these compounds has been determined to be uncoupling of oxidative phosphorylation. A structure–activity relationship for the arylpyrroles has been

Table 11. Bioactivity against selected strains of acaricide-Susceptible, phosphate-resistant and carbamate-resistant strains of *Tetranychus urticae* motiles

	LC_{50}(95% CL) (ppm)[a]
Acaricide-susceptible strain	
CL 303,630	3.2(2.5–3.9)
Phosalone	10.4(7.8–13.5)
Formetanate	4.6(3.8–5.4)
Phosphate-resistant strain	
CL 303,630	4.8(4.1–5.5)
Phosalone	726.1(456.7–1158.2)
Carbamate-resistant strain	
CL 303,630	3.0(2.6–3.4)
Formetanate	823.1(668.4–1011.5)

[a]Leaf-dip assay method.

developed, the activities correlating with the two critical parameters for uncouplers of oxidative phosphorylation, logP and pKa. Extensive biological studies have demonstrated the effectiveness of this series of chemistry.

References

Addor RW (1995) Insecticides. In: Godfrey CRA (ed) Agrochemicals from natural products. Marcel Dekker New York, pp 1–62

Addor RW, Babcock TJ, Black BC, Brown DG, Diehl RE, Jurch JA, Kameswaran V, Kamhi VM, Kremer KA, Kuhn, DG, Lovell JB, Lowen GT, Miller TP, Peevey RM, Siddens JK, Treacy MF, Trotto SH, Wright DP Jr (1992) Insecticidal pyrroles. Discovery and overview. In: Baker DR, Fenyes JG, Steffens JJ (eds) Synthesis and chemistry of agrochemicals III, no 504. American Chemical Society, Washington, DC, pp 283–297

Anonymous (1992) AC 303,630 experimental insecticide–miticide. Technical bulletin. American Cyanamid Company, Princeton, New Jersey, pp 1–11

Benages IA, Albonico SM (1978) 2-Chloroacrylonitrile as a cyclodipolarophile in 1,3-cycloadditions. 3-Cyanopyrroles. J Org Chem 43: 4273–4276

Black BC, Hollingworth RM, Ahammadsahib KI, Kukel CD, Donovan S (1994) Insecticidal action and mitochondrial uncoupling activity of AC 303,630 and related halogenated pyrroles. Pestic Biochem Physiol 50: 115–128

Campanhola C, Plapp FW Jr. (1989) Toxicity and synergism of insecticides against susceptible and pyrethroid-resistant third-instars of the tobacco budworm (*Lepidoptera*: Noctuidae). J Econ Entomal 82: 1495–1501

Carter G, Nietsche J, Goodman J, Tory M, Dunne T, Borders D, Testa R (1987) LL-F42248α, a novel chlorinated pyrrole antibiotic. J Antibiot 40: 233–236

Corbett JR, Wright K, Baillie AC (1984) Pesticides interfering with respiration. In: The biochemical mode of action of pesticides, 2nd edn. Academic Press, London, pp 1–49

Farlow RA, Treacy MF, Miller TP, Burkhart SE, Gard IE (1991) Efficacy of AC 303,630 against insect pests on cotton. In: Herber DJ (ed) Proc Beltwide Cotton Conf, National Cotton Council, Memphis, Tennessee, pp 741–744

Filler R, Kobayashi Y (1986) Biomedical aspects of fluorine chemistry. Elsevier Biomedical Press, New York

Gange DM, Donovan S, Lopata RJ, Henegar K (1995) The QSAR of insecticidal uncouplers. In: Hansch C, Fujita T (eds) Classical and three-dimensional QSAR in agrochemistry, no 606. American Chemical Society, Washington, DC, pp 199–212

Hansch C, Kiehs K, Lawrence G (1965) The role of substituents in the hydrophobic bonding of phenols by serum and mitochondrial proteins. J Am Chem Soc 87: 5770–5773

Henrick CA (1995) Pyrethroids. In: Godfrey CRA (ed) Agrochemicals from natural products. Marcel Dekker, New York, pp 63–146

Hung CF, Kao CH, Liu CC, Lin JG, Sun CN (1990) Detoxifying enzymes of selected insect species with chewing and sucking habits. J Econ Entomol 83: 361–365

Hunt DA (1994) 2-Arylpyrroles: novel uncouplers of oxidative phosphorylation. In: Briggs G (ed) Advances in the chemistry, Cambridge, pp 127–140

Kuhn DG, Addor RW, Diehl RE, Furch JA, Kamhi VM, Henegar KE, Kremer KA, Lowen GT, Black BC, Miller TP, Treacy MF (1993) Insecticidal pyrroles. In: Duke SO, Menn JJ, Plimmer JR (eds) Pest control with enhanced environmental safety, no 524. American Chemical Society, Washington, DC, pp 219–232

Labbe-Bois R, Larvelle C, Godfroid J (1975) Quantitative structure–activity relationships for dicoumarol antivitamins K in the upcoupling of mitochondrial oxidative phosphorylation. J Med Chem 18: 85–90

Loomis WF, Lipmann F (1948) Reversible inhibition of the coupling between phosphorylation and oxidation. J Biol Chem 173: 807–808

Miller TP, Treacy MF, Gard IE, Lovell JB, Wright DP, Addor RW, Kamhi VM (1990) AC

303,630: summary of 1988–1989 field trial results. In: Proc Brighton Crop Prot Conf—Pests and Diseases, British Crop Protection Council, Brighton, Sussex

Mitchell P (1961) Coupling of phosphorylation to electron and hydrogen transfer by a chemiosmotic type of mechanism. Nature 191: 144

Muehleisen DP (1987) Induction and regulation of detoxification enymes in *Heliothis zea* by allelochemicals and insecticides. PhD Diss, Texas A & M University, College Station, 127 pp

Pimprale SS, Besco CL, Bryson PK, Brown TM (1997) Increased susceptibility of pyrethroid-resistant tobacco budworm (Lepidoptera: Noctuidae) to chlorfenopyr. J Econ Entomol 90: 49–54

Tollenarere J (1973) Structure–activity relationships of three groups of uncouplers of oxidative phosphorylation: salicylanilide 2-trifluoromethylbenzimidazoles and phenols. J Med Chem 16: 791–796

Treacy M, Miller T, Black B, Gard I, Hunt D, Hollingworth RM (1994) Uncoupling activity and pesticidal properties of pyrroles. Biochem Soc Trans 22: 244–247

Treacy MF, Miller TP, Gard IE, Lovell JB, Wright DP (1991) Characterization of insecticidal properties of AC 303,630 against tobacco budworm (*Heliothis virescens*) larvae. In: Herber DJ (ed) Proc Beltwide Cotton Conf National Cotton Council, Memphis, Tennessee, pp 738–741

Wilkinson CF (1983) Role of mixed-function oxidases in insecticide resistance. In: Georghiou GP, Sarto T (eds) Pest resistance to pesticides. Plenum, New York, pp 175–206

Yu S (1983) Age variation in insecticide susceptibility and detoxification capacity of fall armyworm (*Lepidoptera: Noctuidae*) larvae. J Econ Entomol 76: 219–222

CHAPTER 9

Avermectins: Biochemical Mode of Action, Biological Activity and Agricultural Importance

R.K. Jansson and R.A. Dybas[1]
Merck Research Laboratories, Agricultural Research and Development, Merck & Co., Inc. P.O. Box 450, Hillsborough Road, Three Bridges, New Jersey 08887, USA
[1]Current address: Lonza Inc., Research and Development, 79 Route 22 East, Annandale New Jersey 08801, USA

1 Introduction

1.1 Discovery

Avermectins are a family of 16-membered macrocyclic lactone natural product homologues produced by the soil microorganism *Streptomyces avermitilis* MA-4680 (NRRL 8165). These compounds are closely related to the milbemycins, another family of 16-membered macrocycles, which also have potent anthelmintic and insecticidal/acaricidal activity. Milbemycins were first reported by Mishima et al. (1975) for their acaricidal activity; however, the potent anthelmintic activity of these compounds was not discovered at that time. Avermectins were isolated at the Merck Research Laboratories from a soil sample collected from a golf course in Kawanas, Ito City, Shizuoka Prefecture, Japan, by researchers at the Kitasato Institute. The Anthelmintic activity of the avermectins was discovered in 1976 by Merck researchers (Burg et al. 1979; Egerton et al. 1979; Miller et al. 1979) in an *in vivo* screen based on the survival of the nematode *Nematospiroides dubius* in mice. Anthelmintic activity of the milbemycins was later discovered at Merck following testing of the individual members of the milbemycin family (Mrozik et al. 1983).

Isolation of the crude fermentation product of *S. avermitilis* yielded a complex of eight closely related avermectin homologues (A_{1a}, A_{1b}, A_{2a}, A_{2b}, B_{1a}, B_{1b}, B_{2a}, and B_{2b}), of which avermectins B_1(a and b) were the major components. Abamectin, the nonproprietary name assigned to avermectins B_1, is a mixture of B_{1a} ($\geq 80\%$) and B_{1b} ($\leq 20\%$). This mixture was found to be very potent against mites and certain insect species (Dybas et al. 1989). Abamectin was subsequently selected for development for crop protection and is currently sold commercially for control of mites and certain insect pests on several ornamental and horticultural crops in nearly 50 countries. Ivermectin, a 22, 23 dihydro semi-synthetic derivative of abamectin, has been developed widely for control of a variety of ecto- and endoparasites of animal health as well as for control of the causative agent of river blindness, *Onchocerca volvulus*, in man (see Lariviere et al. 1985 and chapters therein; Campbell 1989). The structures of the avermectin natural products and semi-synthetic derivatives are shown in Fig. 1 (Albers-Schönberg et al. 1981).

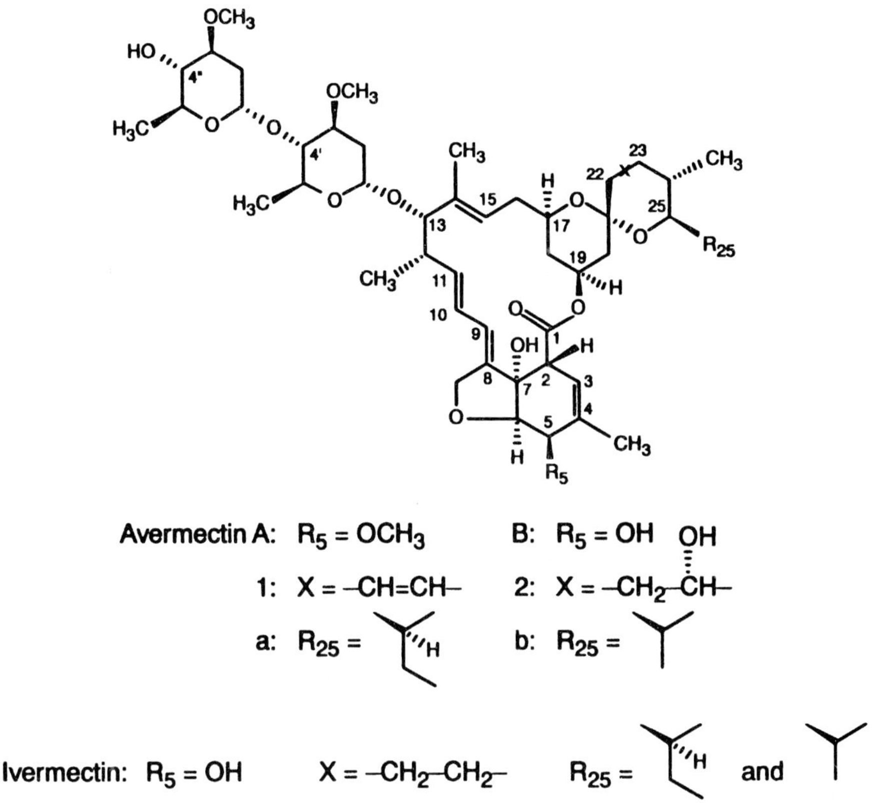

Fig. 1. Avermectin structures

Although abamectin was potent against mites and a select panel of insects, it was markedly less potent against Lepidoptera. This spectrum deficiency for abamectin prompted the initiation of a highly focused, medicinal chemistry and biological testing program resulting in the discovery of a semi-synthetic avermectin, 4″-epi-methylamino-4″-deoxyavermectin B_1 (emamectin) in 1984 (Fig. 2). MK-243 (the hydrochloride salt of emamectin), which was derived from abamectin via a five-step synthesis (Cvetovich et al. 1994), was discovered after screening several hundred avermectin derivatives in an *in vivo* screen using tobacco budworm, *Heliothis virescens*, and southern armyworm, *Spodoptera eridania* (Dybas and Babu 1988; Dybas et al. 1989; Mrozik et al. 1989; Mrozik 1994). This compound was subsequently selected for further development in crop protection. Later studies found that the benzoate salt of emamectin (coded MK-244) had improved thermal stability and greater water solubility compared with the hydrochloride salt. MK-244 was assigned the nonproprietary name emamectin benzoate and is currently being developed as a crop protection insecticide; first registrations in the USA and abroad are expected in 1997.

Since their discoveries, there have been several thousand papers published on the avermectins and milbemycins (Kornis 1995). Several other authors have reviewed the discovery, structure-activity relationships, environmental fate, spectrum

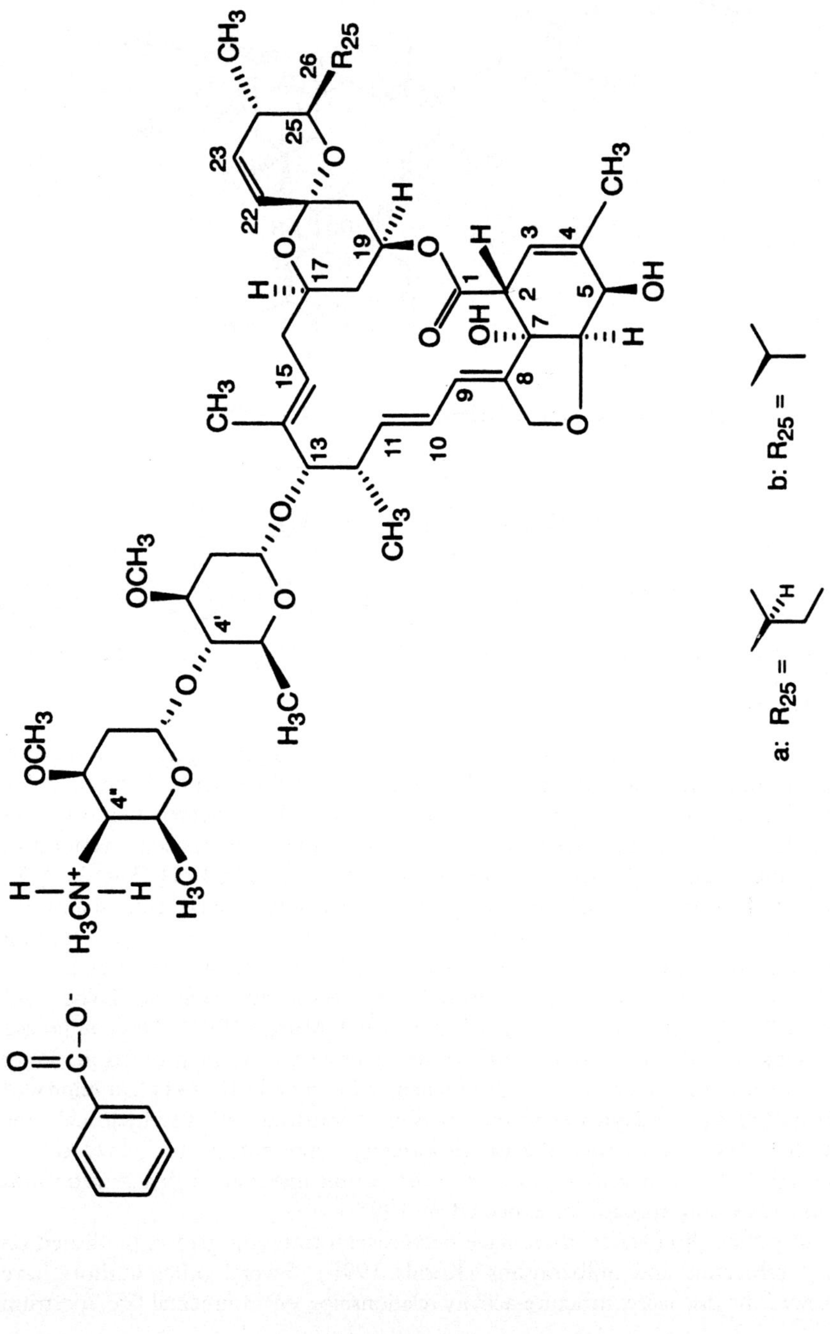

Fig. 2. Structure of emamectin benzoate (MK-244)

of activity, and applications for their anthelmintic, insecticidal, and acaricidal activity (Campbell et al. 1984; Davies and Green 1986; Dybas 1989; Fisher and Mrozik 1984, 1989, 1992; Lasota and Dybas 1991; Mrozik 1994; Kornis 1995; Shoop et al. 1995). The present chapter reviews the mode of action, biological activity, and applications of avermectin insecticides/acaricides in agriculture.

2 Biochemical Mode of Action

It is well known that there are major differences between vertebrate and insect nervous systems. One of the more important differences which can be exploited to lead to insecticide selectivity is the cholinergic nature of the motoneurons regulating vertebrate skeletal muscle, compared with the GABAergic (gamma-aminobutyric acid) and/or glutamanergic nature of motoneurons regulating skeletal muscle activity in insects and other invertebrates. In addition, invertebrates have a class of glutamate-gated chloride channels that are not found in mammalian systems. These differences can lead to the discovery of compounds that reach GABA or glutamate receptors in insect muscle, but do not penetrate the vertebrate CNS, where GABAergic receptors are located. The avermectins are examples of such compounds.

The mode of action of the avermectins has been reviewed by several authors (Turner and Schaeffer 1989; Arena 1994; Rohrer and Arena 1995). All studies suggested that there are few, if any, qualitative differences in the mode of action of those avermectin compounds studied (Fisher and Mrozik 1984); thus, it was suggested that most, if not all, members of the avermectin family have a similar mode of action (Turner and Schaeffer 1989). Milbemycins are also believed to have the same mode of action, in part, as the avermectins (Deng and Casida 1992; Arena et al. 1995).

Avermectins exert various pharmacological effects on different organisms (Turner and Schaeffer 1989). Fritz et al. (1979) were the first to demonstrate that avermectins acted as a chloride channel agonist and opened chloride channels. They showed that the effects of avermectin were reversed by chloride channel blockers, such as picrotoxin, and were specific for tissues that possessed GABA-gated chloride channels. This finding was confirmed by others (Mellin et al. 1983; Albrecht and Sherman 1987). Similarly, Wang and Pong (1982) showed that organisms, such as cestodes and trematodes, which did not contain GABAergic innervation, were not affected by avermectins. Other studies showed that avermectins acted at a site distinct from that of the cyclodienes (Bloomquist 1993 and references therein).

Avermectin B_{1a} and emamectin hydrochloride were shown to bind to a saturable, high affinity site in head membranes of *Drosophila melanogaster* (Deng and Casida 1992). Similar results were reported for abamectin and emamectin benzoate by Rohrer et al. (1995). Rohrer et al. (1995) also demonstrated that abamectin and emamectin benzoate bind to a high affinity site in locust, *Schistocerca americana*, muscle neuronal membranes. Deng and Casida (1992) showed that IC_{50} values for displacement of [^{3}H]-avermectin in housefly head membranes varied among several avermectin analogs, but were correlated with LD_{50} values against adult flies. Ivermectin, avermectin B_{1a}, abamectin, and emamectin hydrochloride were comparable in displacement of [^{3}H]-avermectin and ivermectin in *Drosophila* head membranes (Deng and Casida 1992; Rohrer et al. 1995). LD_{50} values of several avermectin analogs were also correlated with EC_{50} values for displacement of a radiolabeled probe for picrotoxin binding, [^{3}H]-EBOB (4-*n*-propyl-4'-ethynylbicycloorthobenzoate, a noncompetitive blocker of the GABA-gated chloride channel); emamectin hydrochloride was 2.3-fold more potent than abamectin (B_{1a}) at displacing [^{3}H]-EBOB (Deng and Casida 1992). They suggested that avermectin analogs and EBOB appear to act in the same chloride channel at allosteric or closely coupled sites in housefly head membranes. Collectively, these data support the belief that there may be multiple binding sites for avermectins in the same chloride channel, or at other receptor sites.

Studies with locust, *Schistocerca gregaria*, muscle showed that the target of ivermectin was concentration-dependent. At low concentrations (0.000075-0.0075 ppm), ivermectin induced reversible dose-dependent increases in chloride ion permeability and partially blocked GABA-induced chloride ion conductance (Duce and Scott 1985). At higher concentrations (0.01-1.0 ppm), ivermectin potentiated GABA-induced irreversible increases in chloride ion conductance. Ivermectin also induced irreversible increases in chloride ion conductance when applied to GABA-insensitive muscle (Duce and Scott 1985). Their data suggested that GABA was not the site of action for these irreversible increases in conductance. Subsequent studies showed that ivermectin activated a glutamate-sensitive chloride conductance in locust leg muscle (Scott and Duce 1985). Recent studies in nematodes showed that avermectins interact with specific glutamate-gated chloride channels distinct from GABA-sensitive channels (Schaeffer and Haines 1989; Arena et al. 1991, 1992, 1995). The effects of ivermectin on its interaction with glutamate were concentration-dependent. At low concentrations, ivermectin potentiated the response of the channel to glutamate, whereas at high concentrations, ivermectin opened the chloride channel directly (Arena et al. 1992). Thus, in nematodes, avermectins are believed to potentiate or directly open the glutamate-gated chloride channel (Arena et al. 1995).

Cully et al. (1994) recently cloned two functional complementary DNAs encoding an avermectin-sensitive glutamate-gated chloride channel in the nematode *Caenorhabditis elegans*. These two clones were new members of the ligand-gated ion channel superfamily. Cully et al. (1994) demonstrated that the potency of various avermectin compounds (*in vivo*) was correlated with their ability to potentiate glutamate-sensitive current, and to activate ivermectin phosphate-sensitive current, suggesting that these two subunits were the target of avermectins in nematodes. Rohrer et al. (1995) recently identified the binding sites of ivermectin in two arthropods, *Drosophila melanogaster* and *Schistocerca americana*. One subunit, DrosGluCl-alpha, distinct from those cloned from nematodes, was recently cloned from *D. melanogaster*; this subunit is sensitive to both glutamate and avermectins (Cully et al. 1996). They also found transcripts related to DrosGluCl-alpha in other insects, including cat flea, *Ctenocephalides felis*, fall armyworm, *Spodoptera frugiperda*, and cotton bollworm, *Helicoverpa zea*, but not in the twospotted spider mite, *Tetranychus urticae*. It remains to be determined whether or not additional subunits are present in insects.

It has been suggested that avermectins may bind to a common site on all ligand-gated chloride channels, and act in a tissue-specific manner as both positive or negative allosteric modulators similar to those reported for the benzodiazepine site on mammalian GABA-receptors (Arena 1994). This would help to explain some of the additional effects that avermectins exert on chloride channels, such as their inhibition of strychnine binding to the mammalian glycine receptor (Graham et al. 1982); activation of multiple-gated chloride channels (glutamate, GABA, and acetylcholine) in crayfish (Zufall et al. 1989); and both potentiation and blocking of GABA-gated chloride channels in arthropods (Duce and Scott 1985; Bermudez et al. 1991).

In summary, avermectins affect invertebrates by potentiating the ability of neurotransmitters, such as glutamate and/or GABA, to stimulate an influx of chloride ions into nerve cells. It is currently believed that the anthelmintic properties of the avermectins are due predominantly to potentiation and/or direct opening of glutamate-gated chloride channels, whereas in insects it is likely that avermectins bind to multiple sites (including glutamate and GABA) in insect chloride channels. The chloride ion flux produced by the opening of the channel into neurons results in loss of cell function and disruption of nerve impulses. Consequently, invertebrates are paralyzed irreversibly and stop feeding. In insects and mites, maximum mortality is achieved in about 4 days. Although the avermectins do not exhibit rapid knock down effects on insects, paralysis is rapid, and feeding damage to crops is minimal because feeding ceases shortly after ingestion. Avermectins are taken up by arthropods via contact and ingestion, although ingestion is considered to be the primary route whereby arthropods accumulate a lethal dose. The wide margin of safety for avermectin compounds to mammals is attributed to (1) the lack of glutamate-gated chloride channels in mammals; (2) the low affinity of avermectins for other mammalian ligand-gated chloride channels; and (3) their inability to readily cross the blood–brain barrier (Arena et al. 1995).

3
Biological Activity

3.1
Spectrum and Potency of Avermectins against Agricultural Pests

Ostlind et al. (1979) were the first to report on the insecticidal activity of avermectins. Subsequent studies demonstrated the broad spectrum activity of the avermectins, including their potential application in agriculture (Putter et al. 1981). As noted in Section 1.1, avermectin B_1 (abamectin) was selected for use in crop protection based on its high intrinsic toxicity to arthropods (Dybas et al. 1989).

Abamectin has unprecedented potency against a broad spectrum of phytophagous mites, with LC_{90} values ranging between 0.009 and 0.24 ppm for members of Tetranychidae, Eriophyidae, and Tarsonemidae (Lasota and Dybas 1991; Table 1). Abamectin is very toxic to eriophyid mites, such as the citrus

Table 1. Comparative toxicity of avermectin insecticides to different arthropod pests of agricultural importance[a]

Arthropod species	LC_{90} (μg/ml)		PR[b]
	Abamectin	Emamectin benzoate	
Acarina, foliar dip/contact assay			
Aculops lycopersici, adults	0.009[c]	—	—
Tetranychus urticae, adults	0.03 [d]	0.29[c]	0.07
Tetranychus turkestani, adults	0.08 [d]	—	—
Tetranychus pacificus, adults	0.16 [f]	—	—
Phyllocoptruta oleivora, adults	0.02 [d]	—	—
Panonychus citri, adults	0.24 [d]	—	—
Panonychus ulmi, adults	0.04 [d]	—	—
Polyphagotarsonemus latus, adults	0.03 [d]	—	—
Insecta			
Coleoptera, foliar spray			
Leptinotarsa decemlineata, neonates	0.03[d]	0.03[e]	1.0
Epilachna varivestis, neonates	0.2[d]	0.2[e]	1.0
Diptera, plant dip			
Liriomyza trifolii, first instar	0.19[g]	1.45[g]	0.13
Homoptera, foliar spray/contact assay			
Aphis fabae, nymphs and adults	0.2-0.5[h]	19.9[e]	0.01-0.02
Aphis gossypii, nymphs and adults	0.4-1.5[e]	—	—
Acyrthosiphon pisum, nymphs and adults	0.4[d]	—	—
Lepidoptera			
Manducca sexta, neonates, foliar spray	0.02[d]	0.003[e]	7
Plutella xylostella, neonates, foliar spray	0.02[h]	0.002[h]	10
Keiferia lycopersicella, first instar/wet residue	0.03[f]	—	—
Heliothis virescens, neonates, foliar spray	0.13[e]	0.003[e]	43
Trichoplusia ni, neonates, foliar spray	1.0[d]	0.014[e]	71
Helicoverpa zea, neonates, foliar spray	1.5[d]	0.002[e]	750
Spodoptera exigua, neonates, foliar spray	1.97[h]	0.005[e]	394
Spodoptera eridania, neonates, foliar spray	6.0[d]	0.005[e]	1200
Argyrotaenia velutinana, neonates, foliar spray	—	0.009[g]	
Spodoptera frugiperda, neonates, foliar spray	25.0[f]	0.01[e]	2,500
Pseudoplusia includens, neonates, foliar spray	—	0.019[h]	—
Ostrinia nubilalis, neonates, diet	—	0.024[h]	—
Agrotis ipsilon, neonates, diet	—	0.041[h]	
Cydia pomonella, neonates, diet	135.0[i]	0.89[i]	152

[a]Data were generated using either a foliar spray assay, foliar dip assay, or diet assay. In the foliar spray assay, foliage was sprayed with different concentrations of each compound, air dried, infested with target insects, and mortality evaluated 96 h after application. Mite-infested foliage was used in all mite assays. Infested leaves were dipped for 2 seconds in different concentrations of test compounds; mortality was evaluated 96 h after dipping. In diet assays, the surface of an agar-based artificial diet was treated with different concentrations of each compound and then handled as above (as with foliage).
[b]PR, potency ratio = LC_{90} abamectin/LC_{90} emamectin benzoate.
[c]Royalty and Perring (1987).
[d]Dybas and Green (1984).
[e]Dybas et al. (1989).
[f]Dybas (1989).
[g]Cox et al. (1995a)
[h]Merck (unpubl. data).
[i]Cox et al. (1995b).

rust mite, *Phyllocoptruta oleivora*, (LC_{90} = 0.02 ppm) and to various tetranychid mites, such as the twospotted spider mite, *T. urticae* (LC_{90} = 0.03 ppm) (Dybas 1989). However, it is markedly less toxic to other phytophagous mites, such as citrus red mite, *Panonychus citri* (LC_{90} = 0.24) (Dybas 1989).

The toxicity of abamectin to insects is more variable (Table 1). Although the compound is highly toxic to tobacco hornworm, *Manduca sexta*, diamondback moth, *Plutella xylostella*, tomato pinworm, *Keiferia lycopersicella*, tobacco budworm, *Heliothis virescens*, Colorado potato beetle, *Leptinotarsa decemlineata*, and the serpentine leafminer, *Liriomyza trifolii* (LC_{90} values range between 0.02 and 0.19 ppm), it is less potent against certain Homoptera and Lepidoptera (LC_{90} values range between 1 and >25 ppm) (Dybas 1989; Lasota and Dybas 1991). Its lower potency against most Lepidoptera, Homoptera, and certain Coleoptera limited its opportunities for further development. Although abamectin is toxic to certain aphids (e.g., LC_{90} values against black bean aphid, *A. fabae*, are 0.2–0.5 ppm and against cotton aphid, *A. gossypii*, are 0.4–1.5 ppm) (Putter et al. 1981; Dybas and Green 1984), abamectin was not effective at controlling aphids in translaminar assays (Wright et al. 1985). The reduced efficacy at controlling aphids is probably due to sublethal (or lower) concentrations of abamectin in phloem tissue where aphids actively feed. Poor residual control of aphids in the field has been confirmed in several studies (Dybas 1989 and references therein).

The toxicity of abamectin on beneficial arthropods (e.g., honeybees, parasitoids, predators) has been studied extensively (Dybas 1989; Lasota and Dybas 1991 and references therein). In general, abamectin is less toxic to beneficial arthropods, especially when exposure occurs beyond 1 day after application. LC values against most beneficial arthropods are markedly higher than those for the key target pests (Grafton-Cardwell and Hoy 1983; Hoy and Cave 1985; Dybas 1989 and references therein; Zhang and Sanderson 1990). The implications of this differential toxicity are discussed below.

The spectrum and concomitant potency of emamectin benzoate have not been studied as extensively as those for abamectin. In general, emamectin benzoate is highly potent to a broad spectrum of Lepidoptera. It is the most potent insecticide compound registered or in development for agricultural use. LC_{90} values for emamectin benzoate against a variety of Lepidoptera range between 0.002 and 0.89 ppm (Dybas 1989; Cox et al. 1995b; Table 1). Emamectin hydrochloride was up to 1500-fold more potent against armyworm species, e.g., beet armyworm, *S. exigua*, than abamectin (Trumble et al. 1987; Dybas et al. 1989; Mrozik et al. 1989). Emamectin hydrochloride was also 1720-, 884-, and 268-fold more potent against southern armyworm, *S. eridania*, than methomyl, thiodicarb, and fenvalerate, respectively, and 105- and 43-fold more toxic to cotton bollworm, *H. zea*, and *H. virescens*, larvae compared with abamectin (Dybas and Babu 1988). Recent studies have shown that emamectin benzoate was 875- to 2975-fold and 250- to 1300- fold more potent than tebufenozide to *H. virescens* and *S. exigua*, respectively. Emamectin benzoate was also 12.5- to 20-fold and 250- to 500-fold more potent than lambda cyhalothrin, and 175- to 400-fold and 2033- to 8600-fold more potent than fenvalerate to these two Lepidoptera, respectively (Jansson et al. 1997).

Emamectin benzoate is markedly less toxic to other arthropods (Table 1). It is about 8- to 15-fold less toxic to *L. trifolii* and *T. urticae*, respectively, than abamectin (Dybas et al. 1989; Cox et al. 1995a). However, emamectin benzoate and abamectin are comparable in their potency against Mexican bean beetle, *Epilachna varivestis*, and *L. decemlineata* (Dybas 1989). Emamectin benzoate is markedly less toxic to *A. fabae* than abamectin (Table 1).

Similar to abamectin, emamectin benzoate is less toxic to a variety of beneficial organisms than it is to Lepidoptera. Foliar residues of emamectin benzoate were only slightly toxic (< 20% mortality) to most beneficial insects, including honey bee, *Apis mellifera*, and several predators and parasitoids, within 1 day after application and often within a few hours after application (DL Cox et al., unpubl.). The low toxicity was related to the short half-life of emamectin benzoate on foliage. On celery, the half-life of foliar dislodgeable residues was estimated to be approximately 0.66 days (D.M. Dunbar et al., unpubl.). They also found that residues were similar between two crops (celery and alfalfa) 24 after application and averaged about 1.3 ng/cm^2. Although wet residues of emamectin benzoate are generally toxic to many arthropods, several beneficial arthropods, such as common green lacewing, *Chrysopa carnea*, and convergent lady beetle, *Hippodamia convergens*, are tolerant to emamectin benzoate, even when exposed to wet residues (DL Cox et al., unpubl.). Kok et al. (1996) showed that emamectin hydrochloride (MK-243) displayed minimal adverse effects against two hymenopterous parasitoids (*Pteromalus puparum* and *Cotesia orobenae*) in a residual contact bioassay on broccoli. Collectively, these data demonstrate that emamectin benzoate provides ecological selectivity (and in some cases physiological selectivity) to a wide range of beneficial arthropods. For this reason, it is believed to be highly compatible with integrated pest management (IPM) programs.

3.2 Photostability and Translaminar Movement

Avermectin insecticides are very susceptible to photodegradation. MacConnell et al. (1989) showed that the half-life of abamectin was < 10 h in simulated sunlight and that there were marked differences in the half-life of abamectin on petri dishes and on leaves in light and dark environments. Numerous photodegradates have subsequently been identified for both avermectin B_{1a} (Crouch et al. 1991) and emamectin benzoate (Feely et al. 1992).

Despite the short half-life for avermectin insecticides in sunlight, low levels of these compounds are taken up rapidly via translaminar movement into foliage. Translaminar movement of abamectin has been demonstrated in numerous studies (Wright et al. 1985; Dybas 1989 and references therein). MacConnell et al. (1989) showed that the prolonged stability of abamectin in the dark resulted in greater penetrability into leaves and improved efficacy at controlling mites. Wright et al. (1985) showed that translaminar movement of abamectin resulted in good control of mites, but did not control aphids. They suggested that abamectin reservoirs were probably abundant in the parenchyma tissues of leaves where mites feed, whereas little abamectin was probably present in the phloem tissue where aphids feed. Presence of abamectin reservoirs in parenchyma tissue would also account for its excellent efficacy at controlling several dipteran and lepidopteran leafminers.

4 Application in Agriculture

4.1 Crop Applications

Abamectin is currently registered in over 50 countries worldwide for control of mites and certain insect pests on a variety of ornamental and horticultural crops and on cotton. Field use rates for abamectin are currently the lowest among the various insecticides (5.4–27 g ai/ha). Although it is primarily used as an acaricide, abamectin is also effective for controlling agromyzid leafminers, *Liriomyza* spp., *L. decemlineata*, *P. xylostella*, *K. lycopersicella*, citrus leafminer, *Phyllocnistis citrella*, and pear psylla, *Psylla pyricola*, and is sold under the tradenames Avid, Agrimec, Agri-Mek, Dynamec, Epi-Mek, Vertimec, and Zephyr. In addition to ornamental and horticultural crop uses, abamectin has also been incorporated into bait to control red imported fire ant, *Solenopsis invicta*, and cockroaches and is sold under the tradenames Affirm and Avert. These and other urban pest applications have been reviewed previously (Lasota and Dybas 1991).

More recently, studies have shown that avermectins have potential as a component in Africanized honeybee abatement programs aimed at increasing mating of European queens in heavily Africanized areas (Danka et al. 1994). Abamectin was also incorporated into bait to control Caribbean fruit fly, *Anastrepha suspensa* (M.K. Hennessey, pers. comm.). In addition, injection of abamectin into trees controls certain phytophagous arthropods, such as elm leaf beetle, *Pyrrhalta luteola*, for up to 83 days after application (Harrell and Pierce 1994).

The most widespread commercial use of abamectin is as an acaricide. Historically, the compound has demonstrated mite control and residual activity superior to all other acaricides on a variety of ornamental crops, food crops, and cotton (Putter et al. 1981; Dybas and Green 1984; Dybas 1989). Abamectin is also the premier product for control of agromyzid leafminers in ornamental and vegetable crops. Its popularity and success at controlling these pests has often led to its overuse in certain areas. For example, it is not uncommon for certain ornamental growers to apply abamectin once every 5–7 days for control of leafminers and/or mites in the glasshouse. Such high use patterns have led to concerns about resistance development (see Sect. 4.3).

Emamectin benzoate is currently under review at the US Environmental Protection Agency for registration as a broad spectrum insecticide for control of lepidopterous pests on certain vegetable crops. Excellent efficacy of this compound has been demonstrated against numerous lepidopterous pests in a variety of crops (Jansson and Lecrone 1991; Leibee et al. 1995; Jansson et al. 1996; numerous unpubl. reports). Similar to abamectin, the use rates for emamectin benzoate (8.4–16.8 g ai/ha) are among the lowest of all insecticides used in agriculture. It is anticipated that the compound will eventually be used to control lepidopterous pests on a wide variety of vegetable, tree, and row crops. Development programs for this compound are currently focused on the USA, Latin America, Japan and Southeast Asia. In addition to its use for control of lepidopterous pests, the compound is expected to be used for control of thrips on eggplant and tea in Japan. As noted in Section 3.1, because of its ecological selectivity, emamectin benzoate will be very compatible with IPM programs, and may constitute the cornerstone of many IPM programs in certain crops. Results from dozens of field trials have been very consistent. No other insecticide currently available commercially or in development that has been tested by Merck Research Laboratories has outperformed emamectin benzoate at controlling lepidopterous pests on vegetable crops in the field when applied at recommended use rates.

4.2 Selectivity/IPM Compatibility

Selective insecticides that reduce pest abundance and are compatible with natural enemies by allowing their survival and range expansion are central to the design of IPM programs. Compatibility of pesticides with IPM programs is based on the differential toxicity of a compound to pest and beneficial arthropod populations. Differential toxicity can be achieved by exploiting pharmacokinetic or metabolic differences between pests and beneficials (physiological selectivity) or by exploiting a compound's unique qualities which result in toxic exposure to phytophagous pests with concomitant reduced exposure to beneficials (ecological selectivity). The avermectin insecticides possess both of these qualities thereby making them highly selective and compatible with IPM.

It is important to note that compatibility of pesticides with IPM programs may vary among different crop ecosystems. For example, compatibility might be difficult to achieve in an intensively managed crop that receives numerous pesticide applications per year, such as in glasshouse ornamentals. Compatibility might also be difficult to achieve in high value horticultural crops, which also receive numerous pesticide applications per season. Compatibility is easier to achieve in low value crops which receive fewer pesticide applications. Avermectin insecticides have demonstrated compatibility with biological control agents and IPM programs in both cases. For this reason, abamectin has and emamectin benzoate will become an integral part of many IPM programs, indoors and outdoors.

The physiological selectivity (differential toxicity) of abamectin to a variety of pest and predatory mite species is well documented (Lasota and Dybas 1991 and references therein). Hoy and Cave (1985) showed that although abamectin was toxic to the predator *Metaseiulus occidentalis* (Nesbitt), it was more toxic to the twospotted spider mite, *Tetranychus urticae*. Similar results have been found with other predatory mite species (Grafton-Cardwell and Hoy 1983; Zhang and Sanderson 1990). Hoy and Cave (1985) also showed that field-aged residues of abamectin of $\geq$ 48–96 h were safe to *M. occidentalis*. Several other studies demonstrated that abamectin is less toxic to natural enemies (both naturally occurring and introduced), and, for this reason, it selectively kills target pests and conserves natural enemies (Dybas 1989; Lasota and Dybas 1991 and references therein).

Ecological selectivity of the avermectins results from unique qualities of the avermectins when applied in the field. As noted in Section 3.2, avermectin insecticides are readily taken up by plant foliage. Foliar uptake via translaminar movement results in a reservoir of the toxicant (and certain metabolites which may also be toxic to phytophagous insects) inside the leaf (MacConnell et al. 1989). However, while the toxicant is readily taken up by foliage, any remaining compound on the outside of foliage is rapidly broken down by sunlight thereby resulting in minimal residues of the compound less than 1 day after application. These unique qualities of the avermectin insecticides are principally responsible for their compatibility with beneficials and IPM programs.

In glasshouse roses, Sanderson and Zhang (1995) showed that full canopy applications of abamectin (12 ppm) at 3–5-day spray intervals were effective in controlling *T. urticae* populations, but were detrimental to the predatory mite, *Phytoseiulus persimilis.* However, when applications of abamectin were confined to the upper canopy of plants, excellent mite control in the marketable portion of the plant was achieved along with conservation of predatory mite populations in the lower canopy. Their results showed the potential for integrating abamectin with natural enemies for control of a target pest, even under intense abamectin pressure. Similarly, Zchorifein et al. (1994) found that abamectin was compatible with *Encarsia formosa*, an important component of greenhouse IPM programs for greenhouse whitefly, *Trialeurodes vaporariorum.* No parasitoid mortality was found when adult parasitoids were exposed to 24-hour old residues of abamectin. They found that a management program that relied on the combined treatment of abamectin with releases of *E. formosa* maintained lower densities of the whitefly on poinsettia throughout the season with fewer applications of abamectin than did applications of abamectin alone.

In field-grown strawberry, Trumble and Morse (1993) showed that abamectin was also compatible with *P. persimilis.* They found that the highest net return was achieved when several releases of the predatory mite *P. persimilis* were integrated with properly timed applications of abamectin (based on economic threshold). The net return of the combination of these two control strategies was equal to the additive effects of each treatment separately thereby indicating compatibility. Conversely, two other acaricides, hexythiazox and fenbutatin-oxide, were not compatible with releases of the predator.

Other researchers showed that applications of abamectin did not disrupt the leafminer parasitoid complex on a variety of vegetable crops (Dybas 1989 and references therein). Recently, Schuster (1994) showed that abamectin was less detrimental to certain life stages of two leafminer parasitoids, *Diglyphus intermedius* and *Neochrysocharis punctiventris*, than methomyl, permethrin, methamidophos, and endosulfan. Trumble and Alvarado Rodriguez (1993) showed that because of the reduced toxicity of abamectin to natural enemies, it was an important component of a multiple pest IPM program on fresh market tomato that was based on intensive sampling, parasitoid releases, mating disruption, microbial pesticides, and abamectin.

In apple, Cox and coworkers (1994, pers. comm.) found that abamectin was highly compatible with the phytoseiid mites *M. occidentalis* and *Amblyseius fallacis.* They found that abamectin was effective at reducing and maintaining numbers of European red mites, *Panonychus ulmi*, at or below the economic threshold for at least 3 weeks after application. Other studies conducted by Merck researchers found that abamectin controlled mites on apple for up to 6 weeks after application (unpubl.). Although predatory mite numbers declined initially, they quickly resurged, resulting in favorable predator-to-prey ratios which were comparable with or superior to those found in nontreated plots, and aided in mite suppression for the remainder of the growing season (Cox et al. 1994; white et al. 1996).

Because emamectin benzoate is not yet commercially available, there have been fewer studies on its selectivity and compatibility with IPM programs. The majority of the data were collected by Merck researchers. All studies conducted to date indicate that emamectin benzoate is equal or superior to abamectin in its selectivity (ecological and physiological; see above) and will be highly compatible with IPM programs.

4.3 Resistance

Resistance to the avermectins was recently reviewed by Clark et al. (1995). In general, a variety of biochemical and pharmacokinetic mechanisms may contribute to avermectin resistance in arthropods; however, biochemical mechanisms tend to be more important among most arthropod systems studied to date. Target site insensitivity (*kdr*), increased monooxygenase activity, and decreased cuticular penetration were shown to be responsible for resistance to avermectins in the housefly, *Musca domestica* (Scott 1995 and references therein). In twospotted spider mite, *T. urticae*, resistance to abamectin has been attributed to increased excretion and decreased absorption (Clark et al. 1995), monooxygenases (Campos et al. 1996), esterases (Campos et al. 1996; R.K. Jansson et al., unpubl., glutathione-S-transferases (Clark et al. 1995), and decreased penetration (Clark et al. 1995). In *L. decemlineata*, resistance has been attributed to increased enzymatic activity, especially carboxylesterases and monooxygenases, increased excretion, and increased metabolism (Clark et al. 1995 and references therein). In *P. xylostella*, resistance to abamectin was believed to be associated, in part, with monooxygenases, esterases, and glutathione-S-transferases (D.J. Wright, pers. comm.). Cochran (1994) did not find significant levels of resistance to abamectin in 13 strains of German cockroach, *Blattella germanica*, but suggested that small differences in susceptibility to abamectin in at least one strain might be due to decreased feeding, increased metabolism, and/or increased excretion.

In the arthropods examined to date, resistance to avermectins is not stable. Levels of resistance in twospotted spider mite, *T. urticae*, populations collected from the field were shown to drop markedly within 2-6 weeks after collection when maintained in the absence of selection pressure (Campos et al. 1995). Similar results were found by other researchers (R.K. Jansson et al., unpubl.; T Dennehy, pers. comm.). Resistance to abamectin in *M. domestica* and *L. decemlineata* was also unstable (Clark et al. 1995 and references therein). Instability of resistance is undoubtedly related to the inheritance of the gene(s) responsible for resistance. These traits were highly recessive in *M. domestica* and autosomal and incompletely recessive in *L. decemlineata* (Clark et al. 1995). Resistance was believed to be polyfactorial in *L. decemlineata* (Clark et al. 1995).

Cross resistance between avermectin insecticides/acaricides and other classes of chemistry has not been documented widely, nor is it well understood. The first case of cross resistance to abamectin was found in 1986 in a pyrethroid-resistant housefly (Scott and Georghiou 1986). They found that resistance was attributed to microsomal cytochrome P450 monooxygenases, decreased sensitivity of the nervous system (*kdr* type), and decreased cuticular penetration. However, *kdr*-resistant flies were not cross-resistant with abamectin (Scott 1989). Thus, cross resistance was attributed to metabolism and decreased penetration in these flies.

Geden et al. (1990) showed that a wild-type housefly population that was not controlled effectively with permethrin at label rates was less sensitive to abamectin. In another study, a housefly population that was not controlled effectively with tetrachlorvinphos, dimethoate, crotoxyphos, and diazinon at label rates was 3.2-fold less sensitive (LC_{90}) to abamectin (Geden et al. 1992). Roush and Wright (1986) found no evidence for cross resistance between abamectin and pyrethroids in six pyrethroid-resistant strains of the housefly. Parrella (1983) found that abamectin was effective at controlling agromyzid flies that were resistant to pyrethroids.

Lamine (1994) found that abamectin-resistant larvae of *L. decemlineata* were Seven- to ten-fold less susceptible to emamectin benzoate than the baseline colony in a topical bioassay. No cross resistance between abamectin and emamectin benzoate was evident in adult beetles. Recent studies showed that *L. decemlineata* populations which were 15- to 23-fold less sensitive to abamectin in a topical bioassay (Clark et al. 1995) were classified as susceptible to abamectin in a diet-based ingestion assay (G.P. Dively and R.K. Jansson, unpubl.).

Abro et al. (1988) showed that a field-collected population of *P. xylostella* that was resistant to cypermethrin, malathion, and DDT (364- to >2,100-fold) was 26-fold less susceptible to abamectin in a topical bioassay; however, the same population was equal in its susceptibility to the susceptible colony to abamectin in a 10-day foliar ingestion bioassay. Recent studies by Lasota et al. (1996) have suggested that there was no cross-resistance between abamectin, emamectin benzoate, permethrin, and methomyl in *P. xylostella* using an ingestion bioassay. Recent studies conducted at Merck Research Laboratories also suggested that there was little, if any, cross-resistance between abamectin and emamectin benzoate in T. *urticae* and *L. trifolii*, although additional data are needed to confirm this belief (R.K. Jansson et al., unpubl.). Target site insensitivity has not been shown to confer any cross-resistance to avermectin insecticides to date. Rohrer et al. (1995) showed that fipronil, which interacts with GABA-gated chloride channels (as an antagonist), did not affect ivermectin binding in *Drosophila* head membranes or in locust neuronal membranes.

Because of the exceptional control and rapid adoption of abamectin by growers, Merck researchers quickly established pro-active resistance management programs for avermectin insecticides/acaricides. Part of this pro-active strategy includes the development of monitoring systems to detect resistance in high risk populations of arthropods. Monitoring for resistance to abamectin in mites and *Liriomyza* leafminers has been conducted for several years. Genetic resistance has been found in each of these arthropods; however, despite the presence of genetic resistance in monitored populations, abamectin has provided effective control consistently in the field. Higher levels of genetic resistance in mites and leafminers have been found in populations collected from glasshouse-grown ornamentals where the abamectin has been used up to 30-50 times per year (T. Dennehy et al., unpubl.; R.K. Jansson et al., unpubl. For example, mite populations from glasshouse-grown roses tend to have the highest levels of genetic resistance to abamectin; however, abamectin continues to provide effective control of these mites, albeit residual activity appears to have shortened where the product has been overused. Monitoring programs were recently developed to gather baseline data sensitivity to abamectin in *L. decemlineata* and *P. oleivora* (RK Jansson et al., unpubl.). Few monitoring programs for emamectin benzoate have been established to date. As mentioned above, baseline data for susceptibility of *P. xylostella* populations to emamectin benzoate have been generated (Lasota et al. 1996). Monitoring programs for *P. xylostella* and other problematic Lepidoptera are planned.

To minimize the risk of arthropod-resistance to abamectin and emamectin benzoate, development programs for the avermectins in agriculture have been managed carefully. Most labels for avermectin insecticides provide restrictions on the type, number, and sequencing of applications allowed per growing season. Applications have been excluded from transplant nurseries for certain vegetable crops to reduce selection pressure on arthropods that pose the greatest risk. These and other strategies should prolong the life of these insecticides/acaricides in the commercial sector.

5 Summary

The discovery of avermectins and their development as anthelmintics, endectocides, insecticides, and acaricides is unprecedented in terms of their broad spectrum impact on human health, animal health, and crop protection programs. No other family of compounds has found applications as wide as those for the avermectins. In crop protection, avermectins are unprecedented. Their unique mode of action and lack of cross-resistance with any other class of chemistry provides an important tool for crop protection. Avermectin insecticides/acaricides are used at extremely low rates (5.4–27 g a.i./ha) and provide effective control of a broad spectrum of phytophagous pests. Because these compounds are unstable in sunlight, they rapidly degrade, with a half-life of < 1 day after application. However, their translaminar movement ensures that an adequate reservoir of the toxicant is present in parenchyma tissue to provide long-lasting control of phytophagous pests. The combination of their photoinstability and translaminar movement results in ecological selectivity for phytophagous pests and concomitant conservation of beneficial arthropods. These two attributes of the compounds are central to their compatibility with biological control organisms and IPM programs. In addition, avermectins bind tightly to soil and organic matter and are rapidly degraded by soil microorganisms. Thus, their potential to enter surface water or to leach is negligible. If properly managed, these compounds should be a valuable tool in crop protection for many years.

References

Abro GH, Dybas RA, Green ASJ, Wright DJ (1988) Toxicity of avermectin B1 against a susceptible laboratory and an insecticide-resistant strain of *Plutella xylostella* (Lepidoptera: Plutellidae). J Econ Entomol 81: 1575–1580

Albers-Schönberg G, Arison BH, Chabala JC, Douglas AW, Eskola P, Fisher MH, Lusi A, Mrozik H, Smith JL, Tolman RL (1981) Avermectins: structure determination. J Am Chem Soc 103: 4216–4221

Albrecht CP, Sherman M (1987) Lethal and sublethal effects of avermectin B_1 on three fruit fly species (Diptera: Tephritidae). J Econ Entomol 80: 344–347

Arena JP (1994) Expression of *Caenorhabditis elegans* mRNA in *Xenophus* oocytes: a model system to study the mechanism of action of avermectins. Parasitol Today 10: 35–37

Arena JP, Liu KK, Paress PS, Cully DF (1991) Avermectin-sensitive chloride currents induced by *Caenorhabditis elegans* RNA in *Xenopus* oocytes. Mol Pharmacol 40: 368–374

Arena JP, Liu KK, Paress PS, Schaeffer JM, Cully DF (1992) Expression of a glutamate-activated chloride current in *Xenopus* oocytes injected with *Caenorhabditis elegans*: correlation between activation of glutamate-sensitive chloride current, membrane binding and biological activity. Mol Brain Res 15: 339–348

Arena JP, Liu KK, Paress PS, Frazier EG, Cully DF, Mrozik H, Schaeffer J (1995) The mechanism of action of avermectin in *Caenorhabditis elegans*: correlation between activation of glutamate-sensitive chloride current, membrane binding and biological activity. J Parasitol 81: 286–294

Bermudez I, Hawkins CA, Taylor AM, Beadle DJ (1991) Actions of insecticides on the insect GABA receptor complex. J Recept Res 11: 221–232

Bloomquist JR (1993) Toxicology, mode of action and target site-mediated resistance to insecticides acting on chloride channels. Comp Biochem Physiol 106C: 301–314

Burg RW, Miller BM, Baker EE, Birnbaum J, Currie SA, Hartman R, Kong YL, Monaghan RL, Olson G, Putter I, Tunac JB, Wallick H, Stapley EO, Oiwa R, Omura S (1979) Avermectins, new family of potent anthelminthic agents: producing organism and fermentation. Antimicrob Agents Chemother 15: 361–367

Campbell WC (ed) (1989) Ivermectin and abamectin. Springer, Bertin Heidelberg, New York

Campbell WC, Burg RG, Fisher MH, Dybas RA (1984) The discovery of ivermectin and other avermectins. In: Magee PS, Kohn GK, Menn JJ (eds) Pesticide synthesis through rational approaches. ACS Symp Ser no 255. American Chemical Society, Washington, DC, pp 5–20

Campos F, Dybas RA, Krupa DA (1995) Susceptibility of twospotted spider mite, *Tetranychus urticae* (Acari: Tetranychidae) populations in California to abamectin. J Econ Entomol 88: 225–231

Campos FC, Krupa DA, Dybas RA (1996) Susceptibility of populations of twospotted spider mites (Acari: Tetranychidae) from Florida, Holland, and the Canary Islands to abamectin and characterization of abamectin resistance. J Econ Entomol 89: 594–601

Clark JM, Scott JG, Campos F, Bloomquist JR (1995) Resistance to avermectins: extent, mechanisms, and management implications. Annu Rev Entomol 40: 1–30

Cochran, DG (1994) Abamectin resistance potential in the German cockroach (Dictyoptera: Blatellidae). J Econ Entomol 87: 899–903

Cox DL, Lasota JA, Dybas RA, Starner VR, White SM, Fuse M (1994) Compatibility of abamectin with phytoseiid mites for control of European red mite in apple orchards. Proc Giornate Fitopatologiche, May 10–11, 1994 Montesilvano, Italy (Abstr)

Cox DL, Remick D, Lasota JA, Dybas RA (1995a) Toxicity of avermectins to *Liriomyza trifolii* (Diptera: Agromyzidae) larvae and adults. J Econ Entomol 88: 1415–1419

Cox DL, Knight AL, Biddinger DG, Lasota JA, Pikounis B, Hull LA, Dybas RA (1995b) Toxicity and field efficacy of avermectins against codling moth (Lepidoptera: Tortricidae) on apples. J Econ Entomol 88: 708–715

Crouch LS, Feely WF, Arison BH, Vanden Heuvel WJ, Colwell LF, Stearns RA, Kline WF, Wislocki PG (1991) Photodegradation of avermectin B_{1a} thin films on glass. J Agric Food Chem 39: 1310–1319

Cully DF, Vassilatis DK, Liu KK, Paress PS, Van der Ploeg LHT, Schaeffer JM, Arena JP (1994) Cloning of an avermectin-sensitive glutamate-gated chloride channel from *Caenorhabditis elegans.* Nature 371: 707–711

Cully DF, Paress PS, Liu KK, Schaeffer JM, Arena JP (1996) Identification of a *Drosophila melanogaster* glutamate-gated chloride channel sensitive to the antiparasitic agent avermectin. J Biol Chem 271: 20187–20191

Cvetovich RJ, Kelly DH, DiMichele LM, Shuman RF, Grabowski, EJJ (1994) Synthesis of 4″-epi-amino-4″-deoxyavermectins B_1. J Org Chem 59: 7704–7708

Danka RG, Loper GM, Villa JD, Williams JL, Sugden EA, Collins AM, Rinderer TE (1994) Abating feral Africanized honey bees (*Apis mellifera* L.) to enhance mating control of European queens. Apidologie 25: 520–529

Davies GH, Green RH (1986) Avermectins and milbemycins. Nat Prod Rep Engl 3: 87–121

Deng Y, Casida JE (1992) House fly head GABA-gated chloride channel: toxicologically relevant binding site for avermectins coupled to site for ethynylbicycloorthobenzoate. Pestic Biochem Physiol 43: 116–122

Duce IR, Scott RH (1985) Actions of dihydroavermectin B_{1a} on insect muscle. Brit J Pharmacol 85: 395–401

Dybas RA (1989) Abamectin use in crop protection. In: Campbell WC (ed) Ivermectin and abamectin. Springer, Berlin Heidelberg, New York, pp 287–310

Dybas RA, Babu JR (1988) 4″-deoxy-4″-methylamino-4″-epiavermectin B1 hydrochloride (MK-243): a novel avermectin insecticide for crop protection. In: British Crop Prot Conf, Pests and Diseases, British Crop Protection Council, Croydon, pp 57–64

Dybas RA, Green ASJ (1984) Avermectins: their chemistry and pesticidal activity. In: British Crop Prot Conf Pests and Diseases, vol. 9B-3, British Crop Protection Council, Croydon, pp 947–954

Dybas RA, Hilton NJ, Babu JR, Preiser FA, Dolce GJ (1989) Novel second-generation avermectin insecticides and miticides for crop protection. In: Demain AL, Somkuti GA, Hunter-Cevera JC, Rossmoore HW (eds) Novel microbial products for medicine and agriculture. Society of industrial Microbiology, Annandale, Virginia, pp 203–212

Egerton JR, Ostlind DA, Blair LS, Eary CH, Suhayda D, Cifelli S, Riek RF, Campbell WC (1979) Avermectins, new family of potent anthelminthic agents: efficacy of the B_{1a} component. Antimicrob Agents Chemother 15: 372–378

Feely WF, Crouch LS, Arison BH, Vanden Heuvel WJA, Colwell LF, Wislocki PG (1992) Photodegradation of 4″-(epimethylamino)-4″-deoxyavermectin B_{1a} thin films on glass. J Agric Food Chem 40: 691–696

Fisher MH, Mrozik H (1984) The avermectin family of macrolide-like antibiotics. In: Omura S (ed) Macrolide antibiotics. Academic Press, New York, pp 553–606

Fisher MH, Mrozik H (1989) Chemistry. In: Campbell WC (ed) Ivermectin and abamectin. Springer, Berlin Heidelberg, New York, pp 1–23

Fisher MH, Mrozik H (1992) The chemistry and pharmacology of avermectins. Annu Rev Pharmacol Toxicol 32: 537–553

Fritz LC, Wang CC, Gorio AA (1979) Avermectin B_{1a} irreversibly blocks postsynaptic potentials at the lobster neuromuscular junction by reducing muscle membrane resistance. Proc Nat Acad Sci USA 76: 2062–2066

Geden CJ, Steinkraus DC, Long SJ, Rutz DA, Shoop WL (1990) Susceptibility of insecticide-susceptible and wild house flies (Diptera: Muscidae) to abamectin on whitewashed and unpainted wood. J Econ Entomol 83: 1935–1939

Geden CJ, Rutz DA, Scott JG, Long SJ (1992) Susceptibility of house flies (Diptera: Muscidae) and five pupal parasitoids (Hymenoptera: Pteromalidae) to abamectin and seven commercial insecticides. J Econ Entomol 85: 435–440

Grafton-Cardwell EE, Hoy MA (1983) Comparative toxicity of avermectin B_1 to the predator *Metaseiulus occidentalis* (Nesbitt) (Acarina: Phytoseiidae) and the spider mites *Tetranychus urticae* Koch and *Panonychus ulmi* (Koch) (Acari: Tetranychidae). J Econ Entomol 76: 1216–1220

Graham D, Pfeiffer F, Betz H (1982) Avermectin B_{1a} inhibits the binding of strychnine to the glycine receptor of rat spinal cord. Neurosci Lett 29: 173–176

Harrell MO, Pierce PA (1994) Effects of trunk-injected abamectin on the elm leaf beetle. J Arboric 20: 1–3

Hoy MA, Cave FE (1985) Laboratory evaluation of avermectin as a selective acaricide for use with *Metaseiulus occidentalis* (Nesbitt) (Acarina: Phytoseiidae). Exp Appl Acarol 1: 139–152

Jansson RK, Lecrone SH (1991) Efficacy of nonconventional insecticides for control of diamondback moth, *Plutella xylostella* (L.), in 1991. Proc Fla State Hortic Soc 104: 279–284

Jansson RK, Peterson RF, Mookerjee PK, Halliday WR, Dybas RA (1996) Efficacy of solid formulations of emamectin benzoate at controlling lepidopterous pests. Fla Entomol 79: 434–449

Jansson RK, Halliday WR, Argentine JA, Dybas RA (1997) Evaluation of miniature and high volume bioassays for screening insecticides. J Econ Entomol (in review)

Kok LT, Lasota JA, McAvoy TJ, Dybas RA (1996) Residual foliar toxicity of 4″-epimethylamino-4″-deoxyavermectin B_1 hydrochloride (MK-243) and selected commercial insecticides to adult hymenopterous parasites, *Pteromalus puparum* (Hymenoptera: Pteromalidae) and *Cotesia orobenae* (Hymenoptera: Braconidae). J Econ Entomol 89: 63–67

Kornis GI (1995) Avermectins and milbemycins. In: Godfrey CRA (ed) Agrochemicals from natural products. Marcel-Dekker, New York, pp 215–255

Lamine CNG (1994) Biochemical factors of resistance and management of Colorado potato beetle, *Leptinotarsa decemlineata* (Say) (Coleoptera: Chrysomelidae). MS Thesis, University of Massachusetts, Amherst, 91 pp

Lariviere M, Aziz M, Weimann D, Ginoux J, Gaxotte P, Vingtain P, Beauvais B, Derouin F, Schulz-Key H, Basset D, Sarfati C (1985) Double-blind study of ivermectin and diethylcarbamazine in African onchocerciasis patients with ocular involvement. Lancet 2: 174–177

Lasota JA, Dybas RA (1991) Avermectins, a novel class of compounds: implications for use in arthropod pest control. Annu Rev Entomol 36: 91–117

Lasota JA, Shelton AM, Bolognese JA, Dybas RA (1996) Baseline toxicity of avermectins to diamondback moth (Lepidoptera: Plutellidae) populations: implications for susceptibility monitoring. J Econ Entomol 89: 33–38

Leibee GL, Jansson RK, Nuessly G, Taylor J (1995) Efficacy of emamectin benzoate and *Bacillus thuringiensis* for control of diamondback moth on cabbage in Florida. Fla Entomol 78: 82–96

MacConnell JG, Demchak RJ, Preiser FA, Dybas RA (1989) Relative stability, toxicity, and penetrability of abamectin and its 8,9-oxide. J Agric Food Chem 37: 1498–1501

Mellin TN, Busch RD, Wang CC (1983) Postsynaptic inhibition of invertebrate neuromuscular transmission by avermectin B_{1a}. Neuropharmacol 22: 89–96

Miller TW, Chaiet L, Cole DJ, Cole LJ, Flor JE, Goegelman RT, Gullo VP, Joshua H, Kempf AJ, Krellwitz WR, Monaghan RL, Ormond RE, Wilson KE, Albers-Schonberg G, Putter I (1979) Avermectins, new family of potent anthelminthic agents: isolation and chromatographic properties. Antimicrob Agents Chemother 15: 368–371

Mishima H, Kurabayashi M, Tamura C, Sato S, Kuwano H, Saito, A (1975) Structures of milbemycin beta1, beta2, and beta3. Tetrahedron Lett 16: 711–714

Mrozik H (1994) Advances in research and development of avermectins. In: Hedin PA, Menn JJ, Hollingsworth RM (eds) Natural and engineered pest management agents. Am Chem Soc Symp Ser. no 551. American Chemical Society, Washington, DC, pp 54–73

Mrozik H, Chabala JC, Eskola P, Matzuk A, Waksmunski F, Woods M, Fisher MH (1983) Synthesis of milbemycins from avermectins. Tetrahedron Lett 24: 5333–5336

Mrozik H, Eskola P, Linn BO, Lusi A, Shih TL, Tishler M, Waksmunski FS, Wyvratt MJ, Hilton NJ, Anderson TE, Babu JR, Dybas RA, Preiser FA, Fisher MH (1989) Discovery of novel avermectins with unprecedented insecticidal activity. Experientia 45: 315–316

Ostlind DA, Cifelli S, Lang R (1979) Insecticidal activity of the antiparasitic avermectins. Vet Rec 105: 168

Parrella MP (1983) Evaluation of selected insecticides for control of permethrin-resistant *Liriomyza trifolii* (Diptera: Agromyzidae) on chrysanthemums. J Econ Entomol 76: 1460–1464

Putter I, MacConnell JG, Preiser FA, Haidri AA, Ristich SS, Dybas RA (1981) Avermectins: novel insecticides, acaricides, and nematicides from a soil microorganism. Experentia 37: 963–964

Rohrer SP, Arena JP (1995) Ivermectin interactions with invertebrate ion channels. In: Clark JM (ed) Molecular action of insecticides on ion channels. Am. Chem Soc Symp Ser no 591. American Chemical Society, Washington, DC, pp 264–283

Rohrer SP, Birzin ET, Costa SD, Arena JP, Hayes EC, Schaeffer JM (1995) Identification of neuron-specific ivermectin binding sites in *Drosophila melanogaster* and *Schistocerca americana*. Insect Biochem Mol. Biol 25: 11–17

Roush RT, Wright JE (1986) Abamectin toxicity to house flies (Diptera: Muscidae) resistant to synthetic organic insecticides. J Econ Entomol 79: 562–564

Royalty RN, Perring TM (1987) Comparative toxicity of acaricides to *Aculops lycopersici* and *Homeopronematus anconai* (Acari: Eriophydae, Tydeidae). J Econ Entomol 80: 348–351

Sanderson JP, Zhang ZQ (1995) Dispersion, sampling, and potential for integrated control of two-spotted spidermite (Acari: Tetranychidae) on greenhouse roses. J Econ Entomol 88: 343–351

Schaeffer JM, Haines HW (1989) Avermectin binding in *Caenorhabditis elegans*: a two-state model for the avermectin binding site. Biochem Pharmacol 38: 2329–2338

Schuster DJ (1994) Life-stage specific toxicity of insecticides to parasitoids of *Liriomyza trifolii* (Burgess) (Diptera, Agromyzidae). Int J Pest Manage 40: 191–194

Scott JG (1989) Cross-resistance to the biological insecticide abamectin in pyrethroid-resistant house flies. Pestic Biochem Physiol 34: 27–31

Scott JG (1995) Resistance to avermectins in the house fly, *Musca domestica*. In: Clark JM (ed) Molecular action of insecticides on ion channels. Am Chem Soc Symp Ser no 591. American Chemical Society, Washington, DC, pp 284–292

Scott RH, Duce IR (1985) Effects of 22,23-dihydroavermectin B_{1a} on locust (*Schistocerca gregaria*) muscles may involve several sites of action. Pestic Sci 16: 599–604

Scott JG, Georghiou GP (1986) Mechanisms responsible for high levels of permethrin resistance in the house fly. Pestic Sci 17: 195–206

Shoop WL, Mrozik H, Fisher MH (1995) Structure and activity of avermectins and milbemycins in animal health. Vet Parasitol 59: 139–156

Trumble JT, Alvarado Rodriguez B (1993) Development and economic evaluation of an IPM program for fresh market tomato production in Mexico. Agric Ecosyst Environ 43: 267–284

Trumble JT, Morse JP (1993) Economics of integrating the predaceous mite *Phytoseiulus persimilis* (Acari: Phytoseiidae) with pesticides in strawberries. J Econ Entomol 86: 879–885

Trumble JT, Moar WJ, Babu JR, Dybas RA (1987) Laboratory bioassays of the acute and antifeedant effects of avermectin B_1 and a related analogue on *Spodoptera exigua* (Hübner). J Agric Entomol 4: 21–28

Turner MJ, Schaeffer JM (1989) Mode of action of ivermectin. In: Campbell WC (ed) Ivermectin and abamectin. Springer, Berlin Heidelberg, New York, pp 73–88

Wang CC, Pong SS (1982) Actions of ave-mectin B_{1a} on GABA nerves. In: Sheppard JR, Anderson VE, Eaton JW (eds) Membranes and genetic disease. Liss, New York, pp 373–395

White SM, Gillham MC, Norton JA, Starner VR, Dybas RA (1996) The suitability of abamectin 3.4 EC for mite management programs in European apple orchards. Meded Fac Landbouwwet Univ Gent 61/3A: 877–885

Wright DJ, Loy A, Green ASJ, Dybas RA (1985) The translaminar activity of abamectin (MK-936) against mites and aphids. Meded Fac Landbouwwet Rijksuniv Gent 50(2b): 633–637

Zchorifein E, Roush RT, Sanderson JP (1994) Potential for integration of biological and chemical control of greenhouse whitefly (Homoptera: Aleyrodidae) using *Encarsia formosa* (Hymenoptera: Aphelinidae) and abamectin. Environ Entomol 23: 1277–1282

Zhang Z, Sanderson JP (1990) Relative toxicity of abamectin to the predatory mite *Phytoseiulus persimilis* (Acari: Phytoseiidae) and the twospotted spider mite (Acari: Tetranychidae). J Econ Entomol 83: 1783–1790

Zufall F, Franke C, Hatt H (1989) The insecticide avermectin B_{1a} activates a chloride channel in crayfish muscle membrane. J Exp Biol 142: 191–205

CHAPTER 10

Efficacy of Phyto-Oils as Contact Insecticides and Fumigants for the Control of Stored-Product Insects

E. Shaaya and M. Kostjukovsky
Department of Stored Products, Agricultural Research Organization, The Volcani Center, Bet Dagan 50250, Israel

1 Introduction

Insect damage in stored grains and pulses may amount to 10–40% in countries where modern storage technologies have not been introduced. Currently, the measures taken to control pest infestation in grain and dry food products rely heavily upon the use of insecticides, which pose possible health hazards (to warm-blooded animals) and a risk of environmental contamination.

Fumigation is still one of the most effective methods for the protection of stored food, feedstuffs and other agricultural commodities from insect infestation. At present, only two fumigants are still in use: methyl bromide and phosphine. Methyl bromide has been identified as a major contributor to ozone depletion (WMO 1995), which casts doubts on its future use in insect control. There have been repeated indications that certain insects have developed resistance to phosphine, which is widely used today (Nakakita and Winks 1981; Mills 1983; Tyler et al. 1983). In addition, the US Environmental Protection Agency has classified DDVP (2-2-dichlorovinyl dimethyl phosphate) as a possible human carcinogen and is proposing to cancel most uses of this pesticide (Mueller 1995). Thus, there is an urgent need to develop safe alternatives with the potential to replace the toxic fumigants, yet which are simple and convenient to use. Aromatic spices and herbs contain volatile compounds which are known to possess insecticidal activities. These allelo chemical compounds are mainly essential oils (Brattsten 1983; Shaaya et al. 1991; Schmidt et al. 1991). Their high volatility make it possible to extract the oils by water or steam distillation. Most of the essential oil constituents are monoterpenoids, which are secondary plant chemicals and considered to be of little metabolic importance. Among the first observations on the biological activity of essential oils was the resistance of the Scotch pine to flat bugs which was related to the high content of essential oils in the bark (Symelyanets and Khursin 1973).

The toxicity of a large number of essential oils and their constituents has been evaluated against a number of stored-product insects. The essential oils of *Pogostemon heyneanus, Ocimum basilicum* and *Eucalyptus* sp. showed insecticidal activity against *Sitophilus oryzae* (L.), *Stegobium paniceum* (L.), *Tribolium castaneum* (Herbst) and *Callosobruchus chinensis* (L.) (Desphande et al. 1974; Deshpande and Tipnis 1979). Toxic effects of the terpenoids d-limonene, linalool and terpineol were observed on several Coleoptera damaging post-harvest products

(Karr and Coats 1988; Coats et al. 1991; Weaver et al. 1991). Fumigant toxic activity and reproductive inhibition induced by a number of essential oils and their monoterpenoids were also evaluated against the bean weevil *Acanthoscelides obtectus* (Say) and the moth *Sitotroga cerealella* (Klingauf et al. 1983; Regnault-Roger and Hamraoui 1995). Our earlier investigations on the effectiveness of essential oils extracted from various spices and herbs showed great promise for the control of the major stored-product insects and several were found to be active fumigants at low concentrations against these insects (Shaaya et al. 1991, 1993, 1994).

The use of edible oils as contact insecticides to protect grains especially legumes against storage insects is traditional practice in many countries in Asia and Africa. The method is convenient and inexpensive for the protection of stored seeds in households and in small farms. Many different edible oils have been studied as stored grain protectants against insects (Oca et al. 1978; Varma and Pandey 1978; Pandey et al. 1981; Santos et al. 1981; Messina and Renwick 1983; Ivbijaro 1984 a, b; Pierrard 1986; Ahmed et al. 1988; Don Pedro 1989; Pacheco et al. 1995).

The present investigation consisted of two parts: laboratory and field studies to test the efficacy of edible oils as contact insecticides to suppress populations of various stored-product insects; and evaluation of essential oils as fumigants for the control of stored-product insect pests.

2 Studies with Edible Oils and Fatty Acids

2.1 Laboratory Tests

The biological activity of 16 edible oils and straight chain fatty acids which contain carbon from C_5 to C_{19} was evaluated in laboratory tests against two common stored-product insects, *Callosobruchus maculatus* and *S. oryzae*. The first was reared on chick peas and the second on soft wheat. The required amount of the test material was first mixed with the appropriate amount of acetone (50 ml/kg seeds) and the mixture applied to seeds in droplets with continuous hand mixing. The acetone was evaporated under a hood for several hours. Control seeds were treated with acetone alone.

In the case of *S. oryzae*, adult mortality after 10 days of exposure to the treated seeds, the number of adult beetles in F_1, and also food consumption during the life span of the larvae (expressed as milligrans per larva) were recorded. For *C. maculatus*, egg number and percentage of eggs developed to adults were recorded. *S. oryzae* adults live for a number of months while *C. maculatus* live only about 6 days.

All the natural oils tested for the control of the legume pest *C. chinensis* at a concentration of 1 g/kg chickpea seeds were found to have various degrees of activity (Table 1). The most active oils were crude oils of rice, cotton, palm oil and maize. Of the eggs, 90–96% did not develop to larvae, and only 0–1% developed to adults. In the case of *S. oryzae,* a higher concentration of 6 g oil/

Table 1. Activity of edible oils against *Callosobruchus chinensis*

Oil	No. of eggs	Egg Mortality (%)	Adult Emergence (%)
Crude olive oil	175	74	26
Refined olive oil	300	86	6
Crude soya bean	212	85	10
Refined soya bean	262	96	4
Refined sunflower	187	82	16
Crude cottonseed	212	99	1
Refined cottonseed	275	89	8
Crude palm oil	375	99	1
Refined palm oil	237	84	16
Crude coconut	125	95	5
Refined maize	150	97	2
Crude maize	137	96	1
Distilled safflower	287	97	3
Distilled kapok	387	88	10
Distilled peanut	150	95	5
Crude rice	50	100	0
Control287	0	90	

For all the oils tested, 1 g oil/kg chickpeas was used. Five males and females were introduced to 5 g seed. Results are the average of three experiments, each replicated three times.

kg wheat was needed to get effective activity. In all oils tested, the maximum adult mortality after 10 days' exposure to the treated seeds did not exceed 75% (Table. 2). The most active oils were crude soya bean, maize and cotton. The number of adults which emerged (F_1) was 0, 1 and 2, respectively (Table 2). To obtain insight into the nature of the oils' activity, the activity of straight chain fatty acids ranging from C_5 to C_{18} was studied. This study was conducted first on C. chinensis and then on *C. maculatus*. Figure 1 shows the results obtained by applying 4 g/kg seeds againt *C. maculatus*. It was found that C_9 (pelargonic), C_{10} (capric) and C_{11} (undecaoic) acids were the most efficient in preventing oviposition. The activity was remarkably decreased by the lower (C_5-C_7) and higher (C_{17}-C_{18}) acids; the C_{12}-C_{16} acids showed weak activity. At a rate of 1.6 g/kg, C_{11} showed the highest repellency, with no eggs being detected on the treated seeds (Fig. 2). The C_9 and C_{10} acids had a somewhat lesser effect, and some eggs were deposited on the treated seeds, but all died before the larval stage. Lowering the rate to 0.8 and 0.4 g/kg showed that the C_{11} acid was the most active (Fig. 2). At 0.8 g/kg, only two eggs were found on the treated seeds as compared with 300 eggs on the control seeds; and at 0.4 g/kg, 22 eggs were found with no eggs developing to adults. The activity of C_9 and C_{10} fatty acids as repellents was somewhat lower than that of the C_{11} acid. The data presented in Figure 2 show clearly that C_9, C_{10} and C_{11} acids at concentrations of 0.4–1.6 g/kg are strong repellents to *C. maculatus*. These results were confirmed in large-scale laboratory tests (unpubl. results).

To gain insight into the specificity of these molecules, we tested the activity of the methyl esters of the straight chain fatty acids of 6–12 carbon atoms.

Table 2. Activity of edible oils against *Sitophilus oryzae*

Oil	Adult mortality (%)	No. of emerged adults (F_1)
Crude olive oil	20	7
Refined olive oil	63	2
Crude soya bean	56	0
Refined soya bean	41	11
Refined sunflower	76	14
Crude cottonseed	62	2
Refined cottonseed	42	2
Crude palm oil	39	28
Refined palm oil	55	26
Crude coconut	62	10
Refined maize	64	12
Crude maize	67	1
Distilled safflower	55	3
Distilled kapok	3	63
Distilled peanut	4	58
Crude rice	8	56
Control	0	58

For all the oils tested, 6 g oil/kg wheat was used Ten adults aged 10–15 days were introduced to 5 g wheat. After 10-day exposure to the treated seed, adult mortality was recorded and the live insects were discarded. Results are the average of three experiments, each replicated three times.

These compounds were tested at a rate of 4 g/kg seeds against *C. maculatus*. Figure 3 shows clearly that the C_9-C_{11} methyl esters possess no activity as repellents and the number of eggs found on the treated seeds was mostly the same as in the control. The conversion of fatty acids into methyl esters results in total loss of activity. This suggests that the carboxyl group of the fatty acids is a prerequisite for their activity. The correlation between the molecular structure and the biological activity of the acids indicates a specific characteristic effect rather than a non-specific pharmacological one.

The active fatty acids were also found to be active against *S. oryzae* (Fig. 4). Moreover, no new generation of insects emerged. At 8 g/kg seed, C_7-C_{11} acids were lethal to adults, which died after 10 days' exposure to the treated seeds. Acids with a higher (C_{12}-C_{19}) or lower (C_5 and C_6) number of carbon atoms showed no or weak activity. Lowering the rate to 2 and 4 g/kg (Table 3), C_9 was most effective. At the high rate, 80% adult mortality, no adults in F_1, and no loss in seed weight were observed (Table 3).

2.2 Field Tests

The oils were applied to the various seeds by hand mixing with 60 kg of grain in each treatment. Each 60 kg grain was divided into three 20-kg replicates and placed in jute bags which were stored in a warehouse for various periods up to 15 months. Samples of 300 g grain were withdrawn from each bag monthly and the total number of insects was counted.

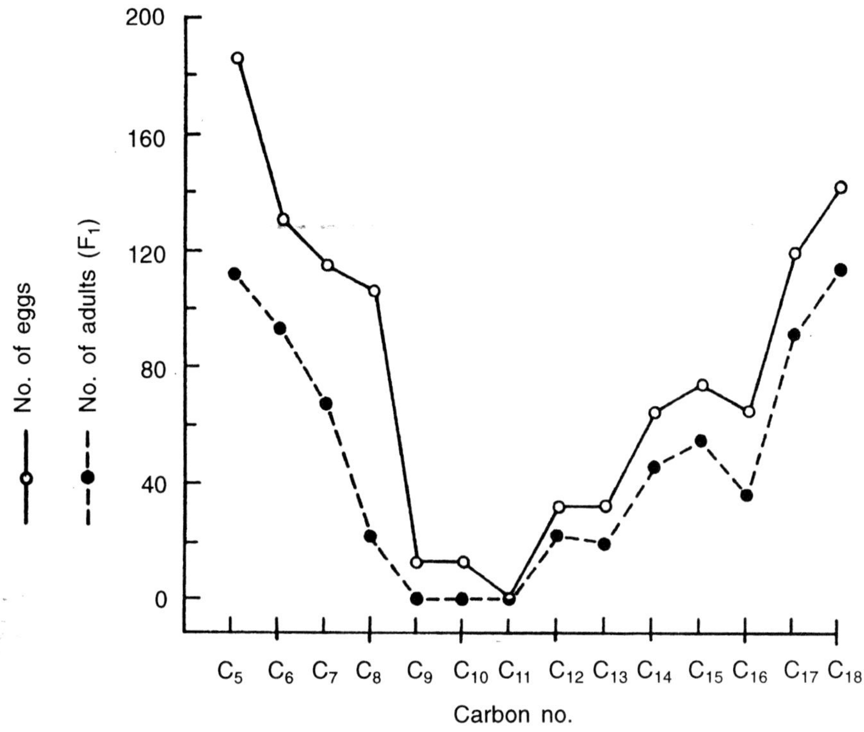

Fig. 1. Activity of straight chain fatty acids, C_5-C_{18}, tested against *Callosobruchus maculatus* at a rate of 4 g/kg Seeds. Results are the average of three experiments, each replicated three times

Crude palm kernel and rice bran oils were tested against *C. maculatus*, a major pest attacking chickpeas. Both oils were effective in controlling *C. maculatus* infestation (Fig. 5), providing full protection for the first 4–5 months when applied at a rate of 1.5 or 3 g/kg seeds. They persisted in controlling insect infestation for up to 15 months, the number of adult insects found in the treated seeds being only approximately 10% of that in the control samples (Fig. 5).

Crude cotton and soya bean oils were tested against *S. oryzae* on wheat. At 10 g/kg, the two oils gave full protection for the first 5 months following oil application (Fig. 6). After 15 months of storage, the number of insects found in the treated seeds was only 6% of that in the control (Fig. 6). At 7.5 g/kg, these oils were somewhat less effective.

3
Studies with Essential Oils

The test insects were laboratory strains of *S. oryzae* and *Rhyzopertha dominica* which were reared on soft wheat; *Oryzaephilus surinamensis* on ground wheat with added glycerin and yeast; and *Tribolium castaneum* on wheat flour. All the insects were cultured at 28 °C and 70% relative humidity (r.h.). Essential oils

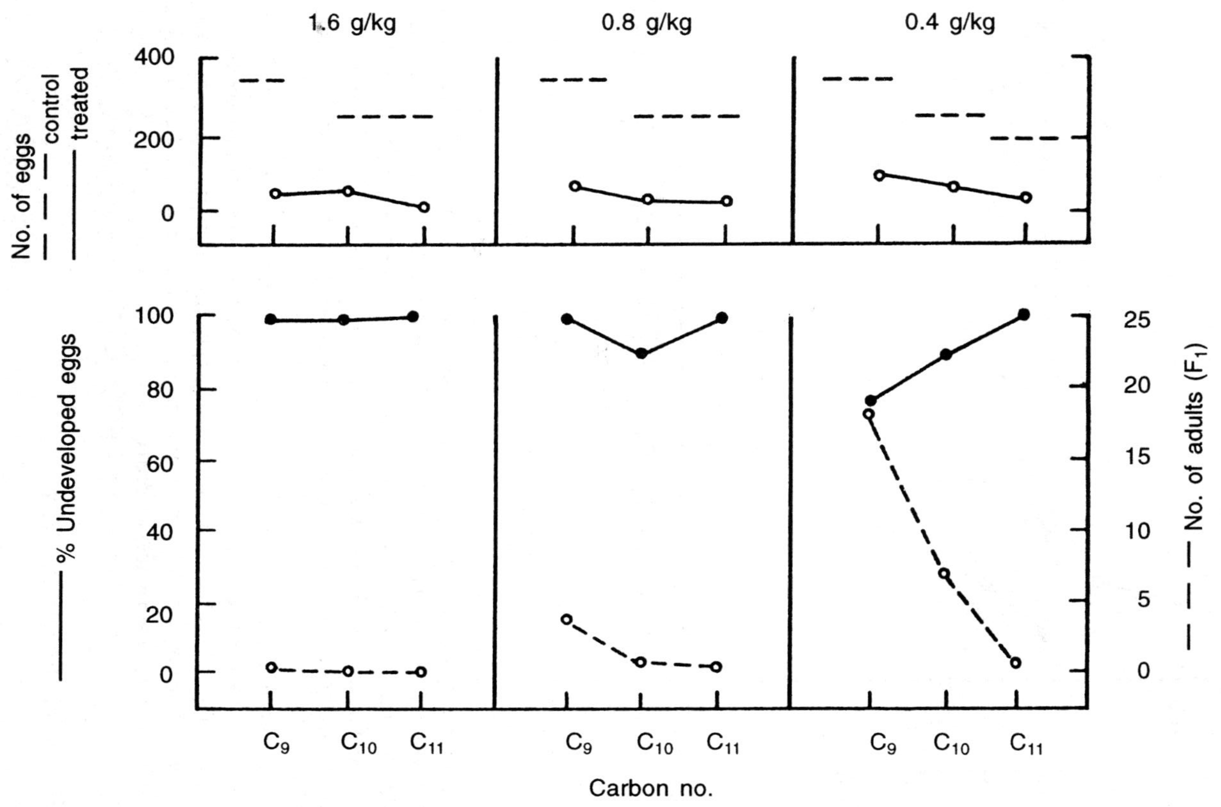

Fig. 2. Activity of straight chain fatty acids, C_9-C_{11}, at a rate of 1.6, 0.8 and 0.4 g/kg seeds, tested against *Callosobruchus maculatus.* Results are the average of three experiments, each replicated three times

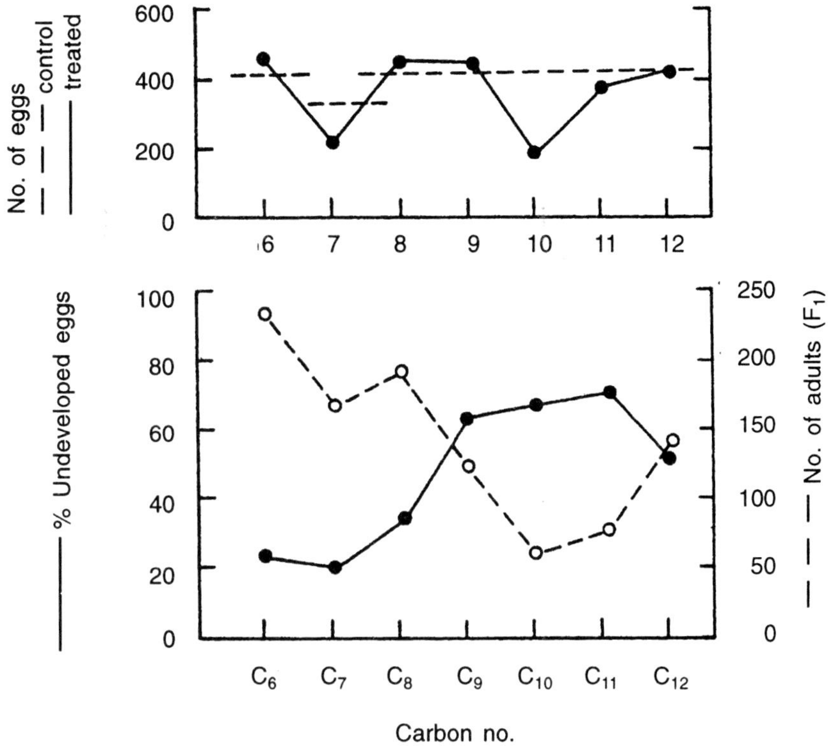

Fig. 3. Activity of C_6-C_{12} methyl derivates of fatty acids at a rate of 4 g/kg seeds, tested against *Callosobruchus maculatus*

were obtained from fresh plants, leaves and stems, by steam distillation (Marcus and Lichtenstein 1979); some of the plants used are listed in Table 4. Two types of tests were performed to evaluate the activity of the oils:

1. The first screening of the essential oils was space fumigation in chambers comprising 3.4-l et glass flasks, as described by Shaaya et al. (1991).
2. The highly active *Labiatae* sp. oil, ZP51 (Table 4), was then assayed in 600-ml glass chambers filled to 20 and 70%, by volume, with wheat (11% moisture content). To each fumigation flask, 20 adults of one of the test insects, *S. oryzae* or *T. castaneum*, were introduced, prior to application of the oil. The test material (3-50 μl/l) was applied on a small piece of filter paper placed on top of the wheat. Exposure times were 1–3 days for *S. oryzae* and 1–7 days for *T. castaneum*.

By screening a large number of essential oils, it was possible to select two oils, ZP51 and SEM76, which were most active against four major stored-product insects (Table 4); with the exception of these two oils, *S. oryzae* and *T. castaneum* showed the highest tolerance to the oils tested. In most cases, a concentration of over 15 μl/l air was needed to obtain LC_{90} of these two

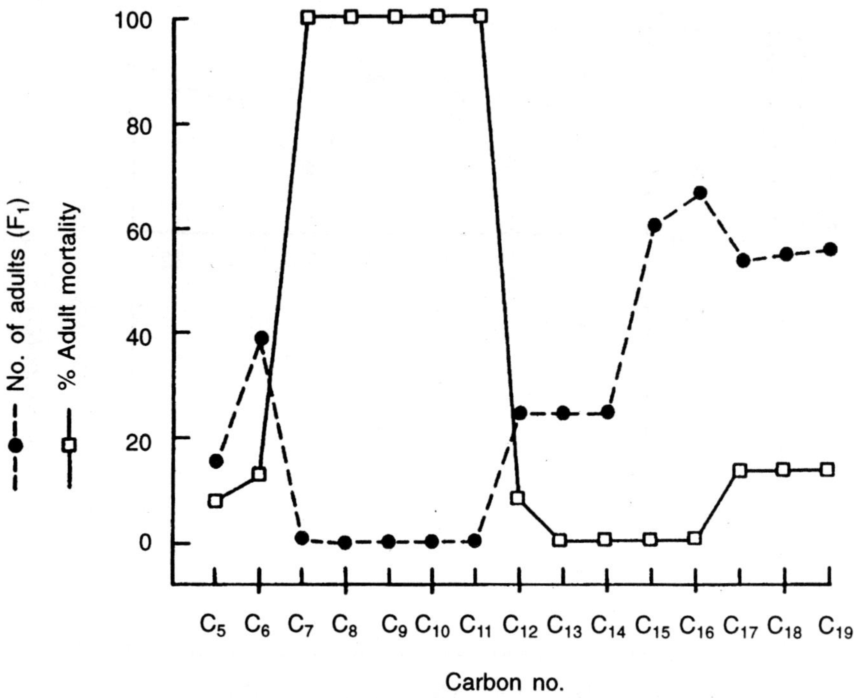

Fig. 4. Activity of straight chain fatty acids, C_5-C_{19}, at a rate of 8 g/kg against *Sitophilus oryzae*

Table 3. Effect of straight chain fatty acids C_9-C_{11} on *Sitophilus oryzae*

Acid tested	Concentration (g/kg)	Adult mortality (%)	No. of Adults (F_1)	Loss in seed weight (%)
C_9	4	80	0	0
	2	54	5	0
C_{10}	4	40	83	2.3
	2	15	375	5.3
C_{11}	2	25	335	4.7
Control	0	0	1150	13.3

For 10 days 200 g wheat seed was infested with 150 beetles. Adult mortality after 10 days, number of adults (F_1) and per cent loss in seed weight were recorded. Other details as in Table 2.

insects. SEM76 was isolated lately and it is the most potent, a concentration of approximately 1 μl/l air being enough to obtain 90% mortality of all insects tested (Table 4). With ZP51, which was also a very effective compound, a concentration of 1.9–5.2 μl/l air was high enough to obtain 90% adult mortality of all four species tested with an exposure time of 1 day (Table 4). This compound was also tested against *S. oryzae* and *T. castaneum* in 600-ml fumigation chambers filled to 20 and 70% volume with wheat. With 20% fill, an exposure time of 1 day and concentrations of 3 and 10 μl/l were enough to cause 100% mortality of *S. oryzae* and *T. castaneum*, respectively (Figs. 7, 8). With 70% fill, concentrations

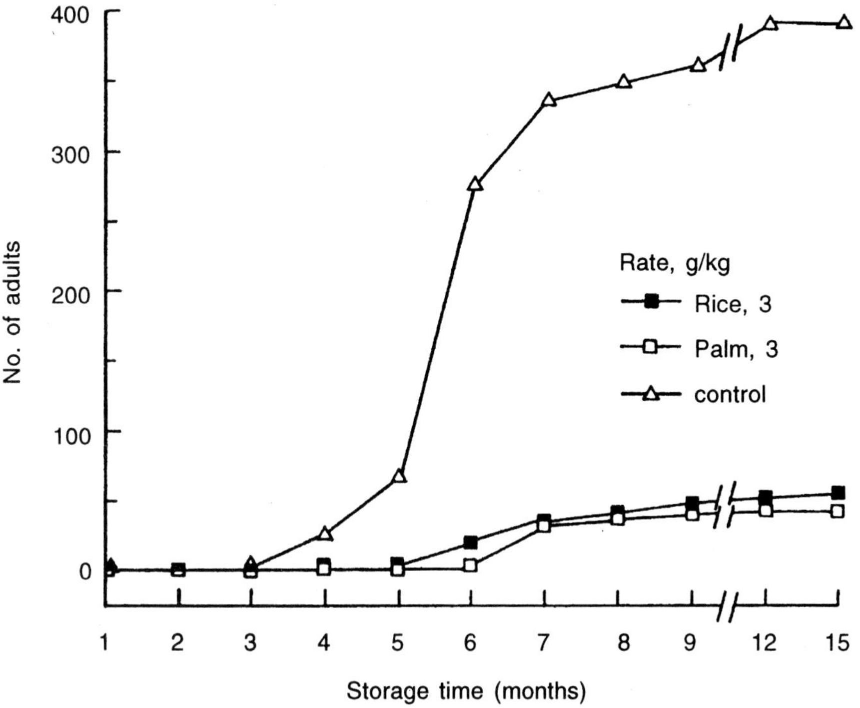

Fig. 5. Field experiment using rice bran and palm kernel oils to protect wheat seeds from infestation by *Callosobruchus maculatus*. There were no significant differences between the two oils ($p > 0.05$) in the t-test

of 30 and 20 μl/l required exposure times of 2 and 3 days, respectively, to obtain 100% mortality for *S. oryzae* (Fig. 7); to obtain 100% mortality of *T. castaneum*, both a higher concentration (40 μl/l) and longer exposure time (4 days) were needed (Fig. 7). The main components of both oils have been isolated and identified and their activity was tested and compared with that of other active monoterpenes. As seen in Table 5, the two components have a higher activity than the oils and they are most effective as compared with all other monoterpenes tested.

4 Discussion

4.1 Studies with Edible Oils and Fatty Acids

Laboratory studies showed that the crude oils of rice bran and palm kernel were the most effective at rates of 1 g/kg seeds of all oils tested, against *C. maculatus*. In field studies, these two oils at rates of 1.5 and 3 g/kg protected chickpeas completely from insect infestation for a period of 4–5 months and partially for

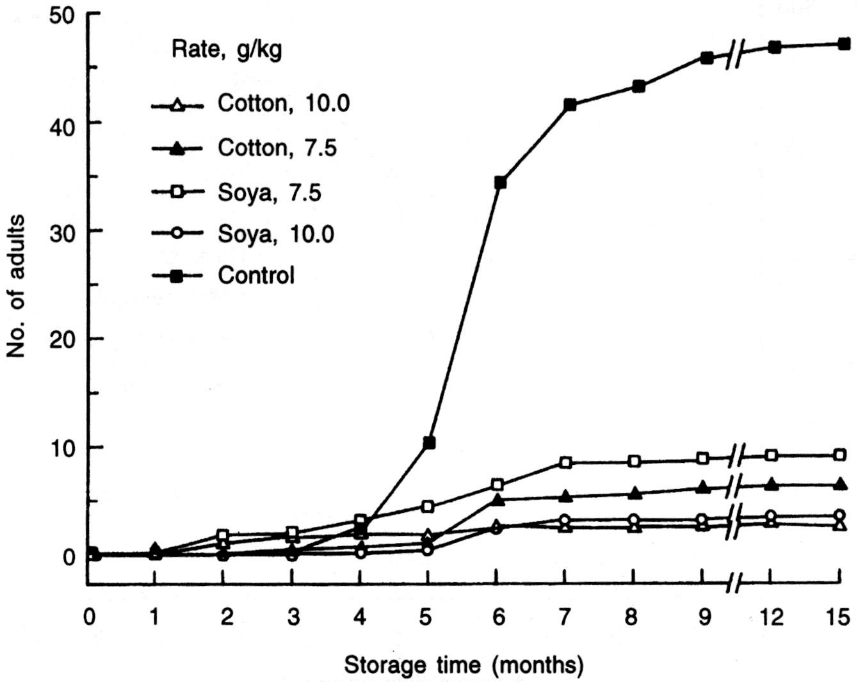

Fig. 6. Field experiment using crude cotton and soya bean oils to protect wheat seeds from infestation by *Sitophilus oryzae*. A significant difference ($p < 0.05$) was found only between the two rates tested with each oil in the t-test

Table 4. Fumigant toxicity of essential oils against stored-product insects in space test

Compound	O. surinamensis LC_{90}	R. dominica LC_{90}	S. oryzae LC_{90}	T. castaneum LC_{90}
SEM76 oil	<1.0	0.8	<1.0	1.4
ZP51 oil	3.6	5.2	1.9	4.3
Peppermint	19.4	16.0	14.9	15.0
Sage	12.7	10.8	23.1	>20
Oregano	8.1	>15	30.4	>20
Basil	11.7	16.7	>20	>20
Thyme	16.4	—	>20	>20
Syrian marjoram	—	15.7	>20	>20
Three-lobed sage	12.9	10.8	>20	>20
Bay laurel	32.0	10.5	>20	>20
Rosemary	13.4	11.6	>20	>20
Lavender	12.8	13.8	>20	>20
Anise	—	21.3	>20	>20

Results are the average of three to five experiments, each replicated three times. LC is expressed in μl/l air.

up to 15 months. In the case of *S. oryzae*, crude cotton oil was the most effective edible oil, 6 g/kg giving effective protection in laboratory tests. In

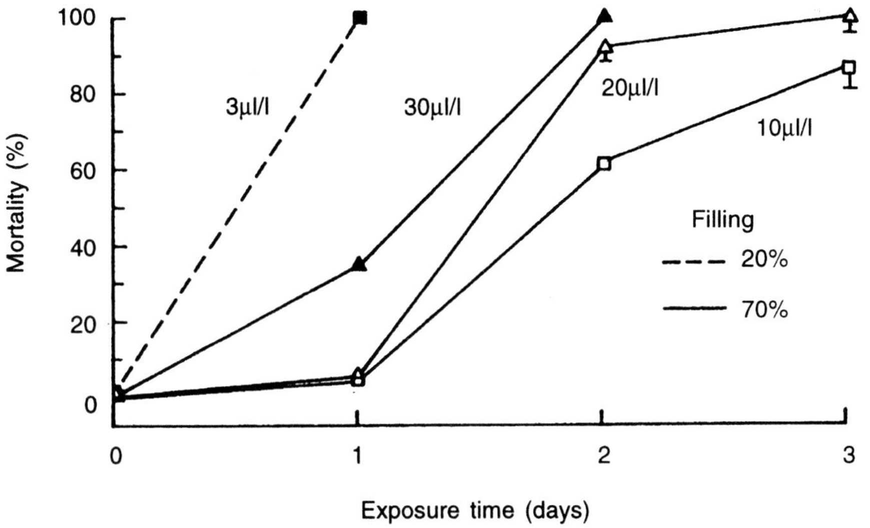

Fig. 7. Fumigant activity of ZP51 against *Sitophilus oryzae* on wheat, at 20 and 70% filling ratio, in 600-ml fumigation chambers (mean ± SE)

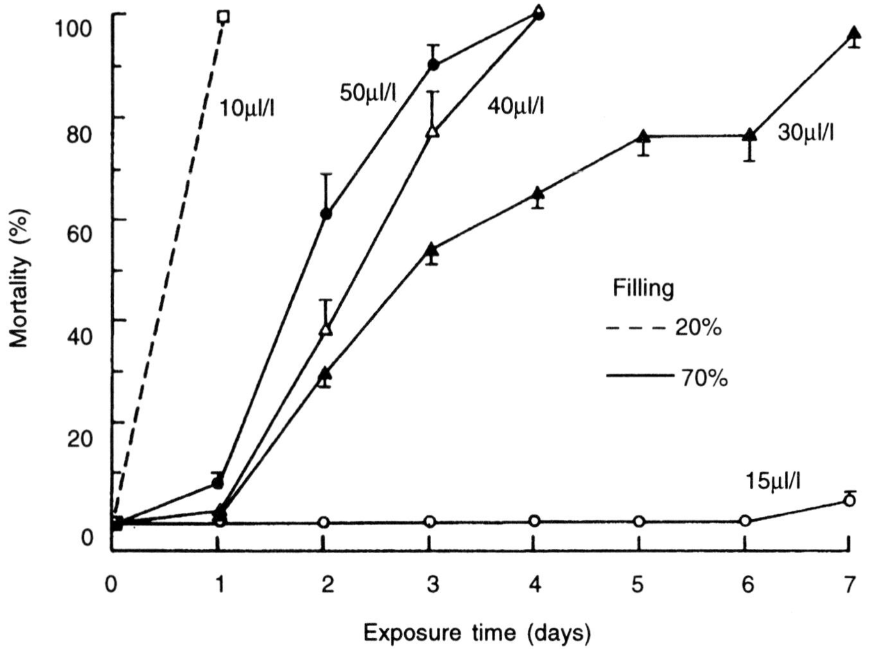

Fig. 8. Fumigant activity of ZP51 against *Tribolium castaneum* on wheat, at 20 and 70% filling ratio, in 600-ml fumigation chambers (mean ± SE)

field studies, 10 g/kg was needed to give full protection for a period of 4–5 months.

Table 5. Fumigant toxicity of essential oils constituents against stored-product insects in space test

Compound	O. surinamensis LC_{90}	R. dominica LC_{90}	S. oryzae LC_{90}	T. castaneum LC_{90}
SEM76	<1.0	0.6	<0.5	1.2
ZP51	2.8	4.5	1.4	3.2
Terpinen-4-ol	11.4	2.0	5.2	3.3
1,8-Cineol	7.3	4.0	14.2	8.5
Carvacrol	—	>20	>20	>20
Terpineol	12.7	>20	>20	>20
Limonene	>20	10.3	>20	8.6
Linalool	6.0	8.5	19.8	>20

Results are the average of three to five experiments, each replicated three times. LC is expressed in μl/l air.

Research, mainly laboratory studies of edible oils protectants, has been conducted on a large number of stored-product insects. The efficacy of a number of vegetable oils as grain protectants of pigenpea against *C. chinensis* was studied by Khaire et al. (1992), who found that adult emergence was completely prevented by 0.5% neem oil and 7.5% karaj oil for up to 100 days after treatment without any adverse effect of the oils on seed germination. Varma and Pandey (1978) and Pandey et al. (1981) showed that groundnut and other oils at a rate of 0.3% w/w gave full protection of greengram against *C. maculatus.* Hill and Schoonhoven (1981) found that crude palm and cotton oils were more effective than crude or refined corn, soya bean and coconut oils for the control of *Zabrotes subfasciatus* (Boh.). Qi and Burkholder (1981) showed that complete control of *Sitophilus. granarius* was achieved for 60 days by mixing the wheat with cotton seed, soya bean, maize or peanut oil at a rate of 10 g/kg seeds. Sighamony et al. (1986) reported that oils of clove, cedar wood and karanja were considerably more effective against *S. oryzae* than *R. dominica.* With a 15-day exposure, the oils produced complete kills of *S. oryzae* at 25–200 ppm. Mortality was reduced to 60 days after treatment.

The mode of action of the edible oils was first attributed mainly to the interference with normal respiration, resulting in suffocation of the animal. In the case of eggs, the oil caused coagulation of the protoplasm by penetrating through the micropyli. The oil's action, however, is more complex, since insects deprived of oxygen survived longer than those treated with oils (Gunther and Jeppson 1960). Wigglesworth (1942) found that droplets of water appear on the surface of the cuticle when an insect is immersed in refined petroleum oil (medicinal liquid paraffin). Studies conducted with termites and scale insects showed that a reduction of the interfacial tension between the solvent and water increased the water loss from the animals. Thus, with the addition of 1% glycerol mono-oleate, the animals lost twice as much weight as those treated with oil alone. It was also noted that the lighter the oil, the higher the rate of water release (Ebeling and Wagner 1959; Ebeling 1976). In a study of the evaluation of the activity of 16 edible oils against *C. maculatus* the source and purity of an oil played a major role in its biological activity. The crude oils were in general more active than the purified ones (Shaaya and Ikan 1979). Moreover, studies of the effect of mineral oils showed that the oil's viscosity might also affect its activity (Calderon 1979).

Form the above findings it may be postulated that the biological activity of the edible oils is attributable to both their physical and chemical properties such as viscosity, volatility, specific gravity and hydrophobicity. It has been postulated also that fatty acid chain length may attribute to the biological activity (Shaaya et al. 1976).

Among the straight chain fatty acids ranging from C_5 to C_{18}, it was found that C_9-C_{11} acids were the most effective in preventing oviposition of *C. maculatus* and were lethal to adults of *S. oryzae*. In our earlier study (Shaaya et al. 1976), we showed that seeds treated with C_{10} fatty acid (at a rate of 4 g/kg) repelled *S. oryzae*, but that forced contact of the beetles with treated seeds had no effect on mortality rate. We postulate, therefore, that fatty acids act as repellents and the beetles die of starvation. This is also apparent from our studies with *C. chinensis*, where the active acids prevent the females from laying eggs on treated seeds but have no lethal effect on adults. It should be noted that *C. chinensis* adults consume no food and live ~6 days, whereas *S. oryzae* adults feed as long as they live, which is approximately 3 months.

4.2
Studies with Essential Oils

A large number of essential oils extracted from various spice and herb plants were screened for activity (Shaaya et al. 1991). Two essential oils, ZP51 and SEM76, were found in the present work to be the most potent fumigants of all the oils tested against four major stored-product insect pests. They had high activity against *S. oryzae* and *T. castaneum*, which were tolerant of all the other oils tested. A concentration of (4.5 μl/l air of ZP51 was enough to obtain 90% kill of all the test insects within 24 h in space tests. In the case of SEM76, which was isolated recently, a concentration of ~1 μl/l was sufficient to achieve 90% kill. ZP51 was also active in studies conducted in fumigation chambers filled either 20 or 70% with wheat. With *S. oryzae*, which was the most susceptible species to this compound, a concentration of 30 μl/l and 2 days' exposure were enough to obtain 100% kill in fumigation chambers 70% filled with wheat. In the case of *T. castaneum*, a less susceptible species, a concentration of 40 μl/l and 4 days' exposure were needed to achieve 100% kill.

The toxicity of a number of essential oils and their constituents, monoterpenes, has been evaluated against stored-product insects. El-Nahal et al. (1989) studied the toxicity of the essential oil *Acorus calamus* (L.) on adults of a number of stored-product insects. The declining order of susceptibility was S. granarius > *S. oryzae* > *Tribolium confusum* > *R. dominica*, the last two being tolerant to all concentrations tested up to 125 μl/l air and an exposure time of 168 h (7 days); 149 h with 25 μl/l air was needed to obtain LC_{50} of *S. granarius*. This oil was also found to cause a reduction in fecundity and regression in the terminal follicle of the vitellarium in treated females of *T. castaneum, S. oryzae, C. chinensis* and *Trogoderma granarium* (Saxena et al. 1976). Coats et al. (1991) evaluated the toxicity of the monoterpenes linalool, d-limonene, myrcene and α-terpineol against *S. oryzae*; 24 h at 14 and 19 μl/l air, respectively, was needed to obtain LC_{50} for the first two compounds, and > 100 μl/l for the other two. A number of essential oils extracted from various spices and medicinal plants of the Mediterranean area were also found to be active against *S. oryzae, R. dominica, O. surinamensis* and *T. castaneum* (Shaaya et al. 1991). Klingauf et al. (1983) studied the fumigant toxicity of 16 essential oils against *Acanthoscelides obtectus*. Rosmarin and caraway oils were the most effective, 100% mortality being obtained after 3 h at 31.4 μl/l air. It should be noted that all these studies were conducted as experimental space fumigations. To the best of our knowledge, there has been no report previously concerning the activity of these compounds as fumigants in grain.

Repellency of essential oils was also observed: *Acorus calamus* was found to repel *T. castaneum* (Jilani et al. 1988), and *Adhatoda vasica* to repel *S. oryzae* and *C. chinesis* (Kokate et al. 1985). The mechanism of the toxic effect of the essential oils is not yet clear. Ryan and Byrne (1988) reported that linalool inhibits the enzyme acetylcholinesterase (AChE). In an earlier study (Greenberg et al. 1993) we found that a number of monoterpenes are competitive inhibitors of AChE of *R. dominica* and *T. confusum*. Moreover, there is a correlation between the relative toxicity of the oils and the relative inhibition of the enzyme activity in the tested insects. The failure of the biologically active terpenes to produce strong enzyme inhibition *in vivo* and *in vitro* leads us to postulate that the enzyme AChE is not the main site of action of the monoterpenes. Neurotoxicity of several terpenes was also tested (Coats et al. 1991).

5 Conclusions

Edible oils can be recommended for the control of *C. maculatus* and other bruchids. In the case of *C. maculatus*, a rate of 1.5–3 g/kg seeds is enough to protect the seeds in a warehouse for a period of up to 5 months. For longer storage times, a second treatment with the oil would be needed. For the control of *S. oryzae, S. zeamais* and *S. cerealella*, a much higher rate is needed; 10–15 g/kg (see also Shaaya and Sukprakarn 1994). At such high dosages, most of the oils tested had severely detrimental effects on seed germination, which makes them unsuitable for insect control in seeds (Shaaya and Sukprakarn 1994). Edible oils could be very useful on the farm level in developing countries and can play an important role in stored-grain protection and reduce the need and risks associated with the use of toxic insecticides.

In this work we reported also on two essential oils which are active as fumigants against major stored-product insects. The low concentration of 40 g oil/m^3 grain is needed to obtain effective control of the insects, as compared with the recommended concentration for methyl bromide of 30 g/m^3. The high activity of the test compounds could render them potential substitutes for methyl bromide use in agriculture.

Acknowledgements

We thank Mrs. Shulamit Atsmy for technical assistance. This work was supported in part by CDR grant C5-077 to E.S. Contribution from the Agricultural Research Organization, The Volcani Center, Bet Dagan 50250, Israel. No. 1948-E, 1996 series.

References

Ahmed K, Khalique F, Afzal M, Malik BA, Malik MR (1988) Efficacy of vegetable oils for protection of greengram from attack of bruchid beetle. Pak J Agric Res 9: 413–416

Brattsten LB (1983) Cytochrome P-450 involvement in the interactions between plant terpenes and insect herbivores. In: Hedin PA (ed) Plant resistance to insects. ACS Symp Ser, 208. American Chemical Society, Washington, DC, pp 173–195

Calderon M (1979) Mixing chickpeas with paraffinoil to prevent *Callosobruchus maculatus* (F.) infestation. Progress Report for the year 1978/79 Stored Products Division, Spec Publ No. 140, Ministry of Agriculture, Israel

Coats JR, Karr LL, Drewes CD (1991) Toxicity and neurotoxic effects of monoterpenoids in insects and earthworms. In: Heiden PA (ed) Naturally occurring pest bioregulators. American Chemical Society, Washington, DC, pp. 305–316

Deshpande RS, Tipnis HP (1979) Insecticidal activity of *Ocimum basilicum* Linn. Pesticides 11: 11–12

Deshpande RS, Adhikary PR, Tipnis HP (1974) Stored grain pest control agents from *Nigella sativa* and *Pogostemon heyneanus*. Bull Grain Technol 12: 232–234

Don Pedro KN (1989) Mechanisms of action of some vegetable oils against *Sitophilus zeamais* Motsch. (Coleoptera: Curculionidae) on wheat. J Stored Prod Res 25: 217–223

Ebeling W (1976) Insect integument: a vulnerable organs system. In: Hepburn HR (ed) The insect integument. Elsevier, Amsterdam, pp 383–399

Ebeling W, Wagner RE (1959) Rapid desiccation of drywood termites with inert sorptive dusts and other substances. J Econ Entomol 52: 190–207

El-Nahal AKM, Schmidt GH, Risha EM (1989) Vapours of *Acorus calmus oil*—a space treatment for stored-product insects. J Stored Prod Res 25: 211–216

Greenberg S, Kostjukovsky M, Ravid U, Shaaya E (1993) Studies to elucidate the effect of monoterpens on acetylcholinesterase in two stored-product insects. Acta Hortic 344: 138–146

Gunter F, Jeppson LR (1960) Modern insecticides and world production. Wiley, New York

Hill JM, Schoonhoven AV (1981) The use of vegetable oils in controlling insect infestations in stored grains and pulses. Recent Adv Food Sci Technol 1: 473–481

Ivbijaro MF (1984a) Groundnut oil as a protectant of maize from damage by the maize weevil *Sitophilus zeamais* Motsch. Prot Ecol 6: 267–270

Ivbijaro MF (1984b) Toxic effects of groundnut oil on the rice weevil *Sitophilus oryzae* (L.). Insect Sci Appl 5: 251–252

Ivbijaro MF, Ligan C, Youdeowei A (1984) Comparative effects of vegetable oils as protectants of maize from damage by rice weevil, *Sitophilus oryzae* (L.). Proc 17th Int Congr Entomol, Hamburg, Germany, 1984, p 643

Jilani G, Saxena RC, Rueda BP (1988) Repellent and growth inhibiting effects of turmeric oil, sweetflag oil, neem oil and Margosan-O on red flour beetle (Coleoptera: Tenebrionidae). J Econ Entomol 81: 1226–1230

Karr L, Coats RJ (1988) Insecticidal properties of d-limonene. J Pestic Sci 13: 287–290

Khaire VM, Kachare BV, Mote UN (1992) Efficacy of different vegetable oils as grain protectants against pulse beetle *Callosobruchus chinensis* L. in increasing the storability of pigeonpea. J Stored Prod Res 28: 153–156

Klingauf F, Bestmann HJ, Vostrowsky O, Michaelis K (1983) Wirkung von atherischen Olen auf Schadinsekten. Mitt Dtsch Ges Allg Angew Entomol 4: 123–126

Kokate CK, D'Cruz JL, Kumar RA, Apte SS (1985) Anti-insect and juvenoidal activity of phytochemicals derived from *Adhatoda vasica* Nees. Indian J Nat Prod 1 (1-2): 7–9

Marcus C, Lichtenstein P (1979) Biologically active components of anise: toxicity and interaction with insecticides in insects. J Agric Food Chem 27: 1217–1223

Messina FJ, Renwick JAA (1983) Effectiveness of oils in protecting stored cowpea from the cowpea weevil (Coleoptera: Bruchidae). J Econ Entomol 76: 634–635

Mills KA (1983) Resistance to the fumigant hydrogen phosphide in some stored-product species associated with repeated inadequate treatments. Mitt Dtsch Ges Allg Angew Entomol 4: 98–101

Mueller D (1995) EPA proposes to cancel Vapona. Fumigation and Pheromones 40: 2

Nakakita H, Winks RG (1981) Phosphine resistance in immature stages of a laboratory selected strain of *Tribolium castaneum* (Herbst). J Stored Prod Res 17: 43–52

Oca GM, Garcia F, Schoonhoven AV (1978) Efecto de cuatro aceites vegetales sobre *Sitophilus oryzae y Sitotroga cerealella* in maiz, sorgo y trigo almacenados. Rev Colomb Entomol 4: 45–49

Pacheco AI, De Castro F, De Paula D, Lourencao A, Bolonhezi S, Barbieri KM (1995) Efficacy of soybean and caster oils in the control of *Callosobruchus maculatus* (F.) and *Callosobruchus phaseoli* (Gyllenhal) in stored chick-peas (*Cicer arietinum* L.). J Stored Prod Res 19: 57–62

Pandey GP, Doharey RB, Varma BK (1981) Efficacy of some vegetable oils for protecting greengram against the attack of *Callosobruchus maculatus* (Fabr.). Indian J Agric Sci 51: 910–912

Pererra J (1983) The effectiveness of six vegetable oils as protectants of cowpeas and bambara groundnuts against infestations by *Callosobruchus maculatus* (F.) (Coleoptera: Bruchidae). J Stored Prod Res 19: 57–62

Pierrard G (1986) Control of the cowpea weevil, *Callosobruchus maculatus*, at the farmer level in Senegal. Trop Pest Manage 32: 197–200

POLO-PC (1987) A user's guide to probit/or logit analysis. Le Ora Software, Berkeley, California

Qi YT, Burkholder WE (1981) Protection of stored wheat from the granary weevil by vegetable oils. J Econ Entomol 74: 502–505

Regnault-Roger C, Hamraoui A (1995) Fumigant toxic activity and reproductive inhibition induced by monoterpenes on *Acanthoscelides obtectus* (Say) (Coleoptera), a bruchid of kidney bean (*Phaseolus vulgaris* L.). J Stored Prod Res 31: 291–299

Ryan MF, Byrne O (1988) Plant insect coevolution and inhibition of acetylcholinesterase. J Chem Ecol 14: 1965–1975

Santos JHR, Beleza MGS, Silva NL (1981) A mortalidade do *Callosobruchus maculatus* em graos de *Vigna sinensis*, tratados com oleo de algodao. Cienc Agron 12: 45–48

Saxena BP, Koul O, Tikku K (1976) Non-toxic protectant against the stored grain insect pests. Bull Grain Technol 14: 190–193

Schmidt GH, Risha EM, Nahal AKM (1991) Reduction of progeny of some stored-products Coleoptera by vapours of *Acorus calamus* oil. J Stored Prod Res 27: 121–127

Schoonhoven AV (1978) The use of vegetable oils to protect stored beans from bruchid attack. J Econ Entomol 71: 254–256

Shaaya E, Ikan R (1979) Insect control using natural products. In: Gessbuhler H (ed) Advances in pesticide science, part 2. Pergamon Press Oxford pp. 303–306

Shaaya E, Sukprakarn C (1994) The use of natural products for the control of stored product insects. Final report submitted to US Agency for International Development, Washington, DC

Shaaya E, Grossman G, lkan R (1976) The effect of straight chain fatty acids on growthy of *Sitophilus oryzae*. Isr J Entomol 11: 81–91

Shaaya E, Kostjukovsky M, Ravid U (1994) Essential oils and their constituents as effective fumigants against stored-product insects. Israel Agresearch 7: 133–139 (in Hebrew)

Shaaya E, Ravid U, Paster N, Juven B, Zisman U, Pissarev V (1991) Fumigant toxicity of essential oils against four major stored-product insects. J Chem Ecol 17: 499–504

Shaaya E, Ravid U, Paster N, Kostjukovsky M, Menasherov M, Plotkin S (1993) Essential oils and their components as active fumigants against several species of stored product insects and fungi. Acta Hortic 344: 131–137

Sighamony S, Anees I, Chandrakala T, Osmani Z (1986) Efficacy of certain indigenous plant products as grain protectants against *Sitophilus oryzae* (L.) and *Rhyzopertha dominica* (F.). J Stored Prod Res 22: 21–23

Symelyanets VP, Khursin LA (1973) Significance of individual terpenoids in the mechanism of population distribution of pests on Scotch pine stands. Zashch Rast (Kiev) 17: 33–44

Tyler PS, Taylor RW, Rees DP (1983) Insect resistance to phosphine fumigation in food warehouses in Bangladesh. Int Pest Control 25: 10–13

Varma BK, Pandey GP (1978) Treatment of stored greengram seed with edible oils for protection from *Callosobruchus maculatus* (Fabr.). Indian J Agric Sci 48: 72–75

Weaver DK, Dunkel FV, Ntezurubaza L, Jackson LL, Stock DT (1991) The efficacy of linalool, a major component of freshly milled *Ocimum canum* Sims (Lamiaceae) for protection against post-harvest damage by certain stored product Coleoptera. J Stored Prod Res 27: 213–220

Wigglesworth VB (1942) Some notes on the integument of insects in relation to the entry of contact insecticides. Bull Entomol Res 33: 205–218

WMO (1995) Scientific assessment of ozone depletion: World Meteorological Organization global ozone research and monitoring project, report no. 37. WMO, Geneva, Switzerland

CHAPTER 11

Novel-Type Insecticides: Specificity and Effects on Non-target Organisms

B. Darvas and L.A. Polgár
Plant Protection Institute, Hungarian Academy of Sciences,
P.O. Box 102, 1525, Budapest, Hungary

1 Introduction

A significant part of the increase in agricultural productivity over the past half century has been due to more efficient and economical pest control. However, there is continuing and growing social and legislative pressure to reduce the ecotoxicological risks of pesticides. One of the most important concerns about a novel type of pesticide is that it should be specific for the target organism(s) (Duke et al. 1993).

1.1 Insect Control Agents

The term "insect control agents" (ICA) involves all kinds of compounds which have direct (on the treated insects) or indirect (not harmful to the treated individuals, but which affects their progeny) actions. A reasonable classification of ICA is as follows (Darvas 1996): (1) neurotoxic zoocides (e.g. neurotoxicants affecting the nervous system of all kinds of organisms); (2) insect behaviour-modifying chemicals (e.g. compounds having no direct effect on the target insect, but reducing indirectly growth of an insect population); and (3) insect development and reproduction disrupters (IDRDs).

For the past 50 years most of the zoocides have been synthesized from petroleum-based sources, as many industrial chemicals. Compounds in the first generation of synthetic zoocides were persistent and resulted in residues of DDT, heptachlor, mirex and other polychlorinated compounds contaminating water and soil, and bioaccumulation and biomagnification in animals (e.g. bioaccumulation through the food chains). Some of them are known to be carcinogenic, teratogenic, deleteriously affecting fertility, growth, enzyme induction and immune response of vertebrates (Murphy 1986). The second generation of synthetic zoocides comprised organophosphorous esters (they were designed as nerve gases during World War II) and carbamate esters. These were more biodegradable, but most of them have broad-spectrum toxicity, with a potential for poisoning beneficial insects, fish, birds, wildlife, livestock and humans. Some showed some selectivity for insects, but the majority of them were not safe enough to inspire a high level of public confidence in their value versus their risks (Coats 1994). The third generation of synthetic zoocides comprised pyrethroids which were developed from natural pyrethrum originating from the pollen of *Chrysanthemum cinerariaefolium* (Trev.) Vis. Pyrethrins are safe to mammals and their rapid photodegradation allowed numerous applications with no appreciable residue problems. The earliest pyrethroids (e.g. allethrin, empenthetrin, pralethrin, tetramethrin) retained most of the beneficial properties of pyrethrins. The development of this group has since yielded a vast array of molecules, some with greater lipophilicity, extremely low water solubility and considerable persistence because of the use of single or multiple halogen atoms (e.g. permethrin) and the introduction of the α-cyano group (e.g. cypermethrin). The most powerful compounds (e.g. deltamethrin, tefluthrin, flucythrinate) are also toxic to non-target insects, fish, birds, wildlife and livestock (Coats 1994; Pap et al. 1996). Most of the pyrethroids are inherently toxic to aquatic organisms including poikilotherm vertebrates; but two recently introduced compounds (cycloprothrin and etofenprox) are less toxic to fish than the earlier ones (Pap et al. 1996).

Neurotoxic zoocides have a direct action on the nervous system; for instance, cyclodienes and avermectins act on the postsynaptic membrane receptors (mostly on the Cl^-channel of γ-aminobutiric acid receptor); organophosphates and zoocide carbamates inhibit acetylcholine esterases; pyrethroids and DDT suppress Na^+ ionpermeability of the nerve membrane; formamidines act through octopaminergic receptors but nicotine, cartap and nitromethylenes act through cholinergic (nicotinic) receptors (Corbett et al. 1984; Voss and Neumann 1992). The similar mode of action, chemical structure and metabolic detoxification of applied neurotoxic zoocides (and benzoyl-phenyl-ureas) is the reason for the zoocide cross-resistance which is so frequently found within this group. Until now, more than 500 insect species have zoocide resistant strains (Georghiou 1990).

At least since the publication of *Silent Spring* (Carson 1962), chemical pest control has been widely viewed as a significant source of environmental pollution. Chlorinated hydrocarbons (e.g. DDT, DDE) eventually reached places as far as Antarctica through atmospheric transport (off-target effect) and biomagnification. Fumigant pesticides such as methyl bromide diminish the ozone layer. Pesticide accidents such as in the Rhine river (by Sandoz AG) and in Bhopal (by Union Carbide) are frequently cited examples of environmental pollution. Groundwater contamination. by pesticides has become critical. Rivers and lakes can also be contaminated by pesticides, e.g. for many years, consumption of fish from Lake Michigan has been discouraged because of the bioaccumulation in their tissues of chlorinated hydrocarbons from agricultural and residential pesticides and industrial uses (Harris and Whalon 1994).

When criticizing the presently used neurotoxic zoocides (organophosphates 36%, pyrethroids 21%, carbamates 20%, cyclodienes 5% of the world sales in the 1990s; Voss and Neumann 1992), we need to remember the following facts:

1. Insufficient information about the benefits of agrochemical products coupled with public campaigns capitalizing on the "fear of the unknown" can have quite dramatic effects. When college students, members of the league of women voters and businessmen in the USA were asked to rank 30 types of

activities or products with a known risk potential, pesticides were placed in positions 4, 9 and 15, respectively, as compared with their real place in rank 28 (based on a survey published in Scientific American in 1982 (Voss and Neumann 1992). In this survey, the indexes of risk factors of smoking, alcohol, and motor vehicles were 150000, 100000, and 50000, respectively, while in the case of pesticides the index was less than 10. We should note that this evaluation is based on chronic effects of smoking and alcohol, but on acute effects of pesticides. It is easy to imagine that the acute effect of smoking is not very strong.

2. Pimentel and Lehman (1993) estimated the cost of pesticide application. The estimated cost of pesticide use in the USA is US$ 8.12 billion per year which originated from bird losses (26%), groundwater contamination (22%), cost of pesticide resistance (17%), crop losses (phytotoxicity) (12%), public health impacts (10%), loss of natural enemies (6%), honeybee and pollination losses (4%) and others (government regulation and enforcement, domestic animal death and contamination, fishery losses) (3%).

3. The reduction of human (and animal) disease vector populations (e.g. Culicidae, Glossinidae, Phlebotomidae, Simuliidae, Tabanidae, Ixodidae) through zoocide treatments is very important in human health. Diseases such as malaria, yellow fever, leishmaniasis, filariasis, onchocerciasis, sleeping sickness and arboviruses have been greatly reduced by zoocide treatments. In India in 1953 there were 75 million cases of malaria. By 1962, 147 million pounds of DDT had been used, so that by 1967 there were fewer than 100000 cases (Anonymous 1973).

4. The natural toxicants found in our food supply may cause considerably more human health problems than pesticide residues. Half of the natural compounds were found to be mutagenic and carcinogenic (Ames et al. 1990).

1.2 Non-Target and Off-Target Impacts

The unintended impacts of pesticides on living creatures are conventionally divided into non-target and off-target impacts. The non-target impacts pertain to living organisms other than the target species that are killed or harmed at the site where and time when the pesticide is applied. Simultaneously, pesticides (as a type of xenobiotic) kill many species of micro- and mesoflora and fauna other than the intended pest(s). These unwanted non-target impacts include species whose role in the ecosystem is well understood (e.g. earthworms have an important function in soil, parasitoids control insect pests) and others whose functions in the trophic structure are not exactly explored. Off-target impacts occur when living organisms are killed or harmed not at the site and time of application. Off-target impacts usually occur when a persistent pesticide is carried by air or water to an unintended location and kills non-target species. These impacts may reduce the diversity of the biota in the agroecosystem; it may thus become less stable and resilient (Harris and Whalon 1994).

The term "non-target organisms" has a very broad meaning (Jepson 1989). For evaluation of the side-effects of zoocides on non-target organisms, various invertebrates and vertebrates representing valuable species in our ecological system are used. The most common grouping is to select—according to a quite anthropomorphic term—"beneficial organisms" existing in the agroecosystem, i.e. natural enemies of pest species and pollinators.

1.3 Toxicity and Specificity

The word "toxic", from the Greek *toxikon* (i.e. poison), means capable of causing injury or death. Toxicity is the degree to which a substance is toxic. In toxicology, we differentiate between acute toxicity, which has a rapid onset and severe course of toxicity, and chronic toxicity, which lasts for a long period of time. The results of acute toxicity tests are expressed as a quantity of the compound which kills half of the subjects (i.e. LC_{50}: lethal concentration, or LD_{50}: lethal dose) during a limited period (e.g. 24, 48 h). Oral LD_{50} (mg/kg body weight) of a pesticide on the female rat (*Rattus norvegicus* Berkenhoud) is one of the most accepted values which represents the acute toxicity in mammals (Fig. 1). Oral LD_{50} (milligrams per kilogram of treated food) of a pesticide on duck (*Anas* sp.) and quail (*Colinus* sp., *Coturnix* sp.) represents the acute toxicity in birds (Fig. 2). LC_{50} (mg/l water) of a pesticide on trout (*Salmon* spp.), carp (*Cyprinus carpio* L.) or bluegill sunfish (*Lepomis macrochir* Rafinesque), which are not easily comparable, represents the acute toxicity in fish (Fig. 3). Organophosphates and organochlorines are usually acutely toxic to vertebrates and beneficial arthropods (e.g. parasitoids, predators, pollinators) as well.

The most important types of chronic toxicity are the mutagenic (i.e. mutagens can induce or increase the frequency of mutation in an organism) and teratogenic (i.e. teratogens cause malformation of an embryo or a foetus) activities, and the bioaccumulation (i.e. accumulation of a substance in various tissues of a living organism) and biomagnification (i.e. accumulation of a substance through the food chains) features of a compound. Some of the widely used neurotoxins have been found recently to be mutagenic (Table 1) and teratogenic (Table 2) on different organisms and they can bioaccumulate (Table 3).

The word "specific" means something particularly fitted to a use or purpose. Originating from the Latin word *species*, the "specificity" of a compound denotes, for a biologist, that it acts at the species level. In ecotoxicology, however, the term is used differently. Perhaps only the sex pheromones act on intraspecific relationships and resulted in a *sensu stricto* specificity called also "superselectivity", which is probably more than we need in plant protection practice.

In a single plant culture, different organisms simultaneously meet and act on each other. In an agroecosystem, cultivated plants, weeds, microorganisms, animal pests and their natural enemies, pollinators, saprophagic organisms, and humans belong to a short-term community. In plant protection, we always treat the whole community, even if the targets of the treatment are only some insect pests. Two different conflicts may arise from this situation: (1) using a non-selective neurotoxic zoocide will kill all kinds of animals independently whether they were the target or not; (2) using a "superselective" compound will solve only a part of the problem by eliminating only one of the pest species.

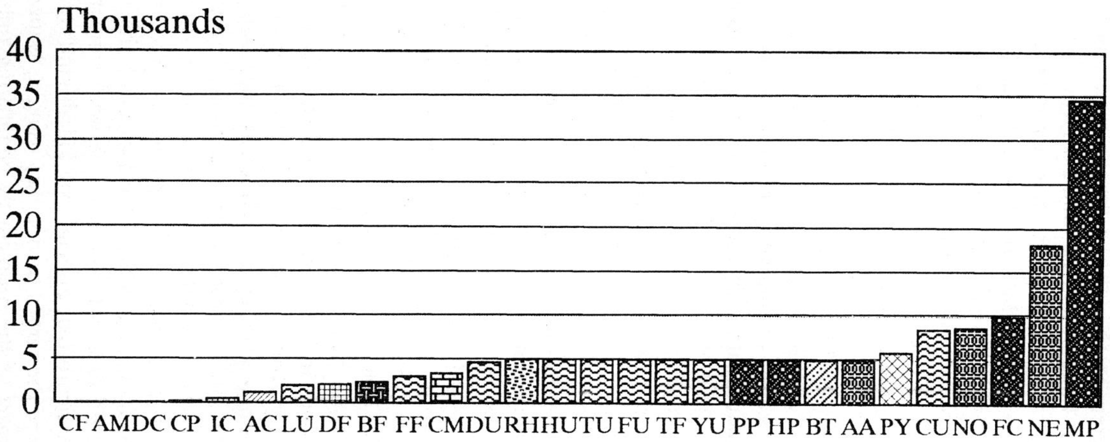

CF: carbofurn
AM: abamectin
DC: diclorvos
CP: cypermethrin
IC: Imidacloprid
AC: AC-303630
LU: lufenuron
DF: diafenthiuron
BF: buprofezin
FF: flufenoxuron
CM: cyromazine
DU: diflubenzuron
RH: tebufenozid
HU: hexaflumuron
TU: teflubenzuron
FU: fluazuron
TF: triflumuron
YU: flucycloxuron
PP: pyriproxifen
HP: hydroprene
BT: Bacillus thuringiensis
AA: azadirachtin A
PY: pymetrozine
CU: chlorfluazuron
NO: neem oil
FC: fenoxycarb
NE: Neem kernel extract
MP: methoprene

Fig. 1. Acute oral LD_{50} values of novel-type insecticides for female rats (mg/kg). (After Tomlin 1995)

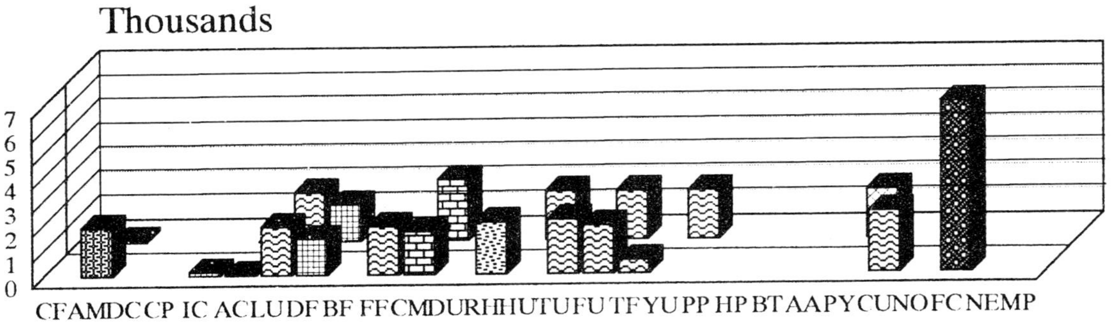

CF: carbofuran
AM: abamectin
DC: diclorvos
CP: cypermethrin
IC: imidacloprid
AC: AC-303630
LU: lufenuron
DF: diafenthiuron
BF: buprofezin
FF: flufenoxuron
CM: cyromazine
DU: diflubenzuron
RH: tebufenozid
HU: hexaflumuron
TU: teflubenzuron
FU: fluazuron
TF: triflumuron
YU: flucycloxuron
PP: pyriproxifen
HP: hydroprene
BT: Bacillus thuringiensis
AA: azadirachtin A
PY: pymetrozine
CU: chlorfluazuron
NO: neem oil
FC: fenoxycarb
NE: neem kernel extract
MP: methoprene

Fig. 2. Acute oral LD_{50} values of novel-type insecticides for birds (mg/kg). *First row* quail; *Second row* duck. (After Tomlin 1995)

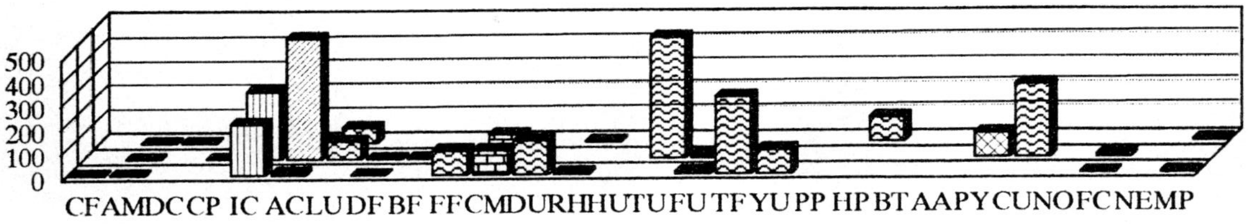

CF: carbofuran
AM: abamectin
DC: diclorvos
CP: cypermethrin
IC: imidacloprid
AC: AC-303630
LU: lufenuron
DF: diafenthiuron
BF: buprofezin
FF: flufenoxuron
CM: cyromazine
DU: diflubenzuron
RH: tebufenozid
HU: hexaflumuron
TU: teflubenzuron
FU: fluazuron
TF: triflumuron
YU: flucycloxuron
PP: pyriproxifen
HP: hydroprene
BT: Bacillus thuringiensis
AA: azadirachtin A
PY: pymetrozine
CU: chlorfluazuron
NO: neem oil
FC: fenoxycarb
NE: neem kernel extract
MP: methoprene

Fig. 3. LD_{50} values of novel-type insecticides for fishes (mg/l). *First row* trout; *Second row* carp; *Third row* bluegill sunfish. (After Tomlin 1995)

Table 1. Mutagenic activities of some neurotoxic zoocides

Active ingredient	Lower organisms	Plant	Invertebrate	Vertebrate	Human
Organochlorines					
Aldrin	H[253]				
Endosulfan	B[187], H[253]				
HCH				o[201]	
Heptachlor	H[255]				
Organophosphates					
Acephate				r[121]	
Azinphos-ethyl	C[66]				
Chlorpyrifos			l[190]	i[9]	
Dichlorvos	C[66]			u[178]	
Dimethoate				u[178]	
Ethion				A[20]	
Fenitrothion	C[66]		I[268]		
Malathion		v[179]		e[299]	
Monocrotophos				c[120],c[21]	
Parathion-methyl		v[179]	D[151]		
Phosmet	j[280,281],k[197,281]			h[234]	g[234]
Phosphamidon		v[179]			
Tetrachlorvinphos				i[9]	
Triazophos			F[277]		
Trichlorfon		v[179]			m[69]
Carbamates					
Carbofuran		E[230]			
Carbosulfan					f[266]
Methomyl				z[30]	s[31]
Propoxur	G[252]				
Bridged biphenyls					
Tetradifon	H[253]				
Pyrethroids					
Bioallethrin	G[252]				
Deltamethrin				A[22]	
Fenvalerate				n[82],p[39],t[81]	
Permethrin					b[17],d[18]
Pyrethroid II		a[157]			

Letters indicate the type of activity (see below) and superscript numbers the reference.
[a]Induced aberrant anaphase and telophase in root tips of *Vicia faba and Hordeum vulgare.*
[b]Induced chromosomal damage in cultured human peripheral blood lymphocytes.
[c]Induced frequency of micronuclei in the erythrocytes of both bone marrow and peripheral blood of chicks.
[d]Characterized as an S-phase independent clastogenic agent in human lymphocyte and CHO (Chinese hamster overies) cultures.
[e]Induced an increase of polychromatic erythrocytes with micronuclei in bone marrow of mice.
[f]Induced chromosomal aberrations and other chromosomal abnormalities in human peripheral lymphocytes.
[g]Induced single-strand breaks in human DNA.
[h]Had a carcinogenic potential in Syrian hamster embryo cells.
[i]Induced a high percentage of metaphases with chromosomal aberrations in cultured mouse spleen cells.
[j]Exhibited an increased frequency of mitotic crossing-over, mitotic gene conversion and reverse mutations on *Saccharomyces cerevisiae.*

[k]Revealed to be a direct mutagen inducing base substitution and frameshift mutations at molecular level on *Salmonella typhimurium.*
[l]Described as genotoxic in somatic and germ cells of *Drosophila melanogaster.*
[m]Caused an increase of the sister chromatid exchanges in the lymphocytes cultured *in vitro* and reduces mitotic activity of lymphocytes.
[n]Increased the incidence of micronuclei in polychromatic erythrocytes in mice.
[o]Caused micronuclei induction *in vivo* in bone marrow cells of rats, hamsters and mice; it was genotoxic in cells of the gastric and nasal mucosa *in vitro* and also *in vivo.*
[p]Induced chromosomal aberrations and sister chromatid exchanges in Chinese hamster ovary cells.
[r]Induced significant increases in micronuclei in both bone marrow and peripheral blood erythrocytes in chicks.
[s]Induced dose-dependent increases in chromosomal aberrations and micronuclei in whole blood human lymphocyte cultures.
[t]Increased the incidence of chromosomal aberrations in germ cells of mice.
[u]Demonstrated mutagenic effects on bone marrow cell chromosomes in rats.
[v]Caused a mutagenic effect in maternal cells of the barley pollen.
[z]Had a clastogenic activity increasing micronucleus frequency in bone marrow of mice.
[A]Inducd chromosomal aberration of mikronuclei in somatic (bone marrow) cells and sperm-shape abnormality in germ cells of mice.
[B]Using a preincubation procedure, showed mutagenic activity with and without metabolic activation with a highly sensitive TA97 *Salmonella* strain.
[C]Was found mutagenic on *S. typhimurium* TA100 strain.
[D]Significant increases in the frequency of micronuclei were observed in mice bone marrow and peripheral blood micronuclei tests.
[E]Higher aberrations are induced by 12-h *Vicia faba* and *Pisum sativum* root tip treatment.
[F]Induced point mutations when assayed in the sex-linked recessive lethal test and induced a weak increase in the non-disjunction frequency on *D. melanogaster.*
[G]Found to be a weak mutagen in the TA98 strains of *S. typhimurium.*
[H]Weak mutagenic on *S. typhimurium* TA100 strain.
[I]Mutagenic in the wing primordial cells of *D. melanogaster* and induces recombination at high doses.

Table 2. Teratogenic activities of some neurotoxic zoocides

Active ingredient	Amphibians	Birds	Mammals
Organochlorines			
DDT		d[196]	
Dieldrin	c[224]		
Heptachlor			a[204]
Organophosphates			
Malathion			e[13]
Pyrethroids			
Deltamethrin			b[3]

Letters indicate type of activity (see below) and superscript numbers the reference.
[a]Accumulation in ovary, uterus and adrenals within 30 minutes of postdosage causes embryotoxicity in rats.
[b]Retardation of growth, hypoplasia of the lungs, dilatation of the renal pelvis of embryo and increase in placental weight in rat.
[c]Gross spinal deformities in frog embryo-larval tests observed.
[d]Mesenchyme differentiation of chick embryo disturbed by Tritox (45% DDT, 3% methoxychlor, 2% HCH).
[e]Significant lag in the development of brain, snout, external pinnae, fore- and hindlimbs and tail while a significant increase in uncovered area of eyeball of mice embryo.

Table 3. Bioaccumulation and biomagnification of some neurotoxic zoocides and diflubenzuron

Active ingredient	Lower organisms	Plant	Invertebrate	Vertebrate	Human
Organochlorines					
Camphechlor				C[186]	
Chlordane					K[259], M[298]
DDE				J[164]	
DDT	H[52]			H[52], R[169]	K[259]
Dicofol			G[254]		
Dieldrin				J[164], P[55]	
Endosulfan			D[173]		
HCB					K[259]
HCH			B[262]	B[262], L[200], N[279]	K[259]
Heptachlor				P[55]	
Polychlorinated biphenyls (PCBs)				J[164]	K[259]
Organophosphates					
Chlorpyriphos				E[287]	
Fenitrotion		S[127]			
Parathion-methyl				F[51], F[52]	
Pyrethroids					
Cypermethrin				A[119], R[169]	
Pyr-Vu-To2				O[183]	
Idrd					
Diflubenzuron				I[7]	

Letters indicate type of activity (see below) and superscript numbers the reference.
[A]Maximum level on 12th day and reached none-detectable level by the 15th day.
[B]Biomagnification through aquatic trophic food chains.
[C]Bioaccumulated significantly in fish.
[D]Bioaccumulation factor for endosulfan II for crayfish tissues was 2; for scallop *Chlamys opercularis*, 26, and for mussel, *Mytilus edulis*, 600.
[E]Bioconcentration factor was calculated of about 1700 for guppy.
[F]In fish it has persistence for uptake and bioaccumulation.
[G]Low biodegradability and capable of accumulating in midgut gland and muscle of prawn.
[H]Ten years after being banned, σ-DDT was present in soil, wetland and lake sediments, surface water and fish.
[I]A low bioaccumulation in fish.
[J]Concentrations of PCBs were 10 or 100 times greater than DDE or dieldrin in Caspian terns (*Sterna capsia*).
[K]PCBs, DDT, HCH, HCB and chlordane were determined in human adipose fat from the provinces of Skierniewice and Gdansk in Poland collected during 1979 and 1990.
[L]Rabbit mothers excreted via the milk to 5-day-old newborns about 30% of the HCH (and pentachlorobenzene) present in tissues during pregnancy.
[M]Chlordane was detected in all 19 human milk samples in Japan (termiticide treated years, 1961–1988).
[N]In the early juvenile stage of a fish, α- and δ-HCH had higher bioconcentration factors than the γ-HCH.
[O]After 6 weeks of Pyr-Vu-To2, a novel cypermethrin-like pyrethroid treatment, an unchanged residue level was measured in heart, lung, muscle, spleen, kidney, liver and brain of sheep.
[P]Partitioning between food in the stomach and body tissues is a major bioaccumulation process in cattle.

[R]Low bioconcentration factors of pyrethroids (cypermethrin, deltamethrin, fenvalerate and permethrin) were primarily due to their rapid depuration: 25- to 50-fold more rapid than DDT in rainbow trout (*Oncorhynchus mykiss*).
[S]Bioconcentration factors for viable *Chlamydomonas reinhardtii* and *C. segnis* algae cells were 293 and 124, respectively, and for dead cells were 1261 and 1025 for the same species, respectively.

Pesticide science looks for chemicals effective against some insect orders with low toxicity to vertebrates (especially humans, their livestock and pets), on beneficial organisms and possibly on saprophagic organisms. These are the desirable characteristics of the specificity of an insecticide. The first desirable character of specificity would be a low toxicity to vertebrates and the second would be selectivity for beneficial arthropods. Standardized test methodology and a sequential procedure (including laboratory, semi-field and field tests) for testing the side-effects of pesticides on different beneficial organisms were developed by the Working Group (WG) Pesticides and beneficial organisms of the International Organization for Biological and Integrated Control (IOBC) of noxious animals and plants, West Palearctic Regional Section (WPRS) (Hassan 1989). This WG was formed in 1974 and its members continuously improve their methodology on parasitoids, predatory insects, mites and spiders as well as on different entomopathogenic organisms. During the evaluation of a pesticide not only the mortality but also the indirect effect (reduction of progeny) is also considered (see Tables 5–8).

This chapter summarizes the results obtained in developing specificity with a low toxicity to vertebrates such as (1) new types of IDRDs: novel moulting inhibitors (benzoyl-phenyl-ureas: chlorfluazuron, flucycloxuron, flufenoxouron, hexaflumuron, lufenuron, teflubenzuron, triflumuron; buprofezin; cyromazine), novel juvenoids (fenoxycarb, pyriproxyfen) and ecdysteroid agonists (tebufenozide); (2) new types of insecticides with natural origin (neem-derived botanical insecticides; avermectins; entomopathogens: *Bacillus thuringiensis*); and (3) new types of insecticides with different modes of action (thiourea insecticides: diafentiuron; pymetrozine; insecticidal pyrroles; nitromethylene insecticides: imidacloprid).

2 Non-Target Effects of Novel-Type Insect Development and Reproduction Disrupters

2.1 Chemicals Interfering with Synthesis and Organization of the Exoskeleton

Insect exoskeleton (i.e. cuticle) consists of chitin, structural proteins, catecholamines and melanin. The cuticles of arthropods, the eggshells of nematodes and cell walls of fungi and green algae all contain chitin. Chitin is based on N-acetylglucosamine polymerization effected by an enzyme named chitin polymerase (or chitin synthetase). Chitin is absent among vertebrates, although several polysaccharides (e.g. glycoproteins, glycosphyngolipids) are built from D-glucosamine or *N*-acetylglucosamine or chondroid tissue is built from

D-galactosamine units. Chitin molecules are subsequently incorporated into complex fibrils. During moulting, the old, rigid cuticle is disintegrated by various enzymes (e.g. chitinase, chitobiose and protease). Sclerotization is the process of cuticular hardening which typically occurs a few hours after each moult, due to the formation of cross-links between catecholamines or other phenolics, and the protein and chitin fibrillar components of the cuticular matrix. It is sometimes accompanied by melanization, although the two processes are separately controlled.

2.1.1 *Benzoyl-Phenyl-Ureas*

The first practically important synthetic representative of novel-type IDRDs was diflubenzuron (Dimilin) discovered by Van Dalen et al. (1972) and Wellinga et al. (1973). There are a number of theories concerning its mode of action (some may be secondary effects): i.e. it (1) inhibits UDP-*N*-acetylglucosamine transport across biomembranes (Mitsui et al. 1984; 1985); (2) inhibits cuticle deposition and fibrillogenesis (Cohen and Casida, 1982; Leopold et al. 1985); (3) inhibits chitin formation and activates chitinases and phenoloxidases (Ishaaya and Casida 1974; Post and Mulder 1974; Leighton et al., 1981), both of which are connected with the catabolism of chitin; and (4) it also inhibits deoxyribonucleic acid (DNA) synthesis (Mitlin et al. 1977; Soltani et al. 1984). For more about mode of action, see Oberlander and Silhacek (this Vol.).

2.1.1.1 Chlorfluazuron

Code No. and Trademark. IKI-7899, CGA 112913, PP145, UC 64644, Aim, Atabron, Helix, Jupiter.

Mode of Action. Chlorfluazuron has no systemic or translaminar effect. It is persistent with transovarial effect. It may reduce the egg-laying rate or hinder the hatching process of embryos.

Uses (Target). Coleoptera, Lepidoptera, Aleyrodidae, Thysanoptera.
Effects on Non-Target Aquatic Organisms
Invertebrates. EC_{50} (48 h) for *Daphnia* sp. 0.91 μg/l (Tomlin 1995).
Vertebrates. LC_{50} (48 h) for carp > 300 mg/l.

Effects on Non-Target Terrestrial Organisms
Invertebrates
Predators. Chlorfluazuron was highly toxic to *Orius* sp. (Het., Anthocoridae), disturbing the moulting process of the larvae (Nagai 1990). Dipping of adult *Chilocorus bipustulatus* (Col., Coccinellidae) in a 0.0125% concentration of chlorfluazuron resulted in 67% adult mortality 7–9 days later. Feeding of adult predators on the host *Aonidiella aurantii* (Maskell) (Hom., Diaspididae) treated with the same concentration resulted in 13% mortality. Chlorfluazuron almost completely prevented egg hatch of the predator (Mendel et al. 1994). Aim was

moderately harmful to *Adalia bipunctata* L. and *Coccinella septempunctata* L. (Col., Coccinellidae) larvae and drastically reduced the fecundity of *C. septempunctata* females (Olszak et al. 1994).
Insect Parasitoids. Chlorfluazuron had no adverse effects on the number of parasitized eggs, female survival, longevity or progeny sex ratio of *Trichogramma pretiosum* Riley (Hym., Trichogrammatidae) (Carvalho et al. 1994a).
Bees: LD_{50} (oral) for honey bee > 100 μg/l (Tomlin 1995).
Vertebrates
Birds. Acute oral LD_{50} for quail and mallard duck (*Anas platyrhynchus* L.) > 2510 mg/kg diet (Tomlin 1995).
Mammals Acute oral LD_{50} for rats > 8500 mg/kg (Tomlin 1995).

2.1.1.2 Flucycloxuron

Code No. and Trademark. PH 70-23, DU 319 722, UBI-A1335, Andalin.

Mode of Action. Incorporation of *N*-acetylglucosamine into chitin was equally inhibited by flucycloxuron and diflubenzuron. Flucycloxuron has ovo-larvicidal activity (Grosscurt et al. 1988). It is non-systemic and has no activity on adults (Tomlin 1995).

Uses (Target). Eriophyidae, Tetranychidae, some lepidopteran pests.

Effects on Non-Target Aquatic Organisms
Invertebrates. EC_{50} (48 h) for *Daphnia* sp. 4.4 μg/l; LC_{50} (96 h) for mysid shrimp 340 ng/l (Tomlin 1995).
Vertebrates: LC_{50} (96 h) for rainbow trout (*Salmo gairdneri* Richardson) and bluegill sunfish > 100 mg/l (Tomlin 1995).

Effects on Non-Target Terrestrial Organisms
Invertebrates
Earthworm. EC_{50} (14 days) for *Lumbricus terrestris* L. (Opisthopora, Lumbricidae) > 1000 mg/kg soil (Tomlin 1995).
Predators. Andalin was harmless to predatory bug *Orius insidiosus* (Say) (Het., Anthocoridae) adults and newly emerged larvae (Van de Veire 1995). Flucycloxuron was harmless to *Amlyseius cucumeris* (Oudemans) (Acari, Phytoseiidae) (Van der Staay 1991). It showed moderately harmful effects on the predatory mite *Phytoseiulus persimilis* A.-H. (Acari, Phytoseiidae), decreasing its egg production (Blümel and Stolz 1993). At 7.5 mg/kg, flucycloxuron was effective on *Teteranychus urticae* Koch (Acari, Tetranychidae), but was safe for *P. persimilis*, although the doubled concentration was toxic to this predator (Stolz 1994a). Andalin (DC-25) was found to be harmless to *Typhlodromus pyri* Scheuten (Acari, Phytoseiidae) (Győ́rffy-Molnár and Polgár 1994; Sterk et al. 1994).
Insect Parasitoids The development of the immature stages of *Trichogramma chilonis* (Ishii) (Hym., Trichogrammatidae), an important egg parasitoid of castor

semilooper, was drastically affected when exposed to flucycloxuron at 4 days after parasitization. Exposure at the Seventh day after parasitization had very little effect on the emergence of adults, but decreased significantly the fecundity of emerged adults (Narayana and Babu 1992). Andalin was harmless to *Encarsia formosa* Gahan (Hym., Aphelinidae), a parasitoid of *Trialeurodes vaporariorum* Westwood (Hom., Aleyrodidae) (Blümel 1990; Van de Veire 1994).
Bees: It has low toxicity; LD_{50} for honeybee > 100 μg per insect (Tomlin 1995).
Vertebrates
Birds. Acute oral LD_{50} for mallard duck > 2000 mg/kg (Tomlin 1995).
Mammals. Acute oral LD_{50} for rats > 5000 mg/kg (Tomlin 1995).

2.1.1.3
Flufenoxuron

Code No. and Trademark. WL 115110, Cascade.

Mode of Action. Flufenoxuron acts in a similar manner to diflubenzuron, reducing chitin incorporation in the cuticle (Clarke and Jewess 1990). It has cuticular and stomach action. It has ovo-larvicidal activity. Treated adults lay non-viable eggs (Tomlin 1995).

Uses (Target). Psyllidae, Tetranychidae, some lepidopteran pests.

Effects on Non-Target Aquatic Organisms:
Vertebrates LC_{50} (96 h) for rainbow trout > 100 mg/l (Tomlin 1995).

Effects on Non-Target Terrestrial Organisms
Invertebrates
Predators. Flufenoxuron had no adverse effects on survival and daily oviposition of *Ceraeochrysa cubana* (Hagen) (Neu., Chrysopidae) adults, but caused total reduction in egg viability (Carvalho et al. 1994b). Cascade was harmful to *Chrysoperla carnea* Steph. (Neu., Chrysopidae) larvae. Larval and nymphal (in cocoons) mortality were found (Vogt 1994). Flufenoxuron showed high toxicity on *Orius* sp. (Het., Anthocoridae), inhibiting the moult of the predator over the long-term (Nagai 1990). Cascade was moderately harmful to *A. bipunctata* larvae and drastically reduced the fecundity of *C. septempunctata* females (Olszak et al. 1994). Flufenoxuron proved to be harmless to the predatory mite *P. persimilis* (Blümel and Stolz 1993) and *Amblyseius stipulatus* A.-H. (Acari, Phytoseiidae) (Miyamoto et al. 1993) at the commercially recommended rates. Cascade was found to be harmless to *T. pyri* (Sterk et al. 1994).
Insect Parasitoids. Flufenoxuron had no negative effects on the number of parasitized eggs, female survival and longevity and sex ratio of the progeny of *T. pretiosum*, but decreased the percentage emergence of progeny (Carvalho et al. 1994a). Cascade was harmless to adults of *E. formosa* (Van de Veire 1994) and on *Opius concolor* Szépligeti (Hym., Braconidae) adults a parasitoid of *Bactrocera oleae* (Gmelin) (Dipt., Tephtritidae) (Jacas and Viñuela, 1994).

Bees:
Vertebrates
Birds. Acute oral LD_{50} for bobwhite quail (*Colinus virginianus* L.) > 2000 mg/kg (Tomlin 1995).
Mammals. Acute oral LD_{50} for rats > 3000 mg technical ingredients/kg (Tomlin 1995). The number of reticulocytes was increased in a flufenoxuron-treated rat group (male Wistar rats for 28 days at oral doses of 100 mg/kg) (Tasheva and Hristeva 1993).

2.1.1.4 Hexaflumuron

Code No. and Trademark. XRD-473, Consult, Trueno.

Mode of Action. Hexaflumuron has systemic properties (Tomlin 1995).

Uses (Target). Coleoptera, Diptera, Homoptera, Lepidoptera.

Effects on Non-Target Aquatic Organisms
Invertebrates EC_{50} (96 h) for *Daphnia* sp. 0.0001 mg/l, which means it is significantly hazardous to it (Tomlin 1995).
Vertebrates. It is not lethal to rainbow trout using maximum solubility of the active ingredient (Tomlin 1995). Hexaflumuron was found to be non-lethal to four species of larvivorous fish, *Poecilia reticulata* Peteres (Cyprinodontiformes, Poeciliidae), *Gambusia affinis holdroucki* Girard (Cyprinodontiformes, Poeciliidae), *Aplocheilus blockii* Arnold (Cyprinodontiformes, Cyprinodontidae) and *Tilapia mossambica* Peters (Perciformes, Cichidae) at 1 mg/l. These species were found to be tolerant with higher LC_{50} values ranging from 2 to 3 mg/l. Survival was observed in all four species at 1 mg/l for more than 8 days. Fish safety factor or suitability index computed for each fish species against three mosquitoes showed maximum value for *P. reticulata* which indicated its higher tolerance than other species. Fish that exhibited decreased swimming activity on exposure at 1 mg/l regained normal activity on withdrawal from exposure. No adverse effect was noticed in the reproduction of *G. affinis* (Vasuki 1992).

Effects on Non-Target Terrestrial Organisms
Invertebrates
Predators. Hexaflumuron generally inhibited the first moulting process of *Poecilus cupreus* L. (Col., Carabidae) (Abdelgader and Heimbach 1992).
Insect Parasitoids. 0.05 μg hexaflumuron per larva applied topically to *Mamestra brassicae* L. (Lep., Noctuidae) resulted in 10% normal *Eulophus pennicornis* (Nees) (Hym., Eulophidae) adults only. Spraying hexaflumuron at a concentration of 3.4 mg/l on the host caused 94% mortality of ectoparasitoid larvae. Female parasitoids did not suffer significant mortality by contact with the chemicals, but there was up to 70% reduction of progeny formation at a concentration of 38 mg/l (Butaye and Degheele 1995).
Bees. LD_{50} (oral and cuticular) for honeybee > 0.1 mg per insect (Tomlin 1995).

Vertebrates
Birds. Acute oral LD_{50} for mallard duck > 2000 mg/kg (Tomlin 1995).
Mammals. Acute oral LD_{50} for rats > 5000 mg/kg (Tomlin 1995). The number of reticulocytes was increased in a hexaflumuron-treated rat group (male Wistar rats for 28 days at oral doses of 100 mg/kg) (Tasheva and Hristeva 1993).

2.1.1.5 Lufenuron

Code No. and Trademark. CGA 184699, Match

Mode of Action. Lufenuron has no systemic or translaminar effect. It is persistent with a transovarial effect. It may reduce the egg-laying rate or hinder the hatching process of embryos.

Uses (Target). Blattidae, Coleoptera, Lepidoptera, Siphonaptera, some Homoptera and Thysanoptera.

Effects on Non-Target Aquatic Organisms
Invertebrates
Vertebrates. LC_{50} (96 h) for rainbow trout > 73, carp > 63, bluegill sunfish > 29, catfish > 45 mg/l (Tomlin 1995).

Effects on Non-Target Terrestrial Organisms
Invertebrates
Predators. Chrysoperla carnea larvae were severely reduced by lufenuron at 80 g a.i./ha (Bourgeois 1994). For the immature stages of the predatory bugs *Anthocoris* spp., *Orius* spp. and *Geocoris* sp., Match was harmful at high rates and with repeated application. At lower rates, the populations recovered soon after application. Under laboratory conditions, Match gave high mortality of Second-instar of *A. nemoralis*, and surviving females laid no eggs (Bourgeois 1994). It was harmless to predatory bug *O. insidiosus* adults and nymphs (Van de Veire 1995). For *C. septempunctata* larvae, Match was moderately harmful at 80 g a.i./ha with three sprays, whereas *Scymnus interruptus* (Goeze) (Col., Coccinellidae) had entirely recovered in a short time after application (Bourgeois 1994). Lufenuron was safe for Syrphidae and *Aphidoletes* sp. larvae (Bourgeois 1994). Match was harmless to *Amblyseius fallacis* (Garman), *Amblyseius addoensis* Van der Merwe et Ryke (Acari, Phytoseiidae), *Typhlodromus* spp. and different spiders (Bourgeois 1994). It was safe on *T. pyri* (Sterk et al. 1994).
Insect Parasitoids. Match was moderately harmful to *Aphelinus mali* Hald. (Hym., Aphelinidae) (Bourgeois 1994). It caused 100% mortality of *E. formosa* adult after 24 h, but it was harmless when the black scales were treated (Van de Veire, pers. comm).
Bees. Oral LC_{50} for honey bee > 38 μg per insect; topical LD_{50} > 8 μg per insect (Tomlin 1995).
Vertebrates
Birds. Acute oral LD_{50} for bobwhite quail and mallard duck > 2000 mg/kg (Tomlin 1995).

Mammals. Acute oral LD_{50} for rats > 2000 mg/kg (Tomlin 1995).

2.1.1.6
Teflubenzuron

Code No. and Trademark. CME 13406, Diaract, Dart, Nomolt.

Mode of Action. Teflubenzuron acts in a similar manner to diflubenzuron, reducing chitin incorporation in the cuticle (Clarke and Jewess 1990). It acts mainly as a stomach poison.

Uses (Target) Aleyrodidae, Diptera, Coleoptera, Hymenoptera, Lepidoptera, Psyllidae.

Effects on Non-Target Aquatic Organisms
Invertebrates. High dosage of teflubenzuron imposed a detrimental effect on the water flea, brine shrimp and freshwater shrimp (Chui et al. 1993). Nomolt was more harmless to aquatic annelid *Tubifex tubifex* Müller (Plesiopora, Tubificidae) (LC_{50}: 10000 mg a.i./kg) than Dimilin (LC_{50}: 850 mg a.i./kg), although both of them belong to the non-earthworm toxic insecticides (Högger and Ammon 1994).
Vertebrates. LC_{50} (96 h) for trout and carp > 500 mg/l (Tomlin 1995). High dosage of teflubenzuron imposed a little or no detrimental effect on the freshwater fish investigated (Chui et al. 1993).

Effects on Non-Target Terrestrial Organisms
Invertebrates
Predators. Teflubenzuron was highly toxic to second-instar of *Forficula auricularia* L. (Dermapt., Forficulidae). Treating the adults, it had no chemosterilizing activity (Blaisinger et al. 1990). Teflubenzuron had no adverse effects on survival and daily oviposition of *C. cubana* adults (Carvalho et al. 1994b). At 0.1–0.2%, Nomolt was highly toxic to *C. carnea* larvae and nymphs under laboratory and semi-field conditions (Bigler and Waldburger 1994). Larval and nymphal (in cocoons) mortality were found (Vogt 1994). At 0.15% a.i., Nomolt killed all treated newly emerged larvae of the predatory bug *O. insidiosus* (Van de Veire 1995). It was moderately harmful to *A. bipunctata* and *C. septempunctata* larvae and drastically reduced the fecundity of *C. septempunctata* females (Olszak et al. 1994). Teflubenzuron was toxic to the pupal stage of *Stethorus punctum* (LeConte) (Col., Coccinellidae), and caused late-season increases of phytophagous mite populations in field trials (Biddinger and Hull 1995).

It was harmless to *P. persimilis, A. cucumeris* and *Amblyseius barkeri* Hughes (Acari, Phytoseiidae) (Van der Staay 1991).
Insect Parasitoids. Teflubenzuron had no negative effects on the number of parasitized eggs, female survival and longevity and sex ratio of the progeny of *T. pretiosum* (Carvalho et al. 1994a). Nomolt had no adverse effects on a lepidopterous egg parasitoid *Trichogramma cacoeciae* Marchal (Hym., Trichogrammatidae) adult and during its postembryonic development (Hassan

1994). Nomolt had cuticular activity on *E. formosa* adults, but it was harmless to pupae (Van de Veire 1994).

Applied topically to *M. brassicae,* 0.05 μg teflubenzuron per larva resulted in 45% normal *E. pennicornis* adults. Spraying teflubenzuron on the host at a concentration of 12.7 mg/l caused 99% mortality in the ectoparasitoid larvae. Female parasitoids did not suffer significant mortality by contact with the chemicals, but there was up to 70% reduction of progeny formation at a concentration of 10 mg/l (Butaye and Degheele 1995). Teflubenzuron was not toxic to *Diadegma eucerophaga* Horstmann (Hym., Ichneumonidae) and *Apanteles plutellae* Kurdjumov (Hym., Braconidae) adults and pupae (Talekar and Yang 1991). When applied on fourth-instar *Plutella xylostella* L. (Lep., Plutellidae), infested with parasitoids, the emergence of adult *Diadegma semiclausum* Hellen (Hym., Ichneumonidae) was significantly reduced, but the compound had no significant effect on another parasitoid, *A. plutellae*, although a small proportion of emerged *A. plutellae* was unable to reproduce. Similarly, different toxicity was observed with teflubenzuron in L_2 hosts, when parasitoids were at the egg or early larval stage. There was no apparent increased effect of teflubenzuron on either species of parasitoid when highly resistant L_2 hosts were treated with concentrations of teflubenzuron two to three orders of magnitude greater than in the equivalent experiments with the susceptible host strain. This suggests that the resistance of host to teflubenzuron confers some protection to parasitoids. Experiments with [^{14}C]-teflubenzuron showed that accumulation of radioactivity was much greater in *D. semiclausum* than in *A. plutellae* and this may account for the differential toxicity observed (Furlong and Wright 1993). In female *D. semiclausum* pretreated with teflubenzuron (50 g a.i./ha), the number of cocoons formed from the first two batches of parasitoid eggs was not significantly different from the untreated control, but the number of cocoons formed from the third batch of parasitoid eggs was significantly reduced. This suggests that teflubenzuron may affect chitin synthesis in developing eggs within a female *D. semiclausum* parasitoid, but has little or no effect on pre-formed eggs which are more likely to be oviposited first (Furlong et al. 1994).

Bees. Non-toxic; LD_{50} (topical) for honey bee > 1000 μ per insect (Tomlin 1995). Bumblebee, *Bombus terrestris* (L.) (Hym., Apiidae), colony was affected and egg development was arrested after feeding on teflubenzuron–sucrose solution. After 5 weeks there was no developing brood in this colony (De Wael et al. 1995).

Vertebrates

Birds. Acute oral LD_{50} for quail > 2250 mg/kg (Tomlin 1995).

Mammals. Acute oral LD_{50} for rats > 5000 mg/kg (Tomlin 1995). The number of reticulocytes was increased in a teflubenzuron-treated rat group (male Wistar rats for 28 days at oral doses of 100 mg/kg) (Tasheva and Hristeva 1993).

Other Effects. Nomolt did not inhibit the entomopathogenic fungus, *Beauveria* sp., which is pathogenic on *Otiorhynchus sulcatus* F. (Col., Curculionidae) (Coremans-Pelseneer 1994).

2.1.1.7
Triflumuron

Code No. and Trademark. SIR 8514, Alsystin.

Mode of Action. Triflumuron acts mainly as a stomach poison.

Uses (Target). Coleoptera, Diptera, Lepidoptera, Psyllidae.

Effects on Non-Target Aquatic Organisms
Invertebrates. EC_{50} (48 h) for Daphnia sp. 0.225 mg/l (Tomlin 1995).
Vertebrates. LC_{50} (96 h) for rainbow trout > 320 mg/l (Tomlin 1995).

Effects on Non-Target Terrestrial Organisms
Invertebrates
Predators. Ladybird larvae were affected by triflumuron treatment. The mortality of mirid bug larvae was 50-95%. Neuroptera (Chrysopa, Hemerobius) larvae respond lightly to moderately to the triflumuron treatment (Boneβ 1983). Triflumuron had no adverse effects on survival and daily oviposition of *C. cubana adults* (Carvalho et al. 1994b). Alsystin was slightly harmful to *A. bipunctata* and *C. septempunctata* larvae and slightly reduced the fecundity of *C. septempunctata* females (Olszak et al. 1994). Triflumuron was harmless to the predatory mite *P. persimilis* at the commercially recommended rates (Blümel and Stolz 1993). On the other hand, a slight mortality of the predacious mites *Typhlodromus athiasae* Parath et Swirski (Acari, Phytoseiidae) (0–12%) and *P. persimilis* (6-10%) with a significant reduction in fecundity was observed. Four days after treatment a reduction of 94% in fecundity of *T. athiasae* and 2 days after treatment a reduction of 78% in fecundity of *P. persimilis* was caused by triflumuron (Mansour et al. 1993a).
Insect Parasitoids. The development of the immature stages of *T. chilonis* was drastically affected when exposed to triflumuron at 4 days after parasitization. Exposure at the seventh day after parasitization had very little effect on the emergence of adults, but decreased significantly the fecundity of adults (Narayana and Babu 1992). Triflumuron had no negative effects on the number of parasitized eggs, female survival and longevity and sex ratio of the progeny of *T. pretiosum*, but decreased the percent emergence of progeny (Carvalho et al. 1994a).
Bees. Toxic to bees (Tomlin 1995).
Vertebrates
Birds. Acute oral LD_{50} for bobwhite quail 561 mg/kg (Tomlin 1995).
Mammals. Acute oral LD_{50} for female rats > 5000 mg/kg (Tomlin 1995). Elevation of methaemoglobin was found in the triflumuron-treated groups (male Wistar rats for 28 days at oral doses of 100 mg/kg). The number of reticulocytes was also increased (Tasheva and Hristeva 1993).

2.1.2 *Buprofezin*

Code No. and Trademark. NNI 750, PP618, Applaud.

Mode of Action. Izawa et al. (1985) found that buprofezin strongly inhibited chitin synthesis from *N*-acetylglucosamine in *Nilaparvata lugens* (Stål) (Hom., Delphacidae). DNA synthesis was weakly reduced by buprofezin. It is persistent with cuticular and stomach actions. [^{3}H]-Prostaglandin biosynthesis from [^{3}H]-arachidonic acid was also inhibited by 84% in the buprofezin-treated *N. lugens* (Uchida et al. 1987). It has ovo-larvicial, activity and treated females lay sterile eggs (Ishaaya et al. 1988). It has translaminar and weak acropetal systemic activities.

Uses (Target). Coleoptera, Heteroptera, Homoptera, Acari.

Effects on Non-Target Aquatic Organisms
Invertebrates. EC_{50} (3 h) for *Daphnia* sp. 50.6 mg/l (Tomlin 1995).
Vertebrates. LC_{50} (48 h) for carp 2.7 and rainbow trout > 1.4 mg/l (Tomlin 1995).

Effects on Non-Target Terrestrial Organisms
Invertebrates
Predators. Buprofezin had no adverse effects on survival and daily oviposition of *C. cubana* adults (Carvalho et al. 1994b). Applaud had no adverse effects on *C. carnea* larvae and nymphs (Bigler and Waldburger 1994). It showed no toxicity to predatory bug *Orius* sp. (Nagai 1990). Buprofezin did not adversely affect egg hatch and larval development of *Elatophilus hebraicus* Pericart (Hem., Anthocoridae) (Mendel et al. 1994). Applaud was harmless to predatory bug *O. insidiosus* adults (Van de Veire 1995). Buprofezin at 250-750 mg a.i./l showed low to moderate toxicity to the predators *Pentilia egena* (Muls.) (Col., Coccinellidae) larvae and adults, and chrysopid larvae (Gravena et al. 1992). Buprofezin did not show toxicity on first-instar of *P. cupreus* (Abdelgader and Heimbach 1992). Opposite to this, none of the larvae of *Rodolia cardinalis* Mulsant (Col., Coccinellidae) developed into adults after application of buprofezin. Buprofezin completely prevented egg hatch of *Chilocorus bipustulatus* L. (Col., Coccinellidae) (Mendel et al. 1994). It was harmless to *P. persimilis* and *A. cucumeris* (Van der Staay 1991).
Insect Parasitoids. The development of the immature stages of *T. chilonis* was drastically affected when exposed to buprofezin at 4 days after parasitization. Exposure at the seventh day after parasitization had very little effect on the emergence of adults, but decreased significantly the fecundity of emerged adults (Narayana and Babu 1992). Buprofezin had no negative effects on the number of parasitized eggs, female survival and longevity and sex ratio of the progeny of *T. pretiosum* (Carvalho et al. 1994a). Applaud had no adverse effects on adults of the lepidopterous egg parasitoid *T. cacoeciae* (Hassan 1994). Buprofezin treatments had a slight adverse effect on the hymenopterous parasitoids *Aphytis mytilaspidis* Le Baron and *Aphytis lepidosaphes* Compere (Hym., Aphelinidae) (Darvas et al. 1994). Buprofezin does not affect *E. formosa* (Garrido et al. 1984). Young *Eretmocerus* sp. (Hym., Encyrtidae) were affected by buprofezin, but young *Encarsia luteola* Howard (Hym., Aphelinidae) were not. The reverse was true for their pupae. No effect on oviposition occurred with females exposed to buprofezin either as immatures or as matures (Gerling and Sinai 1994). Dipping in buprofezin had no appreciable effects on adult mortality, oviposition and development of *Comperiella bifasciata* (Howard) (Hym., Encyrtidae). When

the hosts were treated with buprofezin, mortality of adult *Encyrtus infelix* Embleton (Hym., Encyrtidae) was low. Buprofezin had some detrimental effect on immature stages of *E. infelix* when it was applied prior to parasitization, but not when the host was treated after parasitization (Mendel et al. 1994). Buprofezin had a slight effect on the immature stages of *C. iceryae* (Mendel et al. 1994).
Bees: No direct effect at 2000 mg/l (Tomlin 1995).
Vertebrates
Birds
Mammals: Acute oral LD_{50} for female rats 2355 mg/kg (Tomlin 1995).
Other Effects. Applaud did not inhibit the entomopathogenic fungus, *Beauveria* sp. (Coremans-Pelseneer 1994).

2.1.3
Cyromazine

Code No. and Trademark. CGA 72662, Larvadex, Neoprex, Vetrazine, Trigard.

Mode of Action. The cuticle of cyromazine-treated fly larvae rapidly becomes less extensible and unable to expand to the degree normally seen (Reynolds and Blakey 1989). The cuticle may be stiffer because of increased cross-linking between the various cuticle components, the nature of which remains unknown (Hopkins and Kramer 1992). Cyromazine has strong translaminar activity, and when applied to the soil it is taken up by the roots and translocated acropetally (Tomlin 1995).

Uses (Target). Diptera

Effects on Non-Target Aquatic Organisms
Invertebrates. EC_{50} (48 h) for *Daphnia* sp. > 9.1 mg/l (Tomlin 1995).
Vertebrates. LC_{50} (96 h) for bluegill sunfish > 90, carp and rainbow trout > 100 mg/l (Tomlin 1995).

Effects on Non-Target Terrestrial Organisms
Invertebrates
Predators. Cyromazine (0.1%) applied topically had no detrimental effect on predatory taxa of Histeridae, Staphylinidae (Coleoptera) and Macrochelidae (Acari) (Wills et al. 1990). Cyromazine had no adverse effects on survival and daily oviposition of *C. cubana* adults (Carvalho et al., 1994b). At 0.067%, Trigard was highly toxic to *C. carnea* larvae and nymphs under laboratory and semi-field conditions (Bigler and Waldburger 1994). Trigard was harmless to predatory bug *O. insidiosus* adults (Van de Veire 1995) and was moderately harmful to *C. septempunctata* larvae (Olszak et al. 1994). Cyromazine was harmless to *P. persimilis* and *A. cucumeris* (Van der Staay, 1991). It showed moderately harmful effects to *P. persimilis* egg production in the laboratory test on residual activity (Blümel and Stolz 1993).
Insect Parasitoids. Cyromazine had no negative effects on the number of parasitized eggs, female survival and longevity and sex ratio of the progeny of *T. pretiosum*

(Carvalho et al. 1994a). Trigard had no adverse effects on the adults of lepidopterous egg parasitoid *T. cacoeciae* (Hassan 1994). Cyromazine did not decrease the parasitization of *Liriomyza trifolii* (Burgess) (Dipt., Agromyzidae) by *Chrysocharis parski* Crawford (Hym., Eulophidae) (Parella et al. 1983). Trigard was harmless to *E. formosa* (Van de Veire 1994). It slightly decreased the rate of parasitization caused by *Diglyphus begini* (Ashmead) and *Cirrospilus vittatus* Walker (Hym., Eulophidae) on *Chromatomyia fuscula* (Zetterstedt) (Dipt., Agromyzidae) (Darvas and Andersen 1997). *Muscidifurax raptor* Girault et Sanders, *Spalangia nigroaenea* Curtis, *Spalangia cameroni* Perkins (Hym., Pteromalidae) and *Phygadeuon fumator* Grav. (Hym., Ichneumonidae) were capable of developing in cyromazine-damaged, larviform host puparia as in undamaged puparia. The parasitization rate is, however, about three times and the hatch of flies is twice as high in normal formed puparia compared with larviform puparia. Comparing the attractivity of untreated, normal puparia with cyromazine-damaged, larviform puparia under laboratory conditions showed that the normally formed puparia gave rise to two to three times more parasitoids. Rearing *M. raptor* on larviform host puparia in the laboratory led to a distinct decrease in the development of the parasitoid populations within three to five generations. Parasitoids emerged from larviform and from normal puparia showed no differences in their life cycles in the F_1 generation (Klunker 1991). *Psyttalia incisi* (Silvestri), *Psyttalia fletcheri* (Silvestri), *Diachasmimorpha longicaudata* (Ashmead) and *Diachasmimorpha tryoni* (Cameron) (Hym., Braconidae) endoparasitoid eclosion from cyromazine-treated hosts was not significantly different from that of untreated controls. Cyromazine had no impact on progeny production of either *D. longicaudata* or *D. tryoni* at the concentrations tested (Stark et al. 1992a). Trigard was slightly harmful to *O. concolor* adults and pupae (Jacas et al. 1992; Jacas and Viñuela, 1994). Trigard was moderately harmful to *Aleochara bilineata* Gyll. (Col., Staphylinidae), a parasitoid/predator species of *Delia* spp. (Dipt., Anthomyiidae), reducing the number of offspring (Samsøe-Petersen and Moreth 1994).

Bees. Non-toxic to adult honey bee (no cuticular action up to 5 μg per insect) (Tomlin 1995).

Vertebrates

Birds. Acute oral LD_{50} for bobwhite quail 1785, Japanese quail (*Coturnix japonica* Temminck et Schlegel) 2338, mallard duck > 2510 mg/kg (Tomlin 1995).

Mammals. Acute oral LD_{50} for rats 3387 mg/kg (Tomlin 1995).

Other Effects. Trigard did not inhibit the entomopathogenic fungus, *Beauveria* sp. (Coremans-Pelseneer 1994).

2.2 Chemicals Interfering with Hormonal Regulation

Neurohormones (oligopeptides and bioamines) are secreted by the neurosecretory cells of the insect brain and ganglions, and regulate all vital functions, including further endocrine regulation in insects. Ecdysteroids (secreted by the prothoracic gland, ovaries, testes, fat bodies and oenocytes, and hydroxylated by the cytochrome P-450 isozymes in peripheral tissues) induce moults during postembryonic development and take on a gonadotropic role during adult life. Juvenile hormones (synthesized in the *corpora allata*) are responsible for larval/larval-type moults, for reproduction during adult life and play a role in regulating arthropod dormancy. Vertebrate-type steroids (progestagens, oestrogens, androgens, glucocorticoids, etc.) also occur in insects, but their physiological functions are unknown.

2.2.1 *Juvenoids*

The natural and synthetic juvenile hormone (JH) analogues (Williams 1967), also known as juvenoids, mimic the complex biological effects of JH, causing hyperjuvenilism. Thousands of juvenoids have been synthesized up till now. Chemical structures of most of these compounds are aliphatic-bearing terpenoid chains (they are photo- and termo-labile compounds) or have in some cases an aromatic or alicyclic components.

2.2.1.1 Fenoxycarb

Code No. and Trademark. Ro 13-5223, ACR-2807 B, Insegar, Logic, Pictyl, Torus.

Mode of Action. Juvenile hormone agonist with cuticular and stomach actions.

Uses (Target). Blatellidae, Coccoidea, Culicidae, Lepidoptera, Psyllidae, Siphonaptera, ants.

Effects on Non-Target Aquatic Organisms
Invertebrates. Grass shrimp *Palaemonetes pugio* Holthuis (Decapoda, Palaemonidae) larvae were exposed to fenoxycarb from hatching to postlarval metamorphosis in a chronic, static renewal bioassay. LC_{50}s ranged from 0.92 mg/l at 96 h to 0.35 mg/l at 24 days. In assessing sublethal effects of fenoxycarb, postlarval emergence was significantly reduced in exposed grass shrimp as compared with untreated controls, and the time to reach postlarval status was significantly increased in exposed individuals. Significant differences were not found in other sublethal parameters including postlarval dry weight and intermoult duration. Analysis of fenoxycarb from spiked seawater samples showed concentrations declined by 32 to 42% after 24 h (Key and Scott 1994). EC_{50} (48 h) for *Daphnia* 0.4 mg/l (Tomlin 1995).
Vertebrates. After 24-h exposure, fenoxycarb had 1 mg/kg LC_{50} values on 3–5-days-old *Gambusia affinis* fish (Tietze et al. 1991). LC_{50} (96 h) for rainbow trout 1.07 mg/l (Tomlin 1995).

Effects on Non-Target Terrestrial Organisms
Invertebrates
Predators. Fenoxycarb was not toxic to *F. auricularia*. Treating the adults, it decreased the progeny production near to half (Blaisinger et al. 1990). Opposite

to this, Insegar had no adverse effects on *F. auricularia*, a predator of *Cacopsylla pyri* (L.) (Hom., Psyllidae) (Sauphanor and Stäubli, 1994). At 0.06%, Insegar was highly toxic to *C. carnea* larvae and nymphs under laboratory and semi-field conditions (Bigler and Waldburger 1994). Larval and nymphal (in cocoons) mortality was observed (Vogt 1994). Fenoxycarb revealed no damaging effects on larval or adult anthocorids (mainly *Anthocoris nemoralis* F.) (Solomon and Fitzgerald 1990). Opposite to this statement, Sauphanor and Stäubli (1994) found Insegar highly toxic, at 0.06%, to *A. nemoralis* (Het., Anthocoridae) under laboratory and semi-field conditions. According to the investigations of Peleg (1983a), larvae of *Chilocorus bipustulatus* (L.) (Col., Coccinellidae) fed with fenoxycarb-treated *Chrysomphalus aonidum* (L.) or *Aspidiotus nerii* Bouché (Hom., Diaspididae) were unable to pupate normally, while adults fed with similarly treated scale insects, laid eggs which were unable to hatch. Insegar was harmful to *A. bipunctata* and *C. septempunctata* larvae and drastically reduced the fecundity of both species (Olszak et al. 1994). Fenoxycarb did not show toxicity on first-instar of *P. cupreus* (Abdelgader and Heimbach 1992). None of the larvae of *R. cardinalis* developed into adults after application of fenoxycarb. Fenoxycarb applied either before or after oviposition on pine needles caused total suppression of egg hatch of *Elatophilus hebraicus* (Mendel et al. 1994). Fenoxycarb had ovicidal activity in the laboratory and disrupted the larval–pupal moult of *S. punctum*; it caused late-season increases of phytophagous mite populations (Biddinger and Hull 1995). Fenoxycarb was harmless to the predatory mite *P. persimilis* at the commercially recommended rates (Blümel and Stolz 1993). Opposite to this publication, fenoxycarb caused only slight mortality of the predacious mites *T. athiasae* (0–12%) and *P. persimilis* (6–10%), but a highly significant reduction in fecundity. Four days after treatment, a reduction of 74% in fecundity of *T. athiasae* and 2 days after treatment a reduction of 53% in fecundity of *P. persimilis* was caused by fenoxycarb (Mansour et al. 1993a).

Insect parasitoids. Fenoxycarb did not decrease the parasitization of *L. trifolii* by *C. parski* (Parella et al. 1983). According to Peleg (1983b) fenoxycarb showed good selectivity. It did not damage parasitoids that developed in scale insects, such as *Metaphycus bartletti* Annecke et Mynhardt (Hym., Encyrtidae) in *Saissetia oleae* (Olivier) (Hom., Coccidae), *Aphytis holoxanthus* DeBach (Hym., Aphelinidae) in *C. aonidum, Aphytis chrysomphali* Merc. (Hym., Aphelinidae) and *Comperiella bifasciata* How. (Hym., Encyrtidae) in *Aonidiella aurantii* (Maskell) (Hom., Coccidae), and *Prospaltella inguirenda* Silvestri (Hym., Aphelinidae) in *Parlatoria pergandii* Comstock (Hom., Diaspididae). Darvas and Zsellér (1985) treated first-instars of *Pseudaulacaspis pentagona* Targ. et Tozz. (Hom., Diaspididae) with 0.1% fenoxycarb, and observed that the majority of larvae died during the first moult and the egg production of surviving females was decreased. This treatment had a slight adverse effect on the parasitoid *P. berlesei*. Abd El-Kareim et al. (1988) found that fenoxycarb at 0.1% a.i. did not affect the percentage of parasitization of *Epidiaspis leperii* Sign. (Hom., Diaspididae) by *Aphytis mytilaspidis* (Le Baron) (Hym., Aphelinidae). Fenoxycarb was effective on first- and Second-instars of *Lepidosaphes beckii* Newmann, *Carulaspis juniperi* Bouché (Hom., Diaspididae) and on *Ceroplastes japonicus* Green (Hom., Coccidae) at 0.1% a.i., but had no adverse effect on their hymenopterous parasitoids, *Aphytis mytilaspidis* and *Aphytis lepidosaphes* Comphere (Hym., Aphelinidae) (Darvas et al. 1994). Preimaginal mortality of *Phanerotoma ocularis* Kohl (Hym., Braconidae) was high when fenoxycarb was applied on L_1 of the host (23–46%) and higher when applied to L_6 of the host (40-86%). Fenoxycarb treatment on the last larval instar of the host produced a reduction in the emergence rate of the parasitoid. The treatment of L_1 of the host reduced the parasitism rate by *P. ocularis* (Moreno et al. 1993a). Fenoxycarb had adverse effects on *P. ocularis* pupae as well (Moreno et al. 1993b). Insegar was not harmful to *O. concolor* adults and pupae (Jacas and Viñuela, 1994). Fenoxycarb induced a high mortality of *Pseudoperichaeta nigrolineata* (Walker) (Dipt., Tachinidae) L_2 and L_3 dipping the host into acetone solution at 5 $\mu g/\mu l$. The larval growth was strongly delayed by fenoxycarb. Fifty percent of treated parasitoids were still in L_1 stage after 6 days while all the larvae reached to L_4/L_5 in the untreated control. Larvae stopped their growth in L_2, at a weight near 1 mg, and half of them were still in this stage at 22 days when 92% of the untreated larvae had become puparia (Grenier and Plantevin 1990). Fenoxycarb had marked detrimental effects on parasitization and/or development of the immature stages of *Cryptochetum iceryae* Williston (Dipt., Cryptochetidae) (Mendel et al. 1994). Insegar was harmless to *A. bilineata,* a parasitoid/predator species of *Delia* spp., although it reduced the egg hatch (Samsøe-Petersen and Moreth 1994). Insegar had no adverse effects on adults of a lepidopterous egg parasitoid *T. cacoeciae* (Hassan 1994).

Bees. Non-toxic; oral LC_{50} (24 h) for honey bee adult > 1000 mg/kg (Tomlin 1995). However, the bee workers can carry spoiled pollen, possibly used to feed 4–6-day-old larvae (sensitive stage). Damaged broods are evicted from the hive 10-25 days after the treatment. Abnormal nymphs and newly emerged adults show a white patch on their eyes (Grenier and Grenier 1993). Bumble bee *B. terrestris* colonies developed normally after feeding on fenoxycarb–sucrose solution (De Wael et al. 1995).

Vertebrates

Birds. Acute oral LD_{50} for Japanese quail > 7000 mg/kg (Tomlin ed., 1995).
Mammals. Acute oral LD_{50} for rats > 10 000 mg/kg (Tomlin 1995).
Other Effects. Insegar did not inhibit the entomopathogenic fungus, *Beauveria* sp. (Coremans-Pelseneer 1994).

2.2.1.2
Pyriproxyfen

Code No. and Trademark. S-9318, S-31183, Sumilarv, Admiral.

Mode of Action. Juvenile hormone agonist.

Uses (Target). Blatellidae, Coccoidea, Diptera, Siphonaptera.

Effects on Non-Target Aquatic Organisms

Invertebrates. Crustaceans and aquatic insect larvae are sensitive to pyriproxyfen. *Daphnia magna* Stras and *Daphnia pulex* Leydig (Phyllopoda, Daphniidae), living in treated water, produce fewer number of progeny, but this effect is reversible. *Orthetrum* sp. (Odonata, Libellulidae) and *Chironomus yoshimatsui* Martin et Sublette (Dipt., Chironomidae) last-larval instars are very sensitive to pyriproxyfen causing morphogenetic aberration, although *Chironomus stigmaterus* Say and *Goeldichironomus holoprasinus* Goeldi (Dipt., Chironomidae) were not sensitive to pyriproxifen. At field rates (6-28 g a.i./ha), pyriproxyfen did not exhibit any marked effect on mayfly *Callibaetes* sp.; dragonfly *Tarnetrum* sp., *Anax julius* Brauh (Odonata, Aeschnidae), *Pantala hymenaea* (Say) (Odonata: Libelullidae); ostracods, *Cypridopsis* sp., *Cyprinotus* sp., *Cypris* sp. (Ostracoda, Cypridae); cladocerans, *Simocephalus* sp., *Alona* sp.; copepods, *Cyclops vernalis* (Ficher) (Copepoda, Cyclopidae); beetles, *Tropisternus* sp., *Hydrophilus triangulatus* Say (Col., Hydrophilidae); *Laccophilus* sp., *Copelatus* sp. (Col., Dytiscidae) and further Hydrophilidae and Dytiscidae species. When pyriproxyfen was applied at doubled concentration, minor suppression of the reproductive capacity of daphnoid cladocerans and ostracods was observed. A low degree of induction of morphogenetic aberrations in Odonata at adult emergence was exhibited for 4-10 days after treatment (Miyamoto et al. 1993).

Vertebrates

***Effects on Non-Target Terrestrial organisms*:**

Invertebrates

Predators. Pyriproxyfen caused severe deformities at moult of a predatory bug *Podisus maculiventris* (Say) (Het., Pentatomidae). Treating fifth-instar, it had no adverse effect on reproduction (De Clercq et al. 1995). Admiral was harmless to predatory bug *O. insidiosus* adults and newly emerged larvae (Van de Veire 1995). Pyriproxyfen showed a low toxicity to first-instar of *P. cupreus* (Abdelgader and Heimbach 1992). None of the larvae of *R. cardinalis* developed into adults after application of pyriproxyfen. Pyriproxyfen applied either before or after oviposition on pine needles caused total suppression of egg hatch of *E. hebraicus* (Mendel et al. 1994). Admiral was harmless to *T. pyri* (Sterk et al. 1994) and *P. persimilis* (Van de Veire 1995).

Insect Parasitoids. Admiral (100 EC) was harmless to *E. formosa* adults. Treating the black scales of *E. formosa*, 80-99% mortality was found (Van de Veire, pers. comm.). Pyriproxyfen had marked detrimental effects on the development of the immature stages of *C. iceryae* (Mendel et al. 1994).

Bees. Bumblebee *B. terrestris* colonies developed normally after feeding on pyriproxyfen–sucrose solution (De Wael et al. 1995).

Vertebrates

Birds

Mammals. Acute oral LD_{50} for rats > 5000 mg/kg (Tomlin 1995). ^{14}C-pyriproxyfen was rapidly excreted into faeces and urine of rats, with the former route predominating (approx. 90% of the dose) (Matsunaga et al. 1995).

2.2.2
Ecdysteroid Agonists

Ecdysteroid agonists bind to the ecdysteroid receptor(s) inducing the hyperecdysonism syndrome. A non-steroidal ecdysteroid agonist, RH 5849 (Hsu and Aller 1987), was the first compound acting at the receptor level (Wing 1988).

2.2.2.1 Tebufenozide

Code No. and Trademark. RH 5992, Mimic, Confirm.

Mode of Action. Tebufenozide is a diacylhydrazide ecdysteroid agonist primarily active by ingestion with selective cuticular activity. It mimics the 20-OH ecdysone and works by strong binding on the ecdysteroid receptor protein which initiates the moulting process, especially in Lepidoptera. Treatment leads to precocious induction of lethal moulting which causes synthesis of a new cuticle under the existing one. The caterpillar stops feeding within hours of treatment and dies soon of starvation and dehydration (Smagghe and Degheele 1994, 1995a; Smagghe et al. 1995, 1996a,b). At lower doses, Lepidoptera apparently develop normally but emerge as sterile adults (Smagghe and Degheele 1994).

Uses (Target). Lepidoptera.

Effects on Non-Target Aquatic Organisms
Invertebrates. Tebufenozide does not appear to pose undue risk of direct adverse effects to aquatic macroinvertebrates. Mortality of the amphipod *Gammarus* sp. (Amphipoda, Gammaridae) in one toxicity test was considered an artifact, because there was no significant mortality in subsequent tests at concentrations up to 7 mg/l, or in stream channels treated at 3.5 mg/l (Kreutzweiser et al. 1994). EC_{50} (48 h) for *Daphnia* sp. 3.8 mg/l; LC_{50} (96 h) for mysid shrimp 1.4 mg/l (Tomlin 1995).
Vertebrates. LC_{50} (96 h) for rainbow trout 5.7 and bluegill sunfish 3.8 mg/l (Tomlin 1995).

Effects on Non-Target Terrestrial Organisms
Invertebrates
Earthworm. LC_{50} for *L. terrestris* > 1000 mg/kg (Tomlin 1995).
Predators. Podisus nigrispinus (Dallas) and *Podisus maculiventris* (Say) (Hem., Pentatomidae) showed no sensitivity to RH 5849 and tebufenozide (Smagghe and Degheele 1995b). Likewise, RH 5849 and tebufenozide were ineffective during postembryonic development of *O. insidiosus* (Smagghe and Degheele 1994). Tebufenozide was not toxic to all *S. punctum* stages in the laboratory and field (Biddinger and Hull 1995).
Insect Parasitoids. RH 5849, a structurally similar compound to tebufenozide, does affect the development of 4–6-day-old *Aphidius matricariae* (Hal.) (Hym., Aphidiidae) larvae. The number of diapausing mummies and parasitoid adults with partially developed wings increased (Polgár and Darvas 1991). Host exposure to tebufenozide did not cause apolysis in endo- or ectoparasitic hymenopterans feeding on treated codling moth larvae; however, the endoparasitoid trapped in

the host's haemocoel died as its host's tissue deteriorated. Different results were observed on ectoparasitoids developing on treated hosts. Ectoparasitic *Hyssopus* sp. (Hym., Eulophidae) larvae feeding on tebufenozide-treated hosts pupated in the normal length of time. *Hyssopus* sp. adults which developed from larvae fed tebufenozide-treated hosts were fertile and produced as many progeny as adults reared from solvent controls (Brown 1994). Tebufenozide reduced the survival of the parasitoid *O. concolor* adults, but it has no chemosterilizing effect (Jacas et al. 1996).
Bees. LD_{50} (96 h; cuticular) for honeybees > 234 μg per insect (Tomlin 1995).
Vertebrates
Birds. Acute oral LD_{50} for quail > 2150 mg/kg (Tomlin 1995).
Mammals. Acute oral LD_{50} for rats > 5000 mg/kg (Tomlin 1995).

3
Non-Target Effects of Novel-Type Biological Insecticides

3.1
Neem-Derived Botanical Insecticides: *Azadirachta indica* A. Juss.

Azadirachta indica produce C-seco-meliacins (azadirachtins: azadirachtin A-K, isovepaol, nimbin, nimbinene, salannin, salannol, vepaol, etc.), protomeliacins (meliantriol, etc.) limonoids (gedunin, mahmoodin, meldenin, nimbidinin, nimbinin, nimbocinol, vepinin, vilasinin, etc.) and C-seco-limonoids (margosinolide, salannolide, etc.). In neem seed oil, C-seco-meliacins are the compounds, mainly responsible for the insecticide action; the most important compound is azadirachtin A. Effects of neem-based allelochemicals on Mammalia have been noted, but these are more often therapeutic (used in the Indian folk medicine) than toxic. Useful mammalian effects include anti-inflammatory and anti-ulcer actions (extracts of leaves) and spermicidal birth control (highly volatile compounds of seed oil).

Neem-derived insecticides usually contain a mixture of natural compounds. Thus, two problems have to be borne in mind: (1) the components may vary considerably, depending on origin and the process of their production; and (2) contamination with mycotoxins, such as aflatoxins, may occur, especially in neem oil, extracts of seed kernels and in neem seed cake (Jacobson 1995). In consequence of these, effects of neem-derived pesticides are largely dependent on origin of seeds and formulation of different products.

Code No. and Trademark. Achook, Azatin, Biosol, Green Gold, Jawan, Margocide, Margosan, NeemAzal, Neemark, Neemgold, Neemrich, Nimbecidine, Nimbosol, Pherotech, Repelin, Sukrina, Suneem, Wellgro.

Mode of Action. The mode of action of azadirachtin A lies in (1) effects on deterrent and other chemoreceptors resulting in antifeedancy; (2) effects on ecdysteroid and juvenile hormone titres through a blockage of morphogenetic peptide hormone release (e.g. PTTH , prothoracicotropic hormone; allatotropins); and (3) direct effects on most other tissues resulting in an overall loss of fitness of the insect (Mordue and Blackwell 1993).

Uses (Target). Coleoptera, Diptera, Hemiptera, Homoptera, Lepidoptera, Thysanoptera, Tetranychidae, some Nematoda species.

Effects on Non-Target Aquatic Organisms

Invertebrates NeemAzal was highly toxic to aquatic snails *Lymnaea ovata* Drap. (Basommatophora, Lymnaeidae) and *Planorbella duryi* Wetherby (Basommatophora, Planorbidae) (Schröder three 1992). Stem bark extract of *A. indica* caused 100% mortality when tested against three common snail intermediate host species, *Biomphalaria pfeifferi* Krauss, Bulinus truncatus Audouin (Basommatophora, Physidae), and *Lymnaea natalensis* Krauss, after 24-h exposure. Toxicity test with freeze-dried aqueous extract of the plant gave LC_{50} (96 h) values of 19, 11 and 15 mg/l against *B. pfeifferi, B. truncatus* and *L. natalensis*, respectively (Osuala and Okwuosa 1993). Significant mortality occurred in the mayfly *Isonychia bicolor* (Walker) and *Isonychia rufa* McDunnough (Ephemeroptera, Ecdyonuridae) exposed to Azatin formulation. Pherotech and Azatin were harmless to *Gammarus pseudolimnaeus* Bousfield (Amphipoda, Gammaridae); *Hydatophylax argus* (Harr.), *Hydropsyche* spp. (Trichoptera, Hydropsychidae); *Oligostomis pardalis* (Walker) (Trichoptera, Phryganeidae); *Acroneuria abnormis* (Newman) (Plecoptera: Perlidae); *Pteronarcys dorsata* (Say) (Plecoptera, Pternarcyidae); and *Rhithrogena* sp. (Ephemeroptera, Hetageniidae) (Kreutzweiser 1996).

Vertebrates. Neem products were not toxic at < 0.1% on *Bufo* sp. (Anura, Bufonidae) (Jayaraj 1992). LC_{50} (96 h) of Margosan-O in water with rainbow trout was found to be 8.8 ml/l (Larson 1989). LC_{50} (24 h) value of neem oil for *Oreochromis niloticus* was 1124 mg/kg and for carp was 303 mg/kg (Fernandez et al. 1992). On a fish, *Aphyosemon giardneri* Boulanger (Cyprinodontiformes, Cyprinodontidae), stem bark extract of *A. indica* resulted in LC_{50} (96 h) of 15 mg/l (Osuala and Okwuosa,, 1993). LC_{50} (96 h) value of azadirachtin (49% purity) for juvenile Pacific Northwest salmon is 4 mg/l. The toxicity of the neem-derived products is neem extract > Azatin > Margosan > Pherotech. The toxicity depends on the solvents and emulsifiers used in formulating the material, with LC_{50} (96 h) ranging from 4 mg/l to 72 mg/l. The role of other neem constituent (salannin, salannol, nimbin, nimbadiol, etc.) in fish toxicity is unclear (Wan et al. 1996).

Effects on Non-Target Terrestrial Organisms

Invertebrates

Predators. At 25 mg/kg, none of the treated second-instar of *F. auricularia* could achieve their development. The nymphs exhibited a decrease in food intake and ponderal growth a few days after treatment, as well as an increase in the intermoulting delay. Application of neem-derived insecticides in peach tree orchard resulted in a 70% reduction of nymphal population of earwigs (Sauphanor et al. 1995). High mortality (79%) of the treated *C. carnea* larvae was found after topical application (Kaethner 1990). Neem products are harmful in laboratory tests to larvae of *C. carnea*, disturbing the postembryonic development, although they weve harmless under field conditions probably due to their repellent activity (Vogt et al. 1996). Margosan-O applied on L_3 resulted in a delay of the following moult and morphogenetic defects on *Perillus bioculatus* (F.) (Het., Pentatomidae), a bug predator of the Colorado potato beetle (Hough-Goldstein and Keil 1991). Neem oil acted as a repellent on the beetles *Paederus alfierii* Coiffait (Col., Staphylinidae) and *Coccinella undecimpunctata* L. (Col., Coccinellidae), and the lacewing *C. carnea*, but otherwise had no detrimental effects (Saleem and Matter 1991). Neem oil had a residual activity for up to 6 days efter spraying. There was no effect on survival or behaviour of larvae of *C. undecimpunctata*, except for a prolongation of the fourth larval instar (Matter et al. 1993). Neem seed oil treatment totally prevented adult eclosion of *C. undecimpunctata* (Lowery and Isman 1995). Neem treatment on an aphid host resulted in a strong side-effect on hover flies (Dipt., Syrphidae) (Schauer 1985). Neem seed oil treatment reduced adult eclosion of *Eupeodes fumipennis* (Thompson) (Dipt., Syrphidae), although topical application was ineffective (Lowery and Isman 1995). Larval mortality of *Aphidolethes aphidimyza* (Rondani) (Dipt., Cecidomyiidae) ranged from 30 to 100% for neem-treated L_1, and adult emergence of the treated larvae was significantly lower. Treating the adults, no significant differences occurred (Spollen and Isman 1996).

Karim et al. (1992) recorded that the wolf spider *Oxyopes* sp. (Araneae, Oxyopidae) (LD_{50}: 9.73%) was less sensitive to neem oil 50% EC than the wolf spider *Pardosa pseudoannulata* Boesenberg et Strand (Araneae, Lycosidae) (LD_{50}: 0.18%). Neem extracts at concentractions up to 1% were harmless to the phytoseiid predator *P. persimilis* (Sanguanpong 1992). The effects of different neem extracts, Margosan-O, Azatin-TM and RD9-Repelin on the phytophagous mite *Tetranychus cinnabarinus* (Boisd.) (Acari, Tetranychidae), the predacious mite *T. athiasae* and the predatory spider *Chiracanthium mildei* L. Koch (Araneae, Clubionidae) were studied and compared in laboratory experiments. No toxic effect of any of the mentioned formulations was observed on *C. mildei*. Margosan-OR and Azatin were not toxic to either *T. cinnabarinus* or *T. athiasae*, but RD9-Repelin was highly toxic to both the phytophagous and the predacious mite (Mansour et al. 1993b).

Insect Parasitoids. Emergence of *Anagrus takeyanus* Gordh et Dunbar (Hym., Mymaridae), an egg parasitoid of azalea lace bug , was not affected by azadirachtin (Balsdon et al. 1993). Parasitization of the eggs of diamondback moth by *Trichogramma principium* Sugonjaev et Sorokina (Hym., Trichogrammatidae) was affected by neem treatment. Neem-treated host eggs were less parasitized than untreated host ones. After two treatments, adults of neem-treated *Trichogramma* pupae did not emerge. There was a negative effect on parasitization after a treatment with neem oil. In addition, neem oil reduced the emergence rate of *T. principium* (Klemm and Schmutterer 1993). When larvae of *Henose vigintioctopunctata* (F.) (Col., Coccinellidae) were exposed to adult *Pediobius foveolatus* (Crawford) (Hym., Eulophidae) immediately after their topical application with neem oil (0.05-0.075%), the parasitization rate was strongly reduced, but exposure 1 day later led to no reduction (Tewari and Moorthy 1985). The emergence of adults of the endoparasitic wasp *Tetrastichus howardi* (Olliff) (Hym., Eulophidae) decreased significantly after dipping of parasitized pupae in 1000 mg/kg solution of neem bitters (Lamb and Saxena 1988). *Eretmocerus californicus* Howard (Hym., Aphelinidae) emergence from *Bemisia tabaci* Gernadius (Hom., Aleyrodidae) was strongly (> 50%) reduced by dipping of the parasitized pupae into Margosan-O solution (Hoelmer et al. 1990). A concentration of 10 mg/kg azadirachtin was relatively non-toxic, whereas 20 mg/kg led to a significant reduction (60-70%) in the fitness of *E. formosa* (Feldhege and Schmutterer 1993). Neem seed oil did not reduce the rate of parasitization of *M. persicae* by *Diaeretiella rapae* (McIntosh) (Hym., Aphidiidae) (Lowery and Isman 1995). *Opius incisi* Silvestri (Hym., Braconidae) and *D. longicaudata* developed in and eclosed from oriental fruit flies, *Bactrocera dorsalis* (Hendel) (Dipt., Tephritidae), exposed to azadirachtin concentrations that completely inhibited adult fly eclosion. *Diachasmimorpha tryoni* also eclosed from *Ceratitis capitata* (Wiedemann) (Dipt., Tephritidae) exposed to concentrations of azadirachtin that inhibited fly eclosion. Life spans of parasitoids that emerged from treated flies were not significantly different from untreated controls. Reproduction of *O. incisi*, developed in flies exposed to azadirachtin concentrations of 20 mg/kg, was reduced 63-88%. *Diachasmimarpha longicaudata* and *D. tryoni* reproduction were unaffected (Stark et al. 1992b).

Concentrations of 10 and 20 mg/l of azadirachtin, of an azadirachtin-free fraction and of 50 and 100 mg/l of an enriched product are, under laboratory conditions, only slightly harmful to the hymenopterous parasitoid, and are applied against its host *Pieris brassicae* L. (Lep., Pieridae) L_5. Under these circumstances, numerous larvae of *Apanteles glomeratus* (L.) (Hym., Braconidae) emerge as normal adults. However, higher concentrations (40 mg/l) of azadirachtin and of the azadirachtin-free fraction as well as 50 and 100 mg/l of the enriched formulated neem seed kernel extracts reduce the number of parasitoids considerably (Schmutterer 1992). Neem seed extract was applied orally and topically to *P. brassicae* L_4-L_5 which contained larvae of the parasitoid *A. glomeratus* and also topically to *A. glomeratus* cocoons. As well as heavy parasitoid mortality due to premature host death, *A. glomeratus* suffered increased mortality at all subsequent life stages and deformities in emerged adults. Neem seed extract is capable of having direct negative effects on *A. glomeratus* populations (Osman and Bradley 1993). In residual bioassays on glass with *D. semiclausum* adults of the hymenopteran larval endoparasitoid of *P. xylostella*, neem extract showed little or no activity at rates up to 1000 μg azadirachtin/ml (Verkerk and Wright 1993). Long-term contact (7 days) with residues of the neem product (45 g a.i./ha) slightly reduced (19%) the fecundity of the endoparasitoid *Drino inconspicua* (Meigen) (Dipt., Tachinidae) adult (Beitzen-Heineke and Hofmann 1992).

Bees. Neem seed kernel extract or azadirachtin caused metamorphic disturbances of *Apis mellifera* L. (Hym., Apiidae) larvae. Most of the treated insects died during larval/pupal moult (Rembold et al. 1980). Only a small portion (10%) of the treated colony was affected when flowering white mustard, summer rape or *Phacelia* sp. was treated (Schmutterer and Holst 1987). Foraging honey bee workers were able to discriminate between untainted sugar syrup and syrup containing 0.1 mg/kg azadirachtin, the principal active ingredient of oil-free neem seed extract (NSE). However, there were no significant differences in the numbers of foraging honey bees collected in neem-treated, solvent-treated or untreated canola plots. Other pollinator species present were similarly unaffected. Results suggest that honey bees may be successfully utilized in blooming crops that have been treated with doses of NSE sufficient to control phytophagous insect pests (Naumann et al. 1994). Acute LD_{50} (topical) values of azadirachtin (applying oil-free emulsifiable concentrate) for honey bees were 55 pg/L_1 and 5.9 ng/L_4 (Naumann and Isman 1996), which means that azadirachtin is extremely dangerous for honey bee larvae. Measuring the potential for the azadirachtin transport from sprayed crop to the nest via contaminated nectar or pollen, no such translocation was detected (Naumann and Isman 1996).

Vertebrates

Birds. Forced oral administration of neem seed to red-winged blackbird (*Agelaius phoenicus* L.) resulted in above 1 g/kg lethal dose for expressed seed oil. (Schafer and Jacobson 1983). White Leghorn chickens fed the powdered neem berries water-extract ration looked sluggish soon after administration and, when taken off the ration, showed drooping head and cyanosed combs. Of the latter birds, 60% died within 24 h; post mortem examination of the dead birds showed fragile livers with local congestion, retention of bile in the gall bladder and congested kidneys with localized haemorrhages (Singh et al. 1985). Feeding of neem leaves to chicks for 2 weeks also gave some toxic effects (Ibrahim et al. 1992). In a broiler experiment, 200-day-old (Babcock) broiler chicks were used in a 10-week trial, with the full fat neem seed meal (FFNSM) at 25, 50, 75 and 100 g/kg diet. There was a significant negative correlation between the dietary inclusion of FFNSM, weight gain and food conversion efficiency of the chicks in the starting phase. From 5 to 10 weeks, food intake, weight gain, food conversion and protein efficiency ratio did not differ significantly between the birds on the control diet and diets containing up to 75 g/kg FFNSM (Salawu et al. 1994).

Mammals. LD_{50}s of a single intraperitoneal injection of nimbolide, a major chemical constituent of neem seed oil, on male, female, and weanling mice were 225, 280 and 240 mg/kg body weight, respectively. These values were above 600 and 500 mg/kg body weight in the cases of rat and hamster. The animal died of possible dysfunction of the small intestine, pancreas and liver. A marked drop in arterial blood pressure and respiratory paralysis was also found (Glinsukon et al. 1986). Acute oral LD_{50} of azadirachtin A for rats > 5000 mg/kg (Tomlin 1995).

The effects produced by administration of aqueous suspensions of green or dried neem leaves to goats and guinea pigs were investigated at doses of 50 and 200 mg/kg given orally over a period of up to 8 weeks. A progressive decrease in body weight and condition, and greater weakness, were noted, as well as a reduction in heart, pulse and respiratory rates. Those animals fed green leaves had diarrhoea. In goats, the higher doses produced tremos and ataxia, but no significant haematological changes (Ali 1987). Neem leaf extract produced some hypoglycaemia in normal rats when given in two doses, while in diabetic rats there was a decrease in blood sugar, but it could not alleviate the diabetic state. However, the neem leaf extract produced some toxic effects in rats, as observed in body weight loss and high percentage mortality, especially with a high dose. It was observed that the clotting time of blood was higher than normal. Serum cholesterol level increased with concomitant decrease in the liver fat as compared with normal levels. There was also a drop in liver proteins which was dose-related (El-Hawary and Kholief 1990). In an experiment with 24 rabbits of 3 different breeds used in an 8-week trial, with the FFNSM at 100, 200 and 300 g/kg diet for all the measurements, rabbits on the diet with 100 g/kg FFNSM gave better results than the control. Food intake, weight gain, food conversion and protein efficiency ratios did not differ significantly between rabbits on the control diet and the diet containing 200 g/kg FFNSM. Performance on the diet with 300 g/kg FFNSM was poorest (Salawu et al. 1994).

Humans. Neem oil causes toxic encephalopathy particularly in infants and young children. The usual features are vomiting, drowsiness, tachypnea and recurrent generalized seizures. Leucocytosis and metabolic acidosis are significant laboratory findings (Lai et al. 1990).

Other Effects The activities of carboxymethylcellulase and xylanase were stimulated by cold and hot water extracts of neem seed kernel cake on mixed rumen bacteria from buffalo, and those of α-amylase, protease and urease were inhibited (Agarwal et al. 1991).

Margosan-O was less detrimental than chlorpyrifos to most of the invertebrates living in the soil studied. However, oribatid mites were more sensitive to Margosan-O than to chlorpyrifos. Sminthurid springtails were also susceptible to Margosan-O, although less than to chlorpyrifos. Margosan-O had no significant effect on non-oribatid mites and spiders (Stark 1993). Neem-treated zoosaprophagous blowfly larvae (Dipt., Calliphoridae) died as pharate imago in the puparia (Bidmon et al. 1987).

Entomopathogenic nematodes *Heterorhabditis bacteriophora* Poinar, *Heterorhabditis heliothidis* Khan, Brooks et Hirschmann (Nematoda, Heterorhabditidae), *Steinernema carpocapsae* (Weiser), *Steinernema glaseri* (Steiner) and *Steinernema feltiae* (Filipjev) (Nematoda, Steinernematidae) were affected only by high concentration of aqueous neem extract. The activity of the treated *Steinernema spp.* third juvenile was suppressed. The mortality was 20-30% (Rovesti and Deseő 1991).

Feeding behaviour of the terrestrial slug species (*Deroceras reticulatum* (Müller), *Agriolimax caruanae* Pollonera (Stylommatophora, Limacidae), *Arion distinctus* Mabille (Stylommatophora, Arionidae) and *Maximus* sp. relative to the amount of leaf eaten compared with the controls was not affected by the presence of azadirachtin at those concentrations which deterred aphids from feeding (West and Mordue 1992).

3.2 Avermectins: *Streptomyces avermitilis* Burg et al.

Evaluation of the eight closely related natural products, avermectins (A_{1a}, A_{1b}, A_{2a}, A_{2b}, B_{1a}, B_{1b}, B_{2a}, B_{2b}), and their several hundred semi-synthetic analogues and derivatives showed that avermectin B_{1a} (80%) and B_{1b} (20%), the major components (called abamectin) of the fermentation of *Streptomyces avermitilis* MA-4680 strain (Streptomycetaceae *sectio cinereus serovar. kromaghenes*), were extremely toxic to acarines and to a wide range of insects (Putter et al. 1981; Dybas, 1989). Ivermectin (22,23-dihydroavermectin B_1) is used in animal and human health control.

3.2.1 Abamectin

Code No. and Trademark. MK-0936, Affirm, Agri-Mek, Avid, Dynamec, Vertimec

Mode of Action. Avermectins (1) open neurotransmitter-gated Cl^- channels associated with GABA receptors in vertebrates, GABA and/or glutamate receptors in insect muscle and the multitransmitter-gated channel in crustacean; (2) block GABA-stimulated Cl^- channels in various preparations from mammals, in locust and nematode muscle; and (3) open apparently non-neurotransmitter-gated Cl^- channels in mammals, insects and nematodes (Hawkinson and Casida 1993). Abamectin has cuticular and stomach actions. It has limited systemic and translaminar activities in plants (Tomlin 1995).

Uses (Target). Coleoptera, Diptera, Formicidae, Homoptera, Lepidoptera, Psyllidae, Tetranychidae, Thysanoptera.

Effects on Non-Target Aquatic Organisms

Invertebrates. Invertebrates vary widely in their sensitivity to abamectin. *Mysidopsis* sp. (LC_{50}: 0.022 µg/kg), *Daphia magna* (LC_{50}: 0.34 µg/kg) and *Panaeous duorarum* Burkenroad (LC_{50}: 1.6 µg/kg) were very sensitive species (Wislocki et al. 1989). Ivermectin has been shown to be effective against the sea lice *Lepeophtheirus salmonis* (Kroyer) and *Caligus* sp. (Copepoda, Caligidae). The shrimp *Crangon* sp. (Decapoda, Crangonidae) was exposed to ivermectin in water and via treated salmon food. Shrimp seemed unaffected when exposed to ivermectin in water (maximum concentration is 21.5 µg/l nominal). However, when ivermectin was present in salmon food, it was lethal to the shrimp [nominal LC_{50} (96 h) at 8.5 µg ivermectin/g food]. The NOEC (No Observed Effect Concentration) was approx. 2.6 µg/g. These results suggest that ivermectin, when used at the above concentrations, may present a hazard to non-target organisms during or after oral treatment of fish against parasites (Burridge and Haya 1993). LC_{50} (96 h) for blue crab (*Callinectes* sp.) 153 µg/kg (Tomlin 1995).

Vertebrates. LC_{50} (96 h) for rainbow trout 3.2 and bluegill sunfish 9.6 µg/kg (Tomlin 1995). Ivermectin was slowly absorbed, the highest concentrations being found in lipid-containing organs of Atlantic salmon (*Salmo salar* L.). High concentrations were also found in the central nervous system, indicating that the blood–brain barrier in salmon is poorly developed compared with mammals. The excretion of the drug was very slow. The total amount of radioactivity in blood, muscle, liver and kidney diminished by only 35% from 4 days to 28 days after administration. Excretion was mainly by the biliary route, and enterohepatic circulation of the drug was apparent. The drug was mainly excreted in the unchanged form. Distribution to the central nervous system, and the prolonged excretion period, make the drug unsuitable for the control of parasitic infestations in salmon (Hoy et al. 1990). Cutaneously administered ivermectin was not toxic at dosages up to 20 mg/kg in leopard frog. Nematode infections were eliminated in all ten frogs treated cutaneously with ivermectin at 2 mg/kg (Letcher and Glade 1992).

Effects on Non-Target Terrestrial Organisms

Invertebrates

Earthworm. LC_{50} (28 days) was determined to be 28 mg/kg on *Eisenia foetida* (Savigny) (Opisthopora, Lumbricidae), indicating that the use of abamectin should not have an effect on earthworm (Wislocki et al. 1989).

Predator. Vertimec had a moderate adverse effect on *F. auricularia* and *A. nemoralis* (Sauphanor and Stäubli 1994). It was moderately harmful on predatory bug *O. insidiosus* adults and nymphs (Van de Veire 1995). Abamectin at 30–80 mg a.i./l showed low to moderate action on the predators *Pentilia egena* (Muls.) (Col., Coccinellidae) larvae and adults and chrysopid larvae (Gravena et al. 1992). It was toxic to *S. punctum* larvae and adults in the laboratory (Biddinger and Hull 1995). Dung from the same cattle that received the 200 μg/kg dose of ivermectin was bioassayed for 6 weeks with two species of predaceous staphylinid, *Philonthus flavolimbatus* Erichson and *P. longicornis* Clark (Col., Staphylinidae). The number of *P. flavolimbatus* progeny was reduced for 1 week compared with the untreated dung. There was no apparent effect on the number of *P. longicornis* progeny. Dung from cattle that received the 20 μg/kg dose had no apparent effect on the number of progeny of either predator species (Fincher 1992).

The predatory mites *Typhlodromus occidentalis* Nesbit (Acari, Phytoseiidae) (Sanderson and Barnes 1983) and *A. stipulatus* were not reduced by abamectin treatment under field condition (Morse et al. 1987). Abamectin was harmful to *A. cucumeris* and *A. barkeri* (Van der Staay 1991). Against *T. urticae*, 0.2–0.4 mg abamectin/kg was effective, but it was safe for *P. persimilis*, although at 1 mg/kg it was also harmful to the predator (Stolz 1994a).

Insect Parasitoids. Emergence of *A. takeyanus*, an egg parasitoid of azalea lace bug, was not affected by abamectin (Balsdon et al. 1993). Abamectin applications did not have an adverse impact on the parasitization of *L. trifolii*, on adult parasitoid mortality or emergence of immature parasites of *Diglyphus intermedius* (Ashmead), *Diglyphus begini* (Ashmead), *Neochrysocharis punctiventris* (Crawford), *C. parski*, *Chrysocharis ainsliei* Crawford (Hym., Eulophidae) and *Halticoptera circulus* (Walker) (Hym., Pteromalidae) (Trumble 1985). Abamectin did not reduce the population of *Aphytis melinus* De Bach (Hym., Aphelinidae) and *Cryptolaemus montrouzieri* Mulsant (Col., Coccinellidae) under field condition (Morse et al. 1987). At a concentration of 0.025%, Vertimec was found to be harmless to *E. formosa* pupae (Blümel 1990). Opposite to these earlier results, low adult mortality was found until 24 h, but thereafter mortality was 100% until the third day. Treatment of the black scales resulted in mortality of *E. formosa* at eclosion. Thus, abamectin is definitely toxic to *E. formosa* (Van de Veire, pers. comm). Pteromalid pupal parasitoids of *M. domestica*, as *M. raptor*, *Urolepis rufipes* (Ashmead), *S. cameroni* and *Pachycrepoides vindemmiae* (Rondani) (Hym., Pteromalidae) were sensitive to abamectin (LC_{50}: 0.021-0.036% a.i.), but *Nasonia vitripennis* Walker (Hym., Pteromalidae) was less susceptible (LC_{50}: 0.081% a.i.) (Geden et al. 1992).

Bees. Abamectin was quite toxic to the honey bee; it had cuticular LC_{50} values of 0.002 and 0.017 μg per insect at 24 and 48 h, respectively (Wislocki et al. 1989).

Vertebrates

Birds. Acute oral LD_{50} for mallard duck 84.6 and bobwhite quail > 2000 mg/kg (Tomlin 1995). White Leghorn hens naturally infected with *Ascaridia* spp. (Rhabditidea, Ascarididae), *Heterakis* spp. (Rhabditidea, Heterakidae) and *Capillaria* spp. (Dorylaimida, Capillariidae) were treated with 0.2, 2 or 6 mg/kg intramuscularly or 0.2 or 0.8 mg/kg orally. Faecal samples were collected before treatment and at autopsy, 2, 6 or 16 days after treatment, when the intestines were also examined for helminths. None of the treatments gave satisfactory anthelmintic results (Oksanen and Nikander 1989). Subcutaneously injected ivermectin at a dose of 0.3 mg/kg body weight was found to be 90 and 95% effective against immature and adult *Ascaridia galli* Schrank (Rhabditidea, Ascarididae), respectively. The lower lesion score and post-treatment near-normal haematobiochemical picture in treated white Leghorn chicks confirmed these observations. The treated birds also had a better growth rate than the untreated chickens. The mature worms in the intestinal lumen of the host were more sensitive to the treatment than the immature stages of the parasite in the tissue phase (*Sharma* et al. 1990).

Mammals. Acute oral LD_{50} for rats 10 mg/kg (Tomlin 1995). Ivermectin administered orally to Spanish goats, *Capra hircus* (L.) (Artiodactyla, Bovidae) or to white-tailed deer, *Odocoileus virginianus* (Zimmerman) (Artiodactyla, Cervidae) was highly effective against lone star ticks, *Amblyomma americanum* (L.) (Acari, Ixodidae). For Spanish goats, daily oral doses of 20 µg/kg were sufficient to cause 95% reduction of estimated larvae from feeding ticks (Miller et al. 1989). Single subcutaneous injection of ivermectin at the dose of 200 µg/kg body weight was found to be 100% effective in the treatment of warble fly *Przhevalskiana silenus* (Brauer) (Dipt., Hypodermatidae) in goats (Shahardar et al. 1993).

Humans. A study was carried out in south-eastern Gabon, Africa to evaluate the tolerance and efficacy of single high doses of ivermectin in 31 *Loa loa*-infected humans with low to moderate parasitemia (7-7700 microfilaria/ml). The first group of 16 subjects received 300 µg ivermectin/kg and, 7 days later, a second group of 15 received 400 µg/kg. Complete clinical and biological monitoring was carried out during the first 10 days post-treatment and again after 1 and 3 months. There were no significant changes in blood or urine function test results or in haematologic results, except for a pronounced eosinophil reaction. The 400 µg/kg dose of ivermectin equalled or surpassed in tolerance that of the 300 µg/kg dose. After treatment, *L. loa* microfilaremia decreased rapidly to less than 9% of the pretreatment value by day 10. This decrease was enhanced with the 400 µg/kg dose, although differences between the two groups diminished slightly with time. At 100 days post-treatment, the microfilaremia was less than 10% of the initial values in the two groups, which may indicate an effect of ivermectin on the adult worms (Martin-Prevel et al. 1993). A total of 118 925 individuals in four Nigerian states were treated for onchocerciasis between February and December 1991, using centralized and house-to-house distribution of ivermectin. Pretreatment prevalences of the disease ranged between 28 and 90%. Only 0.7% of those treated reported adverse reactions within 3 days of treatment: 230 individuals (0.19%) had headache, 210 (0.17%) general body pains, 150

(0.12%) pruritis, 120 (0.10%) oedema, 80 (0.06%) fever, 20 (0.02%) dizziness, 15 (0.01%) vomiting, 10 (0.01%) diarrhoea, and 25 individuals (0.02%) passed intestinal worms. Treatment in the endemic communities continues. The results show that mass treatment of onchocerciasis with ivermectin is quite safe and the drug's acceptability increases its potential as the drug of choice for control of onchocerciasis in Nigeria (Ogunba and Gemade 1992). Prothrombin ratios were measured 13–16 days after treatment in 148 subjects from Sierra Leone taking part in a double-blind placebo-controlled trial of ivermectin. Prolonged prothrombin ratios were observed more frequently in the ivermectin group, although this difference was not significant and no patients suffered bleeding complications. Further investigation of these patients failed to reveal any abnormality of liver function, although factor VII and II levels were reduced in most affected individuals, suggesting interference with vitamin K metabolism. Ivermectin has a minimal effect on coagulation and concern about mass treatment for this reason appears to be unjustified (Whitworth et al. 1992).
Other Effects. Abamectin has no effect on entomopathogen fungi *Hirsutella thompsoni* Fish. and *Aschersonia aleyrodes* Webber (McCoy et al. 1982). Horses, donkeys and mules were examined for equine parafilariosis by random sampling from different areas of Iran. The treatment with ivermectin subcutaneously at 0.2 mg/kg was 100% effective with only one injection (Maloufi 1995).

3.3
Entomopathogens: *Bacillus thuringiensis* Berliner

Bacillus thuringiensis (*Bt*) is an aerobic, Gram-positive, spore-forming bacterium found commonly in the environment. It produces a number of toxins, the most distinctive of which are protein crystals formed during sporulation. These crystalline protein (Cry) inclusions, or δ-endotoxins (i.e. MVP), are the principal active ingredients in *Bt* formulations currently in use. The genes encoding δ-endotoxin production have been cloned in other bacteria and transferred into crop plants (transgenic cotton, potato, tomato, etc.) (McGaughey and Whalon 1992). Genetically improved insecticidal crystal protein genes resulted in new *Bt* strains which are not only superior to the natural strain, but also competing effectively with synthetic zoocides (Carlton 1993). Some strains also produce another toxin type as well, called β-exotoxin; however their use, because of their mutagenic activity, is not allowed.

Historically, *Bt* has been classified (De Barjac and Frachon 1990) on the basis of the flagellar or H-antigens of the vegetative cell into 27 serotypes or 34 serovars as follows:

1. *B. thuringiensis serovar. thuringiensis*
2. *serovar.* finitimus

3_a. *serovar. alesti;* $3_{a\text{-}b}$. *serovar. kurstaki* (Agree, Bactospeine, Biobit, Bollgard, Condor, Cutlass, Delfin, Dipel, Ecotech, Foil, Foray, Jackpot, Javelin, Lepinox, Rapax, Steward, Thuricide, Turex, Vault)

$4_{a\text{-}b}$. *serovar. sotto*

$4_{a\text{-}c}$. *serovar. kenyae*

$5_{a\text{-}b}$. *serovar. galleriae*
$5_{a\text{-}c}$. *serovar. canadensis*
6. *serovar. entomocidus*
7. *serovar. aizawai* (Certan, Florbac, Xen Tari)
$8_{a\text{-}b}$. *serovar. morrisoni—pathovar. san diego* (M-One),—*pathovar. tenebrionis* (Novodor)
$8_{a\text{-}c}$. *serovar. ostriniae*
$8_{a\text{-}d}$. *serovar. nigeriensis*
9. *serovar. tolworthi*;
10. *serovar. darmstadiensis*;
$11_{a\text{-}b}$. *serovar. toumanoffi*
$11_{a\text{-}c}$. *serovar. kyushuensis*
12. *serovar. tompsoni*
13. *serovar. pakistani*
14. *serovar. israelensis* (Bactimos, Bactis, Bactoculicid, Gnatrol, Skeetal, Teknar, Vectobac)
15. *serovar. dakota*
16. *serovar. indiana*
17. *serovar. tohokuensis*
18. *serovar. kumamotoensis*
19. *serovar. tochigiensis*
$20_{a\text{-}b}$. *serovar. yunnanensis*
$20_{a\text{-}c}$. *serovar. pondicheriensis*
21. *serovar. colmeri*
22. *serovar. shandongiensis*
23. *serovar. japonensis*
24. *serovar. neoleonensis*
25. *coreanensis*
26. *serovar. silo*
27. *serovar. mexicanensis*

There are some differences between the δ-endotoxins structure of the natural strains which result in different host spectra. The classical nomenclature system fails to consistently reflect the structure or the vast diversity in insect specificity of the inclusion protein. Recently, the crystal proteins and their genes have been classified based on their structure, antigenic properties and activity spectrum, into four major groups: CryI (Lepidoptera-specific: *Bt serovar. kurstaki, thuringiensis, aizawai, entomocidus*, etc.); CryII (Lepidoptera- and Diptera-specific: *Bt serovar. kurstaki*, etc.), CryIII (Coleoptera-specific: *Bt serovar. tenebrionis*), CryIV (Diptera-specific: *Bt serovar. israelensis*) (Höfte and Whiteley 1989). Each of these major groups has been further divided into several toxin types (i.e. $CryIA_a$, $CryIA_b$, $CryIA_c$, CryIB, etc.). Many isolates produce several different Cry proteins, e.g. *Bt serovar. kurstaki* HD-1 strain produces $CryIA_a$, $CryIA_b$, $CryIA_c$, CryIIA and CryIIB (Tabashnik 1994). This heterogeneity in toxin production is responsible for some of the diversity in the activity spectrum among strains.

Mode of Action. After the sporulation, the "mother *Bt* cell" dies and disintegrates, releasing the mature spore into the surrounding medium. During the sporulation, a proteinaceous crystal (parasporal body) is produced. When it is eaten by a susceptible insect, the parasporal body solubilized in the midgut releasing protoxins that range in size from 27 to 140 kDa which then are converted into smaller active toxic polypeptides by a proteolytic activation process. These toxins bind to the midgut epithelium receptors (glycoproteins), generate pores in the cell membrane and cause ion-channel imbalance. Secondary septicaemia is frequent. There is extensive variation in the size and structure of the inclusion protein, the intermediate protoxins and the active toxins that are presumed to relate to insect specificity (McGaughey and Whalon 1992).

Uses (Target). Coleoptera, Diptera, Lepidoptera.

Effects on Non-Target Aquatic Organisms
Invertebrates. High dosage and prolonged application of Vectobac-12AS may impose some stress on the survival of certain small crustaceans (Chui et al. 1993). The toxicity tests indicated that *Bt* was not deleterious to *Mesocyclops* sp. (Copepoda, Cyclopidae) at concentrations exceeding those expected in the field (Tietze et al. 1994). EC_{50} was 0.2 mg/kg for *Chironomus riparius* Meigen (Dipt., Chironomidae). There was no reduction of the major taxa found in the ponds that could be accounted for by the treatments. However, results from laboratory toxicity tests conducted with field-collected Chironomidae indicated that these organisms were affected at the dose 6 kg/ha. This study indicates that chironomids (a major waterfowl food source) are adversely affected by Vectobac-G under a controlled laboratory situation; however, environmental factors reduce the efficacy of the larvicide in the field (Charbonneau et al. 1994). Novodor was harmless to aquatic annelid *T. tubifex* (LC_{50}: 5300 mg a.i./kg) (Högger and Ammon 1994).
Vertebrates. Mortality of brook trout *Salvelinus fontinalis* Mitchill, brown trout Salmo trutta L. and steelhead trout *S. gairdneri*, from eyed embryo to 82-mm fork length, exposed to *Bt serovar. israelensis* in the laboratory increased when dosages exceeded recommended rates by 12·000 times or more. There was generally no toxicity difference between denatured (autoclaved) and non-denatured *Bt* for all three trout species at all *Bt* concentrations tested, indicating that mortality was due to formulation components and not *Bt* toxin. LC_{50} (48 h) values for brown and brook trout alevins ranged from 1561 to 2321 mg/kg for both denatured and non-denatured *Bt*. Scanning electron micrographs showed particle and mucus accumulation on gill surfaces from fish exposed to 2000 mg/kg *Bt* for 4 h. Level of O_2 and CO_2 in blood from exposed (4000 mg/kg *Bt* for 4 h) and unexposed fish were similar. Brown trout of 43-mm fork length fed excess *Bt*-killed or live black flies (*Cnephia dacotensis* Dyar et Shannon; Dipt., Simuliidae) ate similar quantities of each larval type, and both groups experienced similar mortalities and growth rates at 30-day post exposure (Wipfli et al. 1994). Bactoculicid was safe for *P. reticulata* (LC_{50}: 1000 mg/l). Integrated use of larvivorous fish and *Bt* in vector control has been suggested (Mittal et al. 1994). LC_{50}s (96 h) of Teknar for water feeder guppies (*P. retriculata*)

> 156 mg/l and of Thuricide for water gobie *Pomatoschistus minutus* Pallack (Perciformes, Gobidiidae) > 400 mg/l (Tomlin 1995).

Effects on Non-Target Terrestrial Organisms

Invertebrates

Predators. At 1%, Novodor FC had no adverse effects on *C. carnea* larvae and nymphs under laboratory and semi-field conditions (Bigler and Waldburger 1994). In a laboratory trial, *Bt serovar. tenebrionis* adversely affected the survival of *Chauliognathus lugubris* (F.) (Col., Cantharidae), a major predator of the target insect *Chrysophtharta bimaculata* (Olivier) (Col., Chrysomelidae) (Greener and Candy 1994). Over a period of 10 consecutive days, consumption of pollen contaminated with 200 ml M-One/l (5.6 × 10^{-9} CPBIU, Colorado potato beetle international unit/l) had no lethal effects on adult *Coleomegilla maculata lengi* Timberlake (Col., Coccinellidae) (Giroux et al. 1994). Reducing black fly densities in these streams, using *Bt*, indirectly and differentially affected predators. In black-fly-poor environments, feeding habits of specialist predators were most affected and generalist predators least affected because the latter consumed alternative prey. Predator–predator and predator–prey interactions, and prey community structure may be affected indirectly by *Bt* applications reducing food resources and forcing predation onto less preferred prey (Wipfli and Merritt 1994). Bt was harmless to *A. cucumeris* (Van der Staay 1991).

Insect Parasitoids. Although *Bt* did not influence the ability of parasitism of *T. pretiosum*, it did affect the emergence of adults (Marques and Alves 1995). Bactospeine, Dipel, Novodor and Thuricide had no adverse effects on lepidopterous egg parasitoid *T. cacoeciae* adults (Hassan and Krieg 1975; Hassan 1994). Bactospeine and Delfin were not harmful to *O. concolor* adults and pupae (Jacas and Viñuela 1994). *Apanteles fumiferanae* Viereck (Hym., Braconidae) parasitized *Choristoneura fumiferana* (Clemens) (Lep., Tortricidae) larvae are more likely to survive exposure to *Bt* because they feed less than non-parasitized larvae and are thus less likely to acquire a lethal dose of the bacterium. *Bt* nevertheless reduced parasitoid populations by 50-60%, killing their hosts before parasitoid emergence. This negative impact of *Bt* on parasitoid survival was decreased when exposure of budworm larvae to spray deposits was delayed from peak third- to peak fourth-instar (Nealis and Van Frankenhuyzen 1990). *Bt* was not toxic to *D. eucerophaga* and *A. plutellae* adults and pupae (Talekar and Yang 1991).

Bees. Bt serovar. thuringiensis, serovar. kurstaki and *serovar. israelensis* did not show any deleterious effects on *A. mellifera* (Krieg et al. 1980). Non-toxic to bees; LD_{50} (oral) of Delfin WG for honeybee > 0.1 mg per insect (Tomlin 1995).

Vertebrates

Birds. The solubilized parasporal crystalline proteins (SPCP) of *Bt serovar. israelensis* when administered to Japanese quail by intra-abdominal (IA) injection were toxic with an LD_{50} (24 h) of 22.8 mg/kg. No deaths were noted by intranasal, subcutaneous and intravenous administration. SPCP were highly haemolytic on quail red blood cells, but this haemolysis was inhibited by preincubation of SPCP with quail serum. Electrocardiographic analyses of IA-injected quail demonstrated cardiac dysfunction consistent with bradycardia. Hypothermia was also positively correlated with heart rate reduction. SPCP IA injection reduced serum lipid levels and alkaline phosphatase, and increased serum glucose, creatinine phosphokinase and lactate dehydrogenase. Induction of a septic shock-like condition may be the toxic mode of action of IA-injected SPCP in Japanese quail (Kallapur et al. 1992).

Mammals. Mice failed to remove one preparation of *Bt serovar. israelensis* from the spleen, and a constant number of colony-forming units were recovered for 80 days. *Bt* was recovered from the heart blood; its disappearance from heart blood coincided with its clearance from the spleen. *Bt* persisted for 1 week. There was no evidence that *Bt* was infectious (Siegel and Shadduck 1990). *Bt serovar. kyushuensis* synthesizes a mosquitocidal crystalline inclusion containing several proteins ranging from 140 to 14 kDa. Knowles et al. (1992) identified a 25-kDa protein protoxin in this inclusion which is not cytolytic, but when activated proteolytically to 23-22 kDa products is cytolytic to mosquito, lepidopteran and mammalian cells, can release entrapped glucose from liposomes and forms cation-selective channels in a planar lipid bilayer. This broad-spectrum cytolytic toxin is related antigenically to the 23-kDa toxin from *Bt serovar. darmstadiensis* strain 73-E10-2, but not to the 25-kDa CytA toxin of *Bt serovar. israelensis.* The cytolytic activity of these *Bt serovar. kyushuensis* toxins, like that of the latter two toxins, can be neutralized by incubation with liposomes containing phospholipids. The 28000 mol. wt protein of *Bt serovar. israelensis* showed a high degree of toxicity to rat muscle in culture. Application of 1 μg/ml to the culture medium completely inhibited cell fusion. Reversibility of this effect was demonstrated by replacement of the culture medium with fresh medium, and the consequence was that cell fusion was resumed. When differentiated myotubes were treated with 1 μg toxin/ml, the spontaneous contractile activity was abolished within 20 min. Cytotoxic effects were observed 1 h after treatment was initiated, as manifested by creatine kinase release to the medium. Two hours after toxin was applied to the muscle culture, the myotubes were deteriorated whereas the mononucleated cells were not affected. Cultures 6 or 7 days old treated with 1 μg 28000 mol. wt toxin/ml revealed a change in the levels of Na^+ and K^+ within the fibres. Preincubation of the toxin for 20 min with phospholipids before application to the cells reduced the cytotoxic effect. Phosphatidylinositol and phosphatidylserine were the most efficient inhibitors, whereas phosphatidylcholine, sphingomyelin and phosphatidylethanolamine were less effective in protecting cultures from the cytotoxic effects of the 28000 mol. wt protein (Cahan et al. 1994).

Humans. Bt serovar. kurstaki as a microbial pesticide has been widely used for over 30 years in Oregon, USA. Surveillance for human infections caused by *Bt* among Lane County, Oregon, USA residents was conducted during two seasons of aerial *Bt* spraying for gypsy moth control. *Bacillus* isolates from cultures obtained for routine clinical purposes were tested for presence of *Bt.* Detailed clinical information was obtained for all *Bt*-positive patients. About 80000 people lived in the first year's spray area, and 40000 in the second year's area. A total of 55 *Bt*-positive cultures were identified. The cultures had been taken from 18 different body sites or fluids. Fifty-two (95%) of the *Bt* isolates were assessed to be probable contaminants and not the cause of clinical illness. For three patients, *Bt* could neither be ruled in nor out as a pathogen. Each of these three *Bt*-positive patients had pre-existing medical problems. The level of risk for *Bt* and other existing or future microbial pesticides in immunocompromised hosts deserves further study (Green et al. 1990).

Other Effects. A 25-kDa *Bt serovar. israelensis* δ-endotoxin potentiated the antitumour activity of bleomycin (BLM) in both *in vitro* and *in vivo* systems. A characteristic hyperthermic synergy was found toward the cytocidal activity of BLM in the presence of *Bt* δ-endotoxin in cultured L1210 cells. Cytocidal gain, i.e. the ratio of the IC_{50} of BLM in the absence versus the presence of *Bt* δ-endotoxin, reached approx. 20 at 40 °C, suggesting that a more effective heat–BLM combination treatment may be achieved in the presence of *Bt* δ-endotoxin (Yokohama et al. 1993).

Toxicity of *Bt* to species of several invertebrate taxa (Acari, Nematoda, Collembola, Annelida, Hymenoptera) inhabiting the soil has been demonstrated, but only rarely is it possible to relate dosage information to field situations, and in many cases the *Bt serovar* tested is not currently used for pest control. There is an urgent need for further research to elucidate the relationships between *Bt* and the natural soil microflora and fauna (Addison 1993).

Delfin and Novodor did not inhibit the entomopathogenic fungus, Beauveria sp. (Coremans-Pelseneer 1994).

4 Non-Target Effects of Novel-Type Insecticides with Different Mode of Action

4.1 Pyrrole Insecticides

Dioxapyrrolomycin, isolated from a *Streptomyces* strain, was found to have moderate activity against a number of insects and mites. This novel structure led to a search for acidic halogenated nitropyrroles and pyrroles with other substituent combinations, with useful insecticidal activity (Addor et al. 1992; Hunt and Treacy, this Vol.). Certain physiochemical features of the pyrroles such as lipophilicy and acidity suggested that these compounds are acting as uncouplers of the oxidative phosphorylation (Kuhn et al. 1993).

4.1.1 *AC-303630*

Code No. and Trademark. AC-303630, Pirate, Stalker

Mode of Action. AC-303630 is a pro-insecticide that is activated by the oxidative (i.e. by mono-oxygenases) removal of the *N*-ethoxymethyl group. This releases a lipophilic, weakly acidic pyrrole metabolite which exerts its toxicity through mitochondrial uncoupling in insects (Black et al. 1994). It has cuticular and oral activities, although via this latter route it is 75 times more active on Lepidoptera species. It has poor plant-systemic properties, although it has good translaminar activity (Treacy et al. 1994).

Uses (Target). Cicadellidae, Coleoptera, Lepidoptera, Tenuipalpidae, Tetranychidae.

Effects on Non-Target Aquatic Organisms
Invertebrates. EC_{50} (96 h) for *Daphnia* sp. 6.11 μg/l (Tomlin 1995).
Vertebrates: LC_{50} (48 h) for carp 500 μg/l; LC_{50} (96 h) for rainbow trout 7.44 and bluegill sunfish 11.6 μg/l (Tomlin 1995).

Effects on Non-Target Terrestrial Organisms
Invertebrates
Predators. *T.* occidentalis tolerated pyrroles when it was highly active on *T. urticae* (Kuhn et al. 1993).
Insect Parasitoids. AC-303630 was highly toxic to *E. formosa* adults at a dose of 0.83 ml/l, but did not affect the immature stages of the parasitoid inside the host (Van de Veire 1994).
Bees. LD_{50} for honey bee 0.2 μg per insect (Tomlin 1995).
Vertebrates.
Birds. Acute oral LD_{50} for mallard duck 10 and bobwhite quail 34 mg/kg (Tomlin 1995).
Mammals. Acute oral LD_{50} for female rats 1152 mg technical ingredients/kg (Tomlin 1995).

4.2
Pyridine Azomethine Insecticides

Pyridine azomethine, a new structure for an insecticide, was discovered by Kristinsson (1989).

4.2.1
Pymetrozine

Code No. and Trademark. CGA 215944, Chess, Fulfill, Plenum, Reley, Sterling.

Mode of Action. Pymetrozine is not directly toxic to the susceptible insects, but it does cause an immediate cessation of feeding after exposure. It probably acts through the nervous system and on the foregut by stimulating their spontaneous electrical activity and peristalsis (Kayser et al. 1994). It is systemic in plants in both acropetal and basipetal directions. It has translaminar activity (see Fuog et al. this Vol.).

Uses (Target). Aleyrodidae, Aphidae, Cicadellidae.

Effects on Non-Target Aquatic Organisms
Invertebrates. LC_{50} of Plenum for aquatic annelid *T. tubifex* 400 mg/kg (Högger

and Ammon 1994). LC_{50} (48 h) for *Daphnia* sp. > 100 mg/l (Tomlin 1995). *Vertebrates*. LC_{50} (96 h) for rainbow trout and carp > 100 mg/l (Tomlin 1995).

Effects on Non-Target Terrestrial Organisms
Invertebrates
Predators. Pymetrozine was harmless to *Orius majusculus* (Reut.) (Het., Anthocoridae), *C. carnea* and *C. septempunctata* (see Fuog et al. this Vol.). It was also safe to predacious bug *Orius insidiosus*, and *Bembidion obtusum* Audinet-Serville and *Bembidion lampros* (Herbst) (Col., Carabidae) (Sechser et al. 1994). Chess was harmless to *T. pyri* (Sterk et al. 1994) and *A. fallacis* (see Fuog et al. this Vol.).
Insect Parasitoids. In a laboratory residual test, pymetrozine caused a reduction of parasitism of *E. formosa* by 49% and an adult mortality of 80% (Stolz 1994b). Pymetrozine (25 EC) was harmful to *E. formosa* adults. Treating the black scales of *E. formosa*, low mortality was found (Van de Veire, pers. comm.). In the greenhouse, pymetrozine had no negative impact on the parasitism of *E. formosa* and *Aphelinus* sp. (Hym., Aphelinidae) (Sechser et al. 1994; Sechser and Reber 1996).
Bees. Oral LD_{50} (48 h) for honeybee > 117 μg per insect; cuticular LD_{50} (48 h) > 200 μg per insect (Tomlin 1995).
Vertebrates
Birds. Acute oral LD_{50} for bobwhite quail > 2000 mg/kg (Tomlin 1995).
Mammals. Acute oral LD_{50} for rats 5820 mg/kg (Tomlin 1995).
Other Effects. Pymetrozine does afford a valid protection against persistent viruses transmitted by *M. persicae* (Harrewijn and Piron 1994).

4.3 Chloronicotinyl Insecticides

Chloronicotinyl nitromethylene analogues as nicotine and cartap act through cholinergic (nicotinic) receptors (Corbett et al. 1984; Voss and Neumann 1992).

4.3.1 *Imidacloprid*

Code No. and Trademark. NTN 33 893, Admire, Confidor, Gaucho.

Mode of Action. The chloronicotinyl nitromethylene analogue imidacloprid, acts in an agonistic fashion by binding to the nicotinergic acetylcholine receptor in the post-synaptic region of the insect nerve (Shroeder and Flattum 1984; Bai et al. 1991). It is a systemic insecticide, with a very good acropetal translocation and with cuticular and stomach actions (Tomlin 1995).

Uses (Target). Aphidae, Aleyrodidae, Cicadellidae, Thysanoptera, some termites and coleopteran species.

Effects on Non-Target Aquatic Organisms
Invertebrates. Shrimps, amphipods and chironomids are sensitive to imidacloprid

(LC_{50}: 0.034 mg/l). LC_{50} on cladoceran *D. magna* 85 mg/l (Mullins, 1993). *Vertebrates.* LC_{50} (96 h) for golden orfe 237 and rainbow trout 211 mg/kg (Tomlin 1995).

Effects on Non-Target Terrestrial Organisms
Invertebrates
Earthworm. Fourfold overdose (500 g a.i./ha applied twice as a spray) showed only brief influences on *L. terrestris* which had been completely balanced as early as the autumn of the year of application (Pflüger and Schmuck 1991).
Predators. The motility of carabids was affected only for a short time; this effect was almost completely reversible. Semi-field tests confirmed the low toxicity of imidacloprid to *P. cupreus.* There was a noticeable decrease in feeding performance, which can be attributed to the marked repellent effect of imidacloprid (Pflüger and Schmuck 1991). At rates which are relevant in practice, ladybird (Coccinellidae), hover fly (Syrphidae) and lacewing (Chrysopidae) populations were decreased by approx. 50% (Pflüger and Schmuck 1991). Imidacloprid toxicity on *P. maculiventris* decreased in order of topical exposure → ingestion → residual contact (De Cock et al. 1996). Predatory mites are not affected, because imidacloprid does not have acaricidal properties (Elbert et al. 1991). Admire was found to be harmless to *T. pyri* (Sterk et al. 1994) and *P. persimilis* (Van de Veire 1995).
Insect Parasitoids. Penetration of imidacloprid through the insects' cuticle (i.e. cuticular activity) is relatively low (Pflüger and Schmuck 1991). *Cales noacki* Howard (Hym., Aphelinidae) larvae and pupae inside the host *Aleurotrixus floccosus* (Maskell) (Hom., Aleyrodidae) were not harmed (Pflüger and Schmuck 1991). Emergence of *A. takeyanus*, an egg parasitoid of azalea lace bug, was not affected by imidacloprid (Balsdon et al. 1993).
Bees. When applied as a spray in flowering crops during the foraging time, imidacloprid is to be classified as dangerous to honey bee. Imidacloprid has a noticeable repellent effect which, depending on the application rate and crop, can last more than a week (Pflüger and Schmuck 1991).
Vertebrates
Birds. Acute oral LC_{50} (96 h) for Japanese quail 31 and bobwhite quail 152 mg/kg (Tomlin 1995). The reproductive NOEC was 125 mg/kg in duck. Imidacloprid is slightly hazardous to birds. Acute oral toxicity studies resulted in LD_{50} of 41 mg/kg for house sparrow (Mullins 1993). Red-winged blackbirds and brown-headed cowbirds (*Molothrus ater* Boddaert) were strongly deterred from feeding on rice seed treated with imidacloprid at 620 and 1870 mg/kg. When applied to wheat seed, imidacloprid effectively reduced consumption by red-winged blackbirds at rates as low as 165 mg/kg. Treatment-related effects such as ataxia and retching were noted in some birds exposed to the highest treatment levels, but such effects were transitory. Videotapes indicated that imidacloprid was not a sensory repellent or irritant to birds. Avery et al. (1993) conclude that avoidance of imidacloprid-treated food is a learned response mediated by post-ingestional distress.
Mammals. Acute oral LD_{50} for rats approx. 450 mg/kg (Tomlin 1995).
Other Effects. Imidacloprid does not interfere with the activity of soil micro-

organisms (Pflüger and Schmuck 1991). Imidacloprid has no toxicity to the infective juveniles of entomopathogenic nematode *S. carpocapsae* at 100 μg/ml (Zhang et al. 1994).

4.4 Thiourea Insecticides

4.4.1 *Diafenthiuron*

Code No. and Trademark. CGA 106630, Pegasus, Polo.

Mode of Action. Diafenthiuron is a pro-insecticide. It is converted by light, or *in vivo* to the corresponding carbodiimide, which is an inhibitor of mitochondrial ATPase (Ruder and Kayser 1993). It is cuticular and a stomach poison, and also shows some ovicidal action (Tomlin 1995).

Uses (Target). Aleyrodidae, Aphidae, Jasidae, Tarsonemidae, Tetranychidae, some lepidopterous pests.

Effects on Non-Target Aquatic Organisms
Invertebrates. EC_{50} (48 h) for *Daphnia* sp. < 0.5 mg/l (Tomlin 1995).
Vertebrates. LC_{50} (96 h) for carp 0.0038, rainbow trout 0.0007 and bluegill sunfish 0.0013 mg/l (Tomlin 1995).

Effects on Non-Target Terrestrial Organisms
Invertebrates
Predators. Polo was safe for *C. carnea* (Sechser 1994). Diafenthiuron was considered harmless to *Orius niger* Wolff (Het., Anthocoridae) (Van de Veire and Degheele 1993). It was practically non-toxic to *P. maculiventris* when applied topically. However, both ingestion and residual contact caused severe mortality. In general, adults were more sensitive than fifth-instar (De Cock et al. 1996). Polo was harmless to predatory bug *O. insidiosus* adults and nymphs, but caused moderate mortality on the treated newly emerged larvae (Van de Veire 1995). Pegasus was slightly harmful to *C. septempunctata* (Sechser 1994). Diafenthiuron was slightly harmful to *P. persimilis*, *T pyri* and *Amblyseius andersoni* (Chant) (Acari, Phytoseiidae) (Sechser 1994).
Insect Parasitoids. Diafenthiuron and Polo 500 SC were considered to be harmless to *E. formosa* (Van de Veire and Degheele 1993; Van de Veire 1994). In other studies, no mortality was found until 24 h of Polo application, but total mortality occurred at the third day. When treating the "black scales" of *E. formosa*, wasps died at eclosion (Van de Veire, pers. comm.). Polo was harmless to *Dacnusa sibirica* Telenga (Hym., Eulophidae) but it was moderately harmful to *A. plutellae* (Sechser 1994).
Bees. Oral LD_{50} (48 h) for honey bee 2.1 μg per insect; cuticular LD_{50} (48 h) 1.5 μg/l (Tomlin 1995). When diafenthiuron was added to a sugar solution (0.03% a.i.) under semi-field conditions, there were few surviving adults of *B.*

terrestris to be found in the hives. When contaminated honey was fed once under laboratory conditions, there was a mortality of 90% after 12 days (Sechser 1994).

Vertebrates

Birds. Acute oral LD_{50} for bobwhite quail and mallard duck > 1500 mg/kg (Tomlin 1995).

Mammals. Acute oral LD_{50} for rats 2068 mg/kg (Tomlin 1995).

5 Discussion

Voss and Neumann (1992) summarized the contradictions on an "ideal insecticide" as follows:

Highly researched but low-priced; superior to existing products but not more expensive; highly selective but big in market size; long-lived in the market but should not create resistance problems; broadly active on pests but inactive on non-target species; long-lasting on crop and in insect pests but not causing residues in crop and the environment; fast-acting on pest but preferably not as a neurotoxicant; mobile in plants but immobile in soil.

5.1 Effects of Novel-Type Insecticides on Vertebrates

The novel-type insecticides are required to have low acute mammalian toxicity, because of their human applicators and consumers of agricultural products. On the basis of their high acute toxicity (Fig. 1; Table 4) on mammalians, abamectin (although ivermectin is used as a drug) and imidacloprid do not really fit this picture. The values of bird toxicity are usually similar to the mammalian's values (Fig. 2; Table 4). This is very important if we consider the high expenses of bird losses following pesticide treatments (Pimentel and Lehman 1993). Abamectin, AC-303630, imidacloprid and triflumuron unfortunately are harmful to birds. Toxicity of a compound to poikilotherm vertebrates (i.e. fish, amphibians, reptiles) is also considerable, especially because its values usually are quite different (Fig. 3; Table 4) from the values of homoiotherm vertebrates (i.e., birds, mammals). Abamectin, diafenthiuron, buprofezin, fenoxycarb, flucycloxuron, methoprene and tebufenozide are harmful to fish. It is not very easy to qualify the neem-derived insecticides. Botanical insecticides based on natural origin are highly welcomed by environmentally concerned organizations in plant protection. We should note that several very toxic and dangerous compounds [e.g. ricin from *Ricinus communis* (Euphorbiaceae), colhicin from *Colhicum autumnale* (Liliaceae)] are also known to have a natural origin (Ames et al., 1990). Nevertheless, the mode of action of numerous allelochemicals of *Azadirachta indica* is not intensively and strictly researched (azadirachtin A is one of the exceptions) and they may have chronic effects on vertebrates. An additional problem is the changeable chemical profile of a single neem-derived product.

Using a single score system, as in Table 4, we can conclude that *Bacillus*

Table 4. Acute toxicity of some novel-type insecticides on vertebrates

Active ingredient	Score of acute toxicity on vertebrates[a]		
	Fish	Birds	Mammals
Abamectin	4	3	4
AC-303630	2	4	2
Azadirachtin A			1
Bacillus thuringiensis	**1**	**1**	**1**
Buprofezin	3		2
Chlorfluazuron	**1**	2	**1**
Cyromazine	1	2	2
Diafenthiuron	4	2	2
Diflubenzuron	1		1
Fenoxycarb	3	1	1
Fluazuron	2	2	1
Flucycloxuron	3	2	1
Flufenoxuron	1	2	2
Hexaflumuron	**1**	**2**	**1**
Hydroprene	1		1
Imidacloprid	1	3	3
Lufenuron	**1**	**2**	**1**
Methoprene	3		1
Neem insecticides			1
Pymetrozine	**1**	2	**1**
Pyriproxyfen			1
Tebufenozide	3	2	1
Teflubenzuron	**1**	**2**	**1**
Triflumuron	1	3	1

[a]Ingredients in **bold** are safe for vertebrates.
Scores: 1, non-toxic (mammals and birds LD_{50}: > 5000 mg/kg; fish LC_{50}: > 50 mg/l); 2; low risk (mammals and birds: 1001–5000 mg/kg; fish: 5.1–50 mg/l); 3, dangerous (mammals and birds: 101–1000 mg/kg, fish: 0.1–5 mg/l); 4, extremely dangerous (mammals and birds: < 100 mg/kg; fish: < 0.1 mg/l).

thuringiensis δ-endotoxin, chlorfluazuron, hexaflumuron, lufenuron, pymetrozine and teflubenzuron are acutely safe to vertebrates. Surprisingly, some of the important data on bird and fish toxicity are not included in *The Pesticide Manual* (Tomlin 1995) by the manufacturer, and lack of model bird and fish species makes it difficult to compare these values. Nevertheless, it is very difficult to find a world-wide accepted representative test animal.

5.2 Effects of Novel Type Insecticides on Beneficial Insects

An "ideal insecticide" would be broadly active on insect pests but paradoxically inactive on non-target insect species (Voss and Neumann 1992). Specificity of an insecticide includes many different aspects and has insecticide and insect impacts (Elzen 1989; Jepson 1989). We need to emphasize, however independently of the beneficial saving properties of a compound, that insecticides that reduce the number of the host(s) through the food chain always decrease the numbers of their predator(s) and parasitoid(s). Especially obligate endoparasitoids suffer from this dependency (e.g. *B. thuringiensis*; Nealis and Van Frankenhuyzen 1990). The rate of natural enemies and hosts before and after an insecticide treatment would probably be a suitable criterion for the selectivity of a compound (Lowery and Isman 1995).

5.2.1
Insecticide Impacts

5.2.1.1
Specificity Based on the Mode of Action of an Insecticide

We would distinguish two parts of mode of action of a toxicant: (1) the pharmacokinetic part (i.e. the process by which a toxicant is absorbed, distributed, metabolized and/or catabolized and eliminated by an organism); and (2) the pharmacodynamic part (i.e. the distinct action or effects of a toxicant on living organisms). A toxicant finally needs to act on receptor(s) (i.e. structures or sites on the surface or interior of a cell that bind with toxicants). Primary and secondary receptors are distinguished in pharmacology and toxicology, which would suggest more than one receptor site for a toxicant or drug. Pharmacokinetics and pharmacodynamics, originally used in pharmacology related to drugs, were applied to toxicants based on the original meaning of the Greek *pharmakon*, which means poison and drug as well.

Receptor-Based Specificity (i.e. Sensu Stricto Specificity or Pharmacodynamic-Based Specificity). Organization of life, at receptor (i.e. molecular) level, is very similar in a *classis* (class) of organisms. Nature has not developed different biochemical solutions for a single species (Szent-Györgyi 1957, 1963). Insect development and reproduction disrupters (IDRDs) cannot have beneficial insect saving properties based on receptor level simply because both pests and beneficials belong to the same *classis*, called Insecta. There are few known exceptions, e.g. specific inhibitors of enzymes which convert phytosterols to cholesterol [e.g. fucosterol-24(28)-epoxide cleavage enzyme (Clarke et al. 1985)] may have selectivity between herbivorous and carnivorous insects, based on the fact that zoophagous insects do not have this enzyme system. Until now, there were no "herbivoricides" that would act on herbivorous insects only and would be safe for their carnivorous natural enemies at receptor level. This is the principal reason why there are so many specificity problems with the presently used novel insecticides (Tables 5-9), and why so many controversial results were reported in this respect. It is a real challenge for comparative physiology and pesticide research. However, IDRDs and all other insecticides mentioned in this chapter can be safe for beneficial insects based on some secondary mechanisms (see below).

Pharmacokinetic-Based Selectivity. Pharmacokinetic-based selectivity comprises two main parts; the pharmacokinetic properties of a toxicant in plants (contact, translaminar, systemic toxicants) and in insects (respiratory, cuticular or oral toxicants). A toxicant can display contact activity (it does not leave the place which was reached at the application), translaminar activity (it passes through the epidermis of leaves, but it does not move further) and systemic activity. The

Table 5. IOBC rating system for beneficial arthropods

Scoring	Category	Laboratory	Semi-field		Field
		Initial (E%)	Initial (E%)	Persistence (days)	(E%)
1	Harmless	< 30	< 25	< 5 = short-lived	< 25
2	Slightly harmful	30–79	25–50	5–15 = slightly persistet	25–50
3	Moderately harmful	80–99	51–75	16–30 = slightly persistet	51–75
4	Harmful	> 99	> 75	> 30 = persistent	> 75

E%, reduction of efficiency (initial) or days later (persistence). E% = 100% – [(100% – M%) × R], where M = corrected (Abbott) mortality, R = number of eggs per female treated viz. untreated.

systemic action can be acropetal (movement from the roots to the leaves) or basipetal (movement from the leaves to the roots). These properties usually express themselves in the activity spectrum of an insecticide. Systemic insecticides (e.g. imidacloprid, pymetrozine, neem-derived insecticides) usually act on insects with sucking mouth parts (e.g. Homoptera, Hemiptera, Thysanoptera and/or Acari) living on the not directly treated area.

A toxicant may be respiratory (i.e. volatile compounds enter the respiratory system), cuticular (i.e. after contact with the cuticle the compound passes through it) and/or oral (after feeding, it acts through the digestive channel and is also called a "stomach" poison). Respiratory and cuticular toxicants have no specificity for treated insects, but oral poisons without respiratory and cuticular activities may have some special selectivity. Phytophagous insects feeding on the treated leaves may be poisoned by toxicants orally administrated more than predators and exoparasitoids fed on slightly affected hosts. Immatures of endoparasitoids usually escape from the effects of an oral poison but can be affected indirectly inside a treated host. *Bacillus thuringiensis* δ-endotoxin is a typical orally active, protoxic compound with no respiratory and cuticular activity. Probably this is one of the reasons why it has the best specificity among the insecticides mentioned above concerning zoophagous insects.

It is also noticeable that insects, including parasitoids and predators, usually do not immediately respond to an IDRD or a proinsecticide (e.g. diafenthiuron, AC-303630, *Bt* δ-endotoxin) treatment. Effects of IDRDs manifest at the sensitive stage of an insect, i.e. moulting inhibitors kill the insects at moults or a proinsecticide needs to be *in vivo* converted to the active agent (i.e. in the case of diafenthiuron at least 1 day is need for this process). Some frequently applied short time tests generally disregard this effect.

5.2.1.2
Specificity Based on Insecticide Persistence

The natural stability of a compound may also modify its specificity. A quickly degraded (i.e. degraded to non-effective metabolites) insecticide may be used at the time of a non-sensitive period of beneficial insects, but in the case of a

Table 6. Laboratory results according to IOBC scoring system[a] on immatures insects

Species	DU	FF	HU	LU	TU	YU	BF	CM	FC	RH	NE	AM	BT	IC	DF
Dermaptera:															
F. auricularia	4[101]						1[101]	1[101]							
Neuroptera:															
C. carnea	4[282]	4[282]		3[33]	3[282]	2[282]		4[101]	4[219]		4[284]				2[225]
M. tasmaniae									4[219]						
Heteroptera:															
Anthocoridae:															
A. nemoralis	1[101]			4[33]	1[33]		2[225]	2[101]	4[100]						4[225]
O. niger															1[275]
Pentatomidae:															
P. maculiventris	1[63]													1[64]	1[64]
Coleoptera:															
Carabidae:															
P. cupreus			4[1]												
Coccinellidae:															
C. septempunctata				2[33]											2[225]
H. axyridis									1[100]						
S. undecimnotata	3[101]						2[101]	3[101]	1[100]						
Hymenoptera:															
Ichneumonidae:															
C. turionellae[b]	1[99]						1[101]	1[101]	1[100]						
D. semiclausum															1[225]
P. trichops	1[99]						1[101]	2[101]	1[100]						
Braconidae:															
A. plutellae															3[225]
D. sibirica															1[225]
O. concolor		1[114]						2[114]	1[114]	1[115]			1[114]		
Aphelinidae:															
C. noacki	1[101]						1[101]	1[101]							
E. formosa	1[99]						1[101]	1[101]	1[100]			4[274]			1[225]

(Contd)

Species	DU	FF	HU	LU	TU	YU	BF	CM	FC	RH	NE	AM	BT	IC	DF
Aphidiidae:															
A. matricariae	1[101]						1[101]	1[101]	1[100]						
Encyrtidae:															
L. dactylopii	1[99]						1[101]	4[101]	1[100]						
Eulophidae:															
D. iseae															1[225]
Trichogrammatidae:															
T. cacoeciae	1[99]						1[101]	1[101]	1[100]						
Diptera:															
Cecidomyiidae:															
A. aphidimyza	1[101]						2[101]								

DU, diflubenzuron, FF, flufenoxuron, HU, hexaflumuron, LU, lufenuron; TU, teflubenzuron, YU, flucycloxuron, BF, buprofezin, CM, cyromazine, FC, fenoxycarb, RH, tebufenozide, NE, neem extract, AM, abamectin, BT, Bacillus thuringiensis, IC, imidacloprid, DF, diafenthiuron.

[a]For IOBC rating system see Table 5; superscript numbers indicate the reference.

[b]Test on adults.

Table 7. Laboratory results according to IOBC scoring system[a] on nematodes and immature acari

Species	DU	FF	HU	LU	TU	YU	BF	CM	FC	RH	NE	AM	BT	IC	DF
Nematoda															
Steinernematidae:															
S. feltiae							1[101]	1[101]							
Acari															
Phytoseiidae															
A. andersoni	1[99]						1[101]	4[101]							
A. cucumeris															2[225]
A. fallacis															3[225]
A. potentillae									2[100]						
P. persimilis	1[101]						1[101]	4[101]	1[29]						2[225]
T. athiasae									2[147]						
T. pyri	2[101]						1[101]	1[101]	1[100]						
Clubionidae:															
C. mildei	2[101]						1[101]	1[101]	1[100]						

DU, diflubenzuron; FF, flufenoxuron; HU, hexaflumuron; LU, lufenuron; TU, teflubenzuron; YU, flucycloxuron; BF, buprofezin; CM, cyromazine; FC, fenoxycarb; RH, tebufenozide; NE, neem extract; AM, abamectin; BT, *Bacillus thuringiensis*; IC, imidacloprid; DF, diafenthiuron.
[a]For IOBC rating system see Table 5; superscript numbers indicate the reference.

Table 8. Semi-field or field results according to IOBC scoring system[a] on arthropods

Species	DU	FF	HU	LU	TU	YU	BF	CM	FC	RH	NE	AM	BT	IC	DF
Dermaptera:															
F. auricularia	3[214]										1[216]				
Neuroptera:															
C. carnea	4[101]			4[33]				3[101]	3[219]		1[284]				
M. tasmaniae									4[219]						
Heteroptera:															
Anthocoridae:															
A. nemoralis				4[33]					1[100]						
Orius sp.				4[33]											4[225]
Coleoptera:															
Coccinellidae:															
C. septempunctata	2[101]						2[101]	1[101]	1[100]						
Scymnus sp.				1[33]											
Hymenoptera:															
Aphidiidae:															
A. lepidosaphes									1[61]						
A. mytilaspidis									1[61]						
Diptera:															
Syrphidae:															1[225]
Acari															
Phytoseiidae:															
A. andersoni	1[101]			1[33]			1[101]	4[101]	1[100]						3[225]
P. persimilis	1[101]						2[101]		1[100]						
T. pyri	1[101]			1[33]					1[100]						3[225]

DU, diflubenzuron; FF, flufenoxuron; HU, hexaflumuron; LU; lufenuron; TU, teflubenzuron; YU, flucycloxuron; BF, buprofezin; CM, cyromazine; FC, fenoxycarb; RH, tebufenozide; NE, neem extract; AM, abamectin; BT, Bacillus thuringiensis; IC, imidacloprid; DF, diafenthiuron.
[a]For IOBC rating system see Table 5, superscript numbers indicate the reference.

Table 9. Acute toxicity of some novel type insecticides according to IOBC scoring system[a] on arthropods

Active ingredient	Aquatic arthropods		Bees	Predators							Parasitoids				ST	Saproph.	
	PM	CH		DE	NE	HE	CA	CC	SY	AC	HY			TA		CY	SC
											TR	EX	EN				
Abamectin	4		4	2	2	2		3		1	1		1		1	4	4
AC-303630	4		4							1							
Bacillus thuringiensis	**1**	**3**	**2**		**1**			**1**			**1**		**2**				
Buprofezin	**1**		**1**		**1**	**1**		**3**			**2**	**2**	**1**				
Chlorfluazuron																	
Cyromazine	2				4					1	1		2		2	4	
Diafenthiuron																	
Fenoxycarb	3		1	2	4	4	1	4		3	1	1	2	4	2	4	
Fluazuron																	
Flucycloxuron	4		2							2	3		1				
Flufenoxuron					4	4				1	2		1				
Hexaflumuron	4		2				4						4				
Imidacloprid	1		4		3	3	2	3	3	1	1		1				
Lufenuron																	
Neem oil			3	3	1	3		1	4	1	3	3	3	2			
Pymetrozine																	
Pyriproxyfen						4	2	4		1				3			
Tebufenozide	3		2			1		1				1	2				
Teflubenzuron			1	4	4			4		1	1	4	2				
Triflumuron	3		3		2	4		3		3	3						

PH, Phyllopoda; CH, Chironomidae; DE, Dermaptera, NE, Neuroptera; HE, Heteroptera; CA, Carabidae; CC, Coccinellidae; SY, Syrphidae; AC, Acari; HY, Hymenoptera; TR, egg parasitoids; EX, exoparasitoid; EN, endoparasitoid; TA, Tachinidae; ST, Staphylidae; CY, Diptera Cyclorrhapha; SC, Scarabaeidae.

Ingredients in *bold* are safe for arthropods.

For IOBC rating system see Table 5. Scores: 1, Non-toxic (honeybee LD_{50}: > 1000 μg per insect; *Daphnia* sp. LC_{50}: > 50 mg/l); 2, low risk (honeybee: 101 - 1000 μg per insect; *Daphnia* sp.: 5.1 - 50 mg/l); 3, dangerous (honeybee: 10 - 100 μg per insect; *Daphnia* sp.: 0.1 - 5 mg/l); 4, extremely dangerous (honeybee: < 10 μg per insect; *Daphnia* sp.: < 0.1 mg/l).

persistent compound this possibility does not exist, although degradation is strongly dependent on climatic and edaphic factors. Abamectin, most of the benzoyl-phenyl-urea-type insecticides (chlorfluazuron, flucycloxuron, flufenoxuron, hexaflumuron, teflubenzuron, triflumuron), fenoxycarb and imidacloprid appear to be persistent at least in the absence of light and oxygen, e.g. in storage (Eisa and Ammar 1992) and some soil types (Argauer and Cantelo 1980; Liang et al. 1992; Kim and Kim 1993; Mullins 1993). On the other hand, treating the vertebrate hosts, abamectin (Fincher 1992; Strong 1992) and cyromazine (Brake et al. 1991) bioaccumulate for a long time, which is acceptable. We do not know too much about how long *Bt* can survive in nature (Sulaiman et al. 1990; Addison 1993).

5.2.1.3
Specificity Based on Application of an Insecticide

The treatment is usually targeted to soil or to plants. As for soil insecticides, acropetally systemic compounds (e.g. imidacloprid) may be used, although they potentially endanger soil organisms (e.g. micro-organisms, earthworms), but they are safe for the insects living on the aerial parts, because only phytophagous insects are poisoned while predators and parasitoids can be affected only indirectly. Nevertheless, certain predatory bugs can be affected as they feed on plant tissues when prey density is low. The techniques of spraying (i.e. penetration of spray) also carry some "selectivity" because many predators and parasitoids can survive at the minimum required dosage. This is why the results of laboratory (maximal effect, "worst case") and field tests (minimal effect) are often so different (cf. Tables 6 and 8).

5.2.2
Insect Impacts

5.2.2.1
Resistance of Insects

Resistance of insects is based on behavioural and biochemical processes (Brattsten and Ahmad 1986; see details in chapter 12, this Vol. Behavioural resistance includes the learning process of insects. It is a well-known example that after suffering a sublethal pyrethroid poisoning, cockroaches avoid entering pyrethroid treated fields. Compounds may have repellent or deterrent activities. In the case of a repellent agent with volatile properties, the insects turn back before reaching the subject and leave it untouched. In the case of a phagodeterrent, the insects check the subject with their contact receptors on the palpi and reject it after physical contact. Repellent and deterrent activities of an active ingredient are usually insect "group-dependent", e.g. Cu^{2+} is a known phagodeterrent for *L. decemlineata*, but not for the crawlers of *Qudraspidiotus perniciosus* (Comstock) (Hom., Diaspididae). Such kinds of compounds with repellent and/or deterrent activities may have special selectivity for beneficial insects. This is why neem-derived insecticides may be dangerous for beneficial insects under laboratory conditions in "no-choice" situations but are only slightly harmful under field conditions in "free-choice" situations (Naumann et al., 1994; Naumann and Isman 1996; Vogt et al. 1996).

Biochemical resistance is based: (1) on a possible short time sequestration of a poison (e.g. living on plants containing toxic alkaloids some insects can store them for a time). It is also called as physiological resistance; (2) on decrease of the target enzyme sensitivity; and/or (3) on detoxification processes in insects. Living on tobacco, target enzyme insensitivity was found in some insects (e.g. *M. sexta, M. persicae*). The last possibility after consumption of a toxicant is its detoxification. Catabolism of xenobiotics is usually a strongly species- and age-dependent process. Several enzyme systems (i.e. cytochrome P-450-dependent mono-oxygenases, hydrolases, transferases, etc.) are involved in the usually hydrophobic–hydrophilic directed detoxification processes (Brattsten and Ahmad 1986, Darvas, 1988; 1990). The first-instar larvae of insects are usually more sensitive to a poison simply because the detoxifying enzyme systems are induced later by endobiotics of the insects and xenobiotics taken up with their food. In the case of endoparasitoid species, the detoxification processes of the host are additional factors (e.g. teflubenzuron; Furlong and Wright 1993). These may be the reasons that early instars of predators and exoparasitoids are more sensitive, and the sensitivity of an endoparasitoid may be different in various hosts. It is also considerable that there are strong differences between the detoxification capacity of various strains of a single insect species (Georghiou 1990).

5.2.2.2
Specificity Based on Non-sensitive Stage(s) of Insects

There are several reasons why the sensitivity of an insect changes during its lifetime. The egg stadium is usually not very sensitive to toxicants, although some compounds (e.g. buprofezin, fenoxycarb, diafenthiuron) have also ovicidal/embryocidal activity. In many cases, it is hard to differentiate between the embryocid (it kills the individual during the embryonic development) and the early larvicid (it kills the individual during the postembryonic development) activities when a pharate phase larva dies within the eggshell. Works on *Trichogramma* spp. (Narayana and Babu 1992; Klemm and Schmutterer 1993), where the insecticide was applied on host eggs, suggest that many compounds (i.e. flucycloxuron, neem-derived insecticides, triflumuron) can penetrate into the eggs and kill the parasitoid larvae developing inside them.

During the postembryonic development, the age-dependent set of catabolic isozyme systems is one of the reasons why the older larvae may be less sensitive than younger ones. Nevertheless, in the case of juvenoids usually older stadia of Endopterygota insects are more sensitive (Moreno et al. 1993a). Free-living predators and possibly exoparasitoids may be more sensitive to an insecticide simply because they can be contaminated easier by an insecticide than endoparasitoids. Mining insect larvae (e.g. Agromyzidae, Nepticulidae, Lyonetiidae, Elachistidae) hidden in the tissues of a plant can tolerate insecticides which have no at least translaminar activity. The endoparasitoid lifestyle is very similar to the herbivore's. Sometimes the developing larvae (e.g. Aleyrodidae, Diaspididae,

Coccidae) produce a highly hydrophobic "waxy box" (*cunae* and "scale") resulting in difficulties in spray treatment. Other phytophagous insects produce galls (e.g. Aphidae, Cynipidae, etc.) or zoophagous ones produce mummies (e.g. Aphidiidae, Aphelinidae) which defend the larvae against insecticides. Non-feeding and sessile stadia as pupal and any dormant ones are usually the least sensitive to pesticide treatments (Polgár and Darvas 1995), although juvenoids have a marked effect on the pupal stage (Moreno et al. 1993b).

On adults, chemosterilizing activity of a compound may be also measured. In this case, ovo- and spermiogenesis of female and male are disturbed (e.g. tebufenozide; Smagghe and Degheele 1995). After a single adult mortality test, it is frequently "concluded" in the literature that a certain IDRD has no effects on species studied. This type of "evaluation" neglects the mode of action of insect development and reproduction disrupters which have chemosterilizing activity in adults (see Sect. 2.1.1.6; Furlong et al. 1994).

When insecticides are applied, the survival rate of the most sensitive stadia of the natural enemies always limits their population size. Therefore, the most negative results are frosted in Table 9.

Acknowledgements

The authors thank Dr. András Székács (Plant Protection Institute, Hungarian Academy of Sciences, Budapest), Prof. László Papp (Hungarian Natural History Museum, Budapest) and Dr. László Pap (Chinoin Részvénytársaság, Budapest) for critical reading of this manuscript. We also thank Dr. Anneke De Cock and Dr. Guy Smagghe (University of Gent), Dr. David P. Kreutzweiser (Canadian Forest Service, Sault Ste. Marie), Dr. Heidrun Vogt (Biologische Bundesanstalt. für Land- Forstwirtschaft, Institute für Pflanzenschutz in Obstbau, Dossenheim), Dr. Josep-Anton Jacas (Unidad de Protección de Cultivos ETSI Agrónomos, Madrid), Prof. Murray B. Isman (Department of de Plant Science, University of British Columbia, Vancouver) and Dr. Burkhard Sechser (Ciba-Geigy AG, Basel) for sending us their proofs, and Prof. Mark E. Whalon (Department of Entomology and Pesticide Research Center, Michigan State University, East Lansing) and Dr. Marcus Van de Veire (University of Gent) for their unpublished results.

References

Abdelgader H, Heimbach U (1992) The effect of some insect growth regulators (IGRs) on first instar larvae of the carabid beetle *Poecilus cupreus* (Coleoptera: Carabidae) using different application methods. Aspects Appl Biol 31: 171–177

Abd El-Kareim AI, Darvas B, Kozár F (1988) Effects of fenoxycarb, hydroprene, kinoprene and methoprene on first instar larva of *Epidiaspis leperii* (Hom., Diaspididae) and on its ectoparasitoids, *Aphytis mytilaspidis* (Hym., Aphelinidae). Z Angew Entomol 106: 270–275

Abdel-Khalik MM, Hanafy MSM Abdel-Aziz MI (1993) Studies on the teratogenic effects of deltamethrin in rats. Dtsch Tierärztl Wochenschr 100: 142–143

Addison JA (1993) Persistence and nontarget effects of *Bacillus thuringiensis* in soil: a review. Can J For Res 23: 2329–2342

Addor RW, Babcock TJ, Black BC, Brown DG, Diehl RE, Furch JA, Kameswaran V, Kamhi VM, Kremer KA, Kuhn DG, Lovell JB, Lowen GT, Miller TP, Peevey RM, Siddens JK, Treacy MF, Trotto SH, Wright DP (1992) Insecticidal pyrroles. Discovery and overview. In: Baker DR, Fenyes JG, Steffens JJ (eds) Synthesis and chemistry of agrochemicals III. ACS Symp Ser 504: 283–297

Agarwal N, Kewalramani N, Kamra DN, Agarwal DK, Nath K (1991) Effects of water extracts of neem (*Azadirachta indica*) on the activity of hydrolytic enzymes of mixed rumen bacteria from buffalo. J Sci Food Agric 57: 147–150

Ahmed MT, Eid AH (1991) Accumulation of diflubenzuron in bolti fish *Oreochromis niloticus*. Die Nahrung Chem, Biochem Mikrobiol Technol Ernähr 35: 27–31

Ali BH (1987) The toxicity of *Azadiachta indica* leaves in goats and guinea pigs. Vet Hum Toxicol 29: 16–19

Amer SM, Aly FAE (1992) Cytogenetic effects of pesticides: IV. Cytogenetic effects of the insecticides Gardona and Dursban. Mutat Res 279: 165–170

Ames BN, Profet M, Gold LS (1990) Dietary pesticides (99.99% all natural). Proc Natl Acad Sci USA 87: 7777–7781

Anonymous (1973) Agriculture—environmental and consumer protection appropriations for 1974: hearings before a subcommittee on appropriations, House of Representatives, 93rd Congr, US Govt Printing Office, Washington

Argauer RJ, Cantelo WW (1980) Stability of three ureide insect chitin–synthesis inhibitors in mushroom compost determined by chemical and bioassay techniques. J Econ Entomol 73: 671–674

Asmatullah S, Mufti A, Cheema AM, Iqbal J (1993) Embryotoxicity and teratogenicity of malathion in mice. Punjab Univ J Zool 8: 53–61

Avery ML, Decker DG, Fischer DL, Stafford TR (1993) Responses of captive blackbirds to a new insecticidal seed treatment. J Wildl Manage 57: 652–656

Bai D, Lummis SCR, Leicht V, Breer H, Sattelle DB (1991) Actions of imidacloprid and related nitromethylenes on cholinergic receptors of identified insect motor neurone. Pestic Sci 33: 194–204

Balsdon JA, Braman SK, Pendley AF, Espelie KE (1993) Potential for integration of chemical and natural enemy suppression of azalea lace bug (Heteroptera: Tingidae). J Environ Hortic 11: 153–156

Barrueco C, Herrera A, Caballo C, De La Pena E (1992) Cytogenetic effects of permethrin in cultured human lymphocytes. Mutagenesis 7: 433–437

Barrueco C, Herrera A, Caballo C, De La Pena E (1994) Induction of structural chromosome aberrations in human lymphocyte cultures and CHO cells by permethrin. Teratog Carcinog and Mutagen 14: 31–38

Beitzen–Heineke I, Hofmann R (1992) Versuche zur Wirkung des azadirachtinhaltigen, formulierten Neemsamenextraktes AZT–VR–NR auf *Hylobius abietis* L, *Lymantria monacha* Lund *Drino inconspicua* Meig. Z Pflanzenkr Pflanzenschutz 99: 337–348

Bhunya SP, Behera, BC (1989) Evaluation of mutagenicity of a commercial organophosphate insecticide, tafethion, in mice tested *in vivo*. Caryologia 42: 139–146

Bhunya SP, Jena GB (1993) Studies on the genotoxicity of monocrotophos, an organophosphate insecticide, in the chick *in vivo* test system. Mutat Res 292: 231–239

Bhunya SP, Pati PC (1990) Effect of deltamethrin, a synthetic pyrethroid, on the induction of chromosome aberrations, micronuclei and sperm abnormalities in mice. Mutagenesis 5: 229–232

Biddinger DJ, Hull LA (1995) Effects of several types of insecticides on the mite predator, *Stethorus punctum* (Coleoptera: Coccinellidae), including insect growth regulators and abamectin. J Econ Entomol 88: 358–366

Bidmon H–J, Käuser G, Möbus P, Koolman J (1987) Effects of azadirachtin on blowfly larvae and pupae. In: Schmutterer H, Ascher KRS (eds) Natural pesticides from the neem tree and other tropical plants. Proc 3rd Int Neem Conf, Nairobi, Kenya, 1986, pp 253–271

Bigler F, Waldburger M (1994) Effects of pesticides on *Chrysoperla carnea* Steph. (Neuroptera, Chrysopidae) in the laboratory and semi-field. In: Vogt H (ed) Pesticides and beneficial organisms IOBC WPRS Bull, vol 17, Montfavet, France, pp 55–69

Black BC, Hollingworth RM, Ahammadsahib KI, Kukel CD, Donovan S (1994) Insecticidal action and mitochondrial uncoupling activity of AC–303,630 and related halogenated pyrroles. Pestic Biochem Physiol 50: 115–128

Blaisinger P, Sauphanor B, Kielnen JC, Sureau F (1990) Effet des regulateurs de croissance des insectes sur le developpement de *Forficula auricularia*. In: Relations entre les traitements phytosanitaires et la reproduction des animaux. Ann ANPP 1: 75: 82

Blümel S (1990) Results of a direct contact test for evaluation on side-effects of pesticides on the pupal stage of *Encarsia formosa* (Gah.). Pflanzenschutzberichte 51: 139–142

Blümel S, Stolz M (1993) Investigations on the effect of insect growth regulators and inhibitors on the predatory mite *Phytoseiulus persimilis* A.–H. with particular emphasis on cyromazine. Z Pflanzenkr Pflanzenschutz 100: 150–154

Bolognesi C, Peluso M, Degan P, Rabboni R, Munnia A, Abbondandolo A (1994) Genotoxic effects of the carbamate insecticide, methomyl: II. *In vivo* studies with pure compound and the technical formulation "Lannate 25". Environ Mol Mutagen 24: 235–242

Bonatti S, Bolognesi C, Degan P, Abbondandolo A (1994) Genotoxic effect of the carbamate insecticide methomyl. I. *In vitro* studies with pure compound and the technical formulation "Lannate 25". Environ Mol Mutagen 23: 306–311

Boneß M (1983) Peropal, Alsystin and Cropotex: evaluations of their effect on beneficial arthropods. Pflanzenschutz–Nachr Bayer 36: 38–53

Bourgeois F (ed) (1994) Match (lufenuron) IPM fitness and selectivity. Group IPM Services, Ciba-Geigy, Basle

Brake J, Axtell RC, Campbell WR (1991) Retention of larvicidal activity after feeding cyromazine (Larvadex) for the initial 20 weeks of life of single comb white Leghorn layers. Poult Sci 70: 1873–1875

Brattsten LB, Ahmad S (eds) (1986) Molecular aspects of insect–plant associations. Plenum Press, New York

Brown JJ (1994) Effects of a nonsteroidal ecdysone agonist, tebufenozide, on host/parasitoid interactions. Arch Insect Biochem Physiol 26: 235–248

Burridge LE, Haya K (1993) The lethality of ivermectin, a potential agent for treatment of salmonids against sea lice, to the shrimp *Crangon septemspinosa*. Aquaculture 117: 9–14

Butaye L, Degheele D (1995) Benzoylphenyl ureas effect on growth and development of *Eulophus pennicornis* (Hymenoptera: Eulophidae), a larval ectoparasite of the cabbage moth (Lepidoptera: Noctuidae). J Econ Entomol 88: 600–605

Caballo C, Herrera A, Barrueco C, Santa–Maria A, Sanz F, De La Pena E (1992) Analysis of cytogenetic damage induced in CHO cells by the pyrethroid insecticide fenvalerate. Teratog Carcinog Mutagen 12: 243–249

Cahan R, Shainberg A, Malik Z, Nitzan Y (1994) Biochemical and morphological changes in rat muscle cultures caused by 28 000 mol. wt. toxin of *Bacillus thuringiensis israelensis*. Toxicon 32: 1125–1136

Carlton BC (1993) Development of improved bioinsecticides based on *Bacillus thuringiensis*. In: Duke SO, Menn JJ, Plimmer JR (eds) Pest control with enhanced environmental safety. ACS Symp Ser, 524: 258–266

Carson R (1962) Silent spring. Hamish Hamilton, London

Carvalho GA, Tironi P, Rigitano RLO, Salgado LO (1994a) Selectivity of insect growth regulators to *Trichogramma pretiosum* Riley (Hymenoptera: Trichogrammatidae). An Soc Entomol Bras 23: 431–434

Carvalho GA, Salgado LO, Rigitano RLO, Velloso AHPP (1994b) Effects of insect growth regulators on adults of *Ceraeochrysa cubana* (Hagen) (Neuroptera: Chrysopidae). An Soc Entomol Bras 23: 335–339

Charbonneau CS, Drobney RD, Rabeni CF (1994) Effects of *Bacillus thuringiensis* var. *israelensis* on nontarget benthic organisms in a lentic habitat and factors affecting the efficacy of the larvicide. Environ Toxicol Chem 13: 267–279

Chui VWD, Koo CW, Lo WM, Qiu XJ (1993) Laboratory evaluation of Vectobac-12AS and teflubenzuron against Culex and Aedes mosquito larvae under different physical conditions. Environ Int 19: 193–202

Clarke BS, Jewess PJ (1990) The inhibition of chitin synthesis in *Spodoptera littoralis* larvae by flufenoxuron, teflubenzuron and diflubenzuron. Pestic Sci 28: 377–388

Clarke GS, Baldwin BC, Rees HH (1985) Inhibition of the fucosterol–24(28)–epoxide cleavage enzyme of sitosterol dealkylation in *Spodoptera littoralis* larvae. Pestic Biochem and Physiol 24: 220–230

Coats JR (1994) Risks from natural versus synthetic insecticides. Annu Rev of Entomol 39: 489–515

Cohen E, Cassida JE (1982) Properties and inhibition of insect integumental chitin synthetase. Pestic Biochem Physiol 17: 301–306

Cooper CM (1991a) Insecticide concentrations in ecosystem components of an intensively cultivated watershed in Mississippi. J Freshwater Ecol 6: 237–248

Cooper CM (1991b) Persistent organochlorine and current use insecticide concentrations in major watershed components of Moon Lake, Mississippi, USA. Arch Hydrobiol 121: 103–114

Corbett JR, Wright K, Baillie AC (1984) The biochemical mode of action of pesticides. Academic Press, London

Coremans–Pelseneer J (1994) Laboratory tests on the entomopathogenic fungus *Beauveria*. In: Vogt H (ed) Pesticides and beneficial organisms. IOBC WPRS Bull, vol 17, Montfavet, France, pp 147–155

Cullen MC, Connell DW (1994) Pesticide bioaccumulation in cattle. Ecotoxicol Environ Safety 28: 221–231

Darvas B (1988) Induction, organization, functions and inhibition of cytochrome P–450 in insects. Növényvédelem 24: 341–351

Darvas B (1990) Enzyme systems of insects involved in the metabolism of zoocides. Növényvédelem 26: 49–63

Darvas B (1996) Insect development and reproduction disrupters. In: Ben–Dov Y, Hodgson CJ (eds) Soft scale insects: their biology, natural enemies and control. Elsevier, Amsterdam (in press)

Darvas B, Andersen A (1997) Effects of cyromazine on *Chromatomyia fuscula* and its hymenopteran parasitoids. J Appl Entomol (in press)

Darvas B, Zsellér I (1985) Effectiveness of some juvenoids and anti-ecdysones against the mulberry scale, *Pseudaulacaspis pentagona* (Homoptera: Diaspididae). Acta Phytopathol Acad Sci Hung 20: 341–346

Darvas B, Abd El–Kareim AI, Camporese P, Farag AI, Matolcsy Gy, Ujváry I (1994) Effects of some proinsecticide type, fenoxycarb derivatives and related compounds on some scale insects and their hymenopterous parasitoids. J Appl Entomol 118: 51–58

De Barjac H, Frachon E (1990) Classification of *Bacillus thuringiensis* strains. Entomophaga 53: 233–240

De Clercq P, De Cock A, Tirry L, Viñuela E, Degheele D (1995) Toxicity of diflubenzuron and pyriproxifen to the predatory bug *Podisus maculiventris*. Entomol Exp Appl 74: 17–22

De Cock A, De Clercq P, Tirry L, Degheele D (1996) Toxicity of diafenthiuron and imidacloprid to the predatory bug *Podisus maculiventris* (Heteroptera: Pentatomidae). Environ Entomol (in press)

De Wael L, De Greef M, Van Laere O (1995) Toxicity of pyriproxyfen and fenoxycarb to bumble bee brood using a new method for testing insect growth regulators. J Apic Res 34: 3–8

Diril N, Sumer S, Izbirak A (1990) A survey on the mutagenic effects of some organophosphorus insecticides in the *Salmonella*/microsome test system. Doga Muhendislik ve Cevre Bilimleri 14: 272–279

Duke SO, Menn JJ, Plimmer JR (1993) Challenge of pest control with enhanced toxicological and environmental safety. In: Duke SO, Menn JJ, Plimmer JR (eds) Pest control with enhanced environmental safety. ACS Symp Ser 524: 1–13

Dybas RA (1989) Abamectin use in crop protection. In: Campbell WC (ed) Ivermectin and abamectin. Springer, Berlin Heidelberg New York, pp 287–310

Dzwonkowska A, Hubner H, Zajac E (1991) The influence of trichlorfon, an organophosphorus insecticide, on SCE and dynamics of *in vitro* human lymphocyte divisions. Genet Pol 32: 185–188

Eisa AA, Ammar IMA (1992) Persistence of insect growth regulators against the rice weevil, *Sitophilus oryzae*, in grain commodities. Phytoparasitica 20: 7–13

Elbert A, Becker B, Hartwig J, Erdelen C (1991) Imidacloprid—ein neues systemishes Insektizid. Pflanzenschutz-Nachr Bayer 44: 113–136

El-Hawary ZM, Kholief TS (1990) Biochemical studies on hypoglycemic agents: (I) Effect of *Azadirachta indica* leaf extract. Arch Pharm Res (Seoul) 13: 108–112

Elzen GW (1989) Sublethal effects of pesticides on beneficial parasitoids. In: Jepson PC (ed) Pesticides and non-target invertebrates. Intercept, Wimborne, Dorset, pp 129–150

Feldhege M, Schmutterer H (1993) Investigations on side-effects of Margosan-O on *Encarsia formosa* Gah. (Hym., Aphelinidae), parasitoid of the greenhouse whitefly, *Trialeurodes vaporariorum* Westw. (Hom., Aleyrodidae). J Appl Entomol 115: 37–42

Fernandez NJ, Palanginan EL, Soon LL, Botterell DG (1992) Impact of neem on nontarget organisms. Proc Final Worksh, Botanical Pest Control Proj Phase 2 IRRI, Los Baños, Philippines, pp 117–121

Fincher GT (1992) Injectable ivermectin for cattle: effects on some dung-inhabiting insects. Environ Entomol 21: 871–876

Furlong MJ, Wright DJ (1993) Effect of the acylurea insect growth regulator teflubenzuron on the endo-larval stages of the hymenopteran parasitoids *Cotesia plutellae* and *Diadegma semiclausum* in a susceptible and an acylurea-resistant strain of *Plutella xylostella*. Pestic Sci 39: 305–312

Furlong MJ, Verkerk RHJ, Wright DJ (1994) Differential effects of the acylurea insect growth regulator teflubenzuron on the adults of two endolarval parasitoids of *Plutella xylostella, Cotesia plutellae* and *Diadegma semiclausum*. Pestic Sci 41: 359–364

Garrido A, Beitia F, Gruenholz P (1984) Effects of PP618 on immature stages of *Encarsia formosa* and *Cales noacki* (Hymenoptera: Aphelinidae). Proc Brit Crop Prot Conf, Brighton, pp 305–310

Geden CJ, Rutz DA, Scott JG, Long SJ (1992) Susceptibility of house flies (Diptera: Muscidae) and five pupal parasitoids (Hymenoptera: Pteromalidae) to abamectin and several commercial insecticides. J Econ Entomol 85: 435–440

Geetha KY, Devi KR (1992) Induction of micronuclei in bone marrow erythrocytes of mice by fenvalerate. Agric Biol Res 8: 57–61

Geetha KY, Devi KR (1993) Cytogenetic effects of fenvalerate in germ cells of mice. Agric Biol Res 9: 37–40

Georghiou GP (1990) Overview of insecticide resistance. In: Green MB, Moberg WK, LeBaron H (eds) Managing resistance to agrochemicals: from fundamental research to practical strategies. ACS Symp Ser 421: 18–41

Gerling D, Sinai P (1994) Buprofezin effects on two parasitoid species of whitefly (Homoptera: Aleyrodidae). J Econ Entomol 87: 842–846

Giroux S, Cote JC, Vincent C, Martel P, Coderre D (1994) Bacteriological insecticide M-One effects on predation efficiency and mortality of adult *Coleomegilla maculata lengi* (Coleoptera: Coccinellidae). J Econ Entomology 87: 39–43

Glinsukon T, Somjaree H, Piyachatirawat P, Thebtaramonth Y (1986) Acute toxicity of nimbolide and nimbic acid in mice, rats and hamsters. Toxicol Lett (Amit.) 30: 159–166

Gravena S, Fernandes OD, Santos AC, Pinto AS, Paiva PB (1992) Effect of Buprofezin and abamectin on *Pentilia egena* (Muls) (Coleoptera: Coccinellidae) and chrysopids in citrus. An Soc Entomol Bras 21: 215–222

Green M, Heumann M, Sokolow R, Foster LR, Bryant R, Skeels M (1990) Public health implications of the microbial pesticide *Bacillus thuringiensis*: an epidemiological study, Oregon, (USA) 1985–86. Am J Public Health 80: 848–852

Greener A, Candy SG (1994) Effect of the biotic insecticide Bacillus thuringiensis and a pyrethroid on survival of predators of *Chrysophtharta bimaculata* (Olivier) (Coleoptera: Chrysomelidae). J Aust Entomol Soc 33: 321–324

Grenier S, Grenier A–M (1993) Fenoxycarb, a fairly new insect growth regulator: a review of its effects on insects. Ann Appl Biol 122: 369–403

Grenier S, Plantevin G (1990) Development modifications of the parasitoid *Pseudoperichaeta nigrolineata* (Diptera, Tachinidae) by fenoxycarb, an insect growth regulator, applied onto its host *Ostrinia nubilalis* (Lepidoptera, Pyralidae). J Appl Entomol 110: 462–470

Grosscurt AC, ter Haar M, Jongsma B, Stoker A (1988) PH 70–23: a new acaricide and insecticide interfering with chitin deposition. Pestic Sci 22: 51–59

Győrffy-Molnár J, Polgár L (1994) Effects of pesticides on the predatory mite *Typhlodromus pyri* Scheuten—a comparison of field and laboratory results. In: Vogt H (ed) Pesticides and beneficial organisms. IOBC WPRS Bull, vol 17, Montfavet, France pp 21–26

Harrewijn P, Piron PGM (1994) Pymetrozine, a novel agent for reducing virus transmission by Myzus persicae. Brighton Crop Prot Conf—Pests and Diseases, vol 1: pp 737–742

Harris CK, Whalon ME (1994) Mapping the middle road for Michigan pest management policy. In: Grummon PTH, Mullan, PB (eds) Policy choices: framing the debate for Michigan's future. Institute for Public Policy Studies and Research, Michigan State University, East Lansing, Michigan, pp 103–137

Hassan SA (1989) Testing methodology and the concept of the IOBC/WPRS working group. In: Jepson PC (ed) Pesticides and non-target invertebrates. Intercept, Wimborne, Dorset, pp 1–18

Hassan SA (1994) Comparison of three different laboratory methods and one semi-field test method to assess the side effects of pesticides on Trichogramma cacoeciae. In: Vogt H (ed) Pesticides and beneficial organisms. IOBC WPRS Bull, vol 17, Montfavet, France, pp 133–141

Hassan SA, Krieg A (1975) Über die schonende Wirkung von *Bacillus thuringiensis* Präparaten auf den Parasiten *Trichogramma cacoeciae* (Hym., Trichogrammatidae). Z Pflanzenkr Pflanzenschutz 82: 515–521

Hassan SA, Bigler F, Bogenschütz H, Brown JU, Firth SI, Huang P, Leidieu MS, Naton E, Oomen PA, Overmeer WPJ, Rieckmann W, Samsøe–Petersen L, Viggiani G, Van Zon AQ (1983) Results of the second joint testing programme by the IOBC/WPRS working group "Pesticides and beneficial arthropods". Z Angew Entomol 95: 151–158

Hassan SA, Bigler F, Bogenschütz H, Boller E, Brun J, Calis JNM, Chiverton P, Coremans–Pelseneer J, Duso C, Lewis GB, Mansour F, Moreth L, Oomen PA, Overmeer WPJ, Polgár L, Rieckmann W, Samsøe–Petersen L, Stäubli A, Sterk G, Tavares K, Tuset JJ, Viggiani G (1991) Results of the fifth joint pesticide testing programme carried out by the IOBC/WPRS working group "Pesticides and beneficial organisms". Entomophaga 36: 55–67

Hassan SA, Bigler F, Bogenschütz H, Boller E, Brun J, Calis JNM, Coremans–Pelseneer J, Duso C, Grove A, Heimbach U, Helyer N, Hokkanen H, Lewis GB, Mansour F, Moreth L, Polgár L, Samsøe–Petersen L, Sauphanor B, Stäubli A, Sterk G, Vainio A, Van de Veire M, Viggiani G, Vogh H (1994) Results of the sixth joint pesticide testing programme of the IOBC/WPRS working group "Pesticides and beneficial organisms". Entomophaga 39: 107–119

Hawkinson JE, Casida JE (1993) Insecticide binding sites on γ-aminobutyric acid receptors of insects and mammals. In: Duke SO, Menn JJ, Plimmer JR (eds) Pest control with enhanced environmental safety. ACS Symp Ser 524: 126–143

Hoelmer KA, Osborne LS, Yokomu RK (1990) Effects of neem extracts on beneficial insects in greenhouse culture. Proceedings of USDA Neem Workshop on Neem's potential in pest management programs, Beltsville, Maryland, pp 100–105

Hopkins TL, Kramer KJ (1992) Insect cuticle sclerotization. Ann Rev Entomol 37: 273–302

Hough–Goldstein J, Keil CB (1991) Prospects for integrated control of the Colorado potato beetle (Coleoptera: Chrysomelidae) using *Perillus bioculatus* (Heteroptera: Pentatomidae) and various pesticides. J Econ Entomol 84: 1645–1651

Hoy T, Horsberg TE, Nafstad I (1990) The disposition of ivermectin in Atlantic salmon (*Salmo salar*). Pharmacol Toxicol 67: 307–312

Höfte H, Whiteley HR (1989) Insecticidal crystal proteins of *Bacillus thuringiensis*. Microbiol Rev 53: 242–255

Högger CH, Ammon HU (1994) Testing the toxicity of pesticides to earthworms in laboratory and field tests. In: Vogt H (ed) Pesticides and beneficial organisms. IOBC WPRS Bull, vol 17, Montfavet, France, pp 157–178

Hsu ACT, Aller HE (1987) European patent application No 236,61816, Sept

Ibrahim IA, Omer SA, Ibrahim FA, Kahlid SA, Adam SEI (1992) Experimental *Azadirachta indica* toxicosis in chicks. Vet Hum Toxicol 34: 221–224

Ishaaya I, Casida JE (1974) Dietary TH 6040 alters composition and enzyme activity of housefly larval cuticle. Pestic Biochem Physiol 4: 484–490

Ishaaya I, Mendelson Z, Melamed-Madjur V (1988) Effects of buprofezin on embryogenesis and progeny formation of the sweetpotato whitefly, *Bemisia tabaci* (Homoptera: Aleyrodidae). J Econ Entomol 81: 781–784

Izawa Y, Uchida M, Sugimoto T, Asai T (1985) Inhibition of chitin biosynthesis by buprofezin analogs in relation to their activity controlling *Nilaparvata lugens* Stål. Pestic Biochem Physiol 24: 343–347

Jacas JA, Viñuela E (1994) Side-effects of pesticides on *Opius concolor* Szépl. (Hymenoptera, Braconidae), a parasitoid of the olive fly. In: Vogt H (ed) Pesticides and beneficial organisms. IOBC WPRS Bull, vol 17. Montfavet, France, pp 143–146

Jacas JA, Viñuela E, Adán A, Budia F, del Estal P, Marco V (1992) Laboratory evaluation of selected pesticides against *Opius concolor* Szépl. (Hymenoptera, Braconidae). Ann Appl Biol 120: 140–141

Jacas JA, Gonzáles M, Viñuela E (1996) Influence of the application method on the toxicity of the moulting accelerating compound tebufenozide on adults of the parasitic wasp *Opius concolor* Szépl. Meded Fac Landbouwwet Rijkuniv Gent 60 (in press)

Jacobson M (1995) 5.1 Toxicity to vertebrates. In: Shmutterer H (ed) The neem tree *Azadirachta indica* A. Juss and other meliaceous plants. VCH Verlagsgesellschaft mbH, Weinheim, pp 484–495

Jayaraj S (1992) Studies on IPM in rice-based cropping systems with emphasis on use of botanicals, their safety, and socioeconomic consideration. Proc of Final Works, Botanical Pest Control Proj Phase 2, IRRI, Los Baños, Philippines, pp 63–78

Jebakumar SRD, Flora SDJ, Ganesan RM, Jagathessan G, Jayaraman J (1990) Effect of short-term sublethal exposure of cypermethrin on the organic constituents of the freshwater fish *Lepidocephalichthys thermalis*. J Environ Biol 11: 203–210

Jena GB, Bhunya SP (1992) Thirty days genotoxicity study of an organophosphate insecticide, monocrotophos, in a chick *in vivo* test system. *In vivo* (Athens) 6: 527–530

Jena GB, Bhunya SP (1994) Mutagenicity of an organophosphate insecticide, acephate: an *in vivo* study in chicks. Mutagenesis 9: 319–324

Jepson PC (1989) The temporal and spatial dynamics of pesticide side-effects on non-target invertebrates. In: Jepson PC (ed) Pesticides and non-target invertebrates. Intercept, Wimborne, Dorset, pp 95–128

Kaethner M (1990) Wirkung von Niemsamenprodukten auf die Reproduktionsfähigkeit und Fitneß von *Leptinotarsa decemlineata* Say, *Melolontha hippocastani* Fand *Melolontha melolontha* L. sowie Nebenwirkungen auf die aphidophagen Nützlinge *Coccinella septempunctata* L. und *Chrysoperla carnea* (Stephens). Doctoral Thesis, Universität Giessen, Germany

Kallapur VL, Mayes ME, Edens FW, Held GA, Dauterman WC, Kawanishi CY, Roe RM (1992) Toxicity of the crystalline polypeptides of *Bacillus thuringiensis* ssp. *israelensis* in Japanese quail. Pestic Biochem Physiol 44: 208–216

Karim RANM, Chowdhury MMA, Hoque MNM (1992) Current research on neem in rice in Bangladesh. Proc of Final Worksh, Botanical Pest Control Proj Phase 2, IRRI, Los Baños, Philippines, pp 30–34

Kayser H, Kaufmann L, Schürmann F, Harrewijn P (1994) Pymetrozine (CGA 215'944): a novel compound for aphid and whitefly control. An overview of its Mode of Action. Proc Brighton Crop Prot Conf–Pests and Diseases, vol 2. pp 737–742

Kent RA, Caux PY (1995) Sublethal effects of the insecticide fenitrothion on freshwater phytoplankton. Can J Bot 73: 45–53

Key PB, Scott GI (1994) The chronic toxicity of fenoxycarb to larvae of the grass shrimp,

Palaemonetes pugio. J Environ Sci Health Part B Pestic Food Contam Agric Wastes 29: 873–894

Kim TH, Kim JE (1993) Behavior of benzoylurea insecticide teflubenzuron and flucycloxuron in soil environment. J Korean Agric Chem Soc 36: 510–516

Klemm U, Schmutterer H (1993) Effects of neem preparations on *Plutella xylostella* Land its natural enemies of the genus *Trichogramma*. Z Pflanzenkr Pflanzenschutz 100: 113–128

Klunker R (1991) On side-effects of Neporex on puparium parasitoids of stable flies Angew Parasitol 32: 205–218

Knowles BH, White PJ, Nicholls CN, Ellar DJ (1992) A broad-spectrum cytolytic toxin from *Bacillus thuringiensis* var. *kyushuensis*. Proc R Soc Lond Ser B Biol Sci 248: 1–7

Kreutzweiser DP (1996) Non-target effects of neem-based insecticides on aquatic invertebrates. Ecotoxicol Environ Safety (in press)

Kreutzweiser DP, Capell SS, Wainio-Keizer KL, Eichenberg DC (1994) Toxicity of a new molt-inducing insecticide (RH-5992) to aquatic macroinvertebrates. Ecotoxicol Environ Safety 28: 14–24

Krieg A, Hassan S, Pindorf W (1980) Wirkungsvergleich der Varietät *israelensis* mit anderen Varietäten de *Bacillus thuringiensis* an Nicht–Zielorganismen der Ordung Hymenoptera: *Trichogramma cacoeciae* und *Apis mellifera*. Anz Schädlingsk, D Planzenschutz Umweltschutz 53: 81–83

Kristinsson H (1989) European patent application No EP–A2–0314615, May 3

Kuhn DG, Addor RW, Diehl RE, Furch JA, Kamhi VM, Henegar KE, Kremer KA, Lowen GT, Black BC, Miller TP, Treacy MF (1993) Insecticidal pyrroles. In: Duke SO, Menn JJ, Plimmer JR (eds) Pest control with enhanced environmental safety. ACS Symp Ser 524: 219–232

Lai SM, Lim KW, Cheng HK (1990) Margosa oil poisoning as a cause of toxic encephalopathy. Singapore Med J 31: 463–465

Lamb R, Saxena RC (1988) Effects of neem seed derivatives on rice leaffolders (Lepidoptera: Pyralidae) and their natural enemies. Proc of Final Worksh IRRI–ADB–EWC Projon botanical pest control in rice-based cropping systems, IRRI, Los Baños, Philippines, pp 47

Larson RO (1989) The commercialization of neem. In: Jacobson M (ed) Focus on Phytochemical Pesticides. The Neem Tree. vol 1, CRC Press, Boca Raton, pp 155–168

Leighton T, Marks E, Leighton F (1981) Pesticides, insecticides and fungicides are chitin synthesis inhibitors. Science 213: 905–907

Leopold RA, Marks EP, Eaton JK, Knoper J (1985) Ecdysial failures in the cotton boll weevil: synergism of diflubenzuron with juvenile hormone. Pestic Biochem Physiol 24: 267–283

Letcher J, Glade M (1992) Efficacy of ivermectin as an anthelmintic in leopard frogs. J Am Vet Med Assoc 200: 537–538

Liang DX, Shang HL, Qian CF, Ming JX (1992) The degradation and residues of chlorfluazuron in Chinese cabbage. Acta Phytophylacica Sin 19: 187–191

Lowery DT, Isman MB (1995) Toxicity of neem to natural enemies of aphids. Phytoparasitica 23: 297–306

Maloufi F (1995) Equine parafilariosis in Iran. Vet Parasitol 56: 189–197

Mansour FA, Cohen H, Shain Z (1993a) Integrated mite management in apples in Israel: augmentation of a beneficial mite and sensitivity of tetranychid and phytoseiid mites to pesticides. Phytoparasitica 21: 39–51

Mansour FA, Ascher KRS, Abo-Moch F (1993b) Effects of Margosan-O, Azatin and RD9–Repelin on spiders, and on predacious and phytophagous mites. Phytoparasitica 21: 205–211

Marques IMR, Alves SB (1995) Influence of *Bacillus thuringiensis* Berliner var. *kurstaki* on *Scrobipalpuloides absoluta* (Meyer) (Lepidoptera: Gelechiidae) parasitism for *Trichogramma pretiosum* R. (Hymenoptera: Trichogrammatidae). Arq Biol Tecnol (Curitiba) 38: 317–325

Martin-Prevel Y, Cosnefroy JY, Tshipamba P, Ngari P, Chodakewitz JA, Pinder M (1993) Tolerance and efficacy of single high-dose ivermectin for the treatment of loiasis. Am J Trop Med Hyg 48: 186–192

Mathew G, Rahiman MA, Vijayalaxmi KK (1990) *In vivo* genotoxic effects in mice of Metacid 50, an organophosphorus insecticide. Mutagenesis 5: 147–150

Matsunaga H, Yoshino H, Isobe N, Kaneko H, Nakatsuka I, Yamada H (1995) Metabolism of pyriproxyfen in rats: 1. absorption, disposition excretion, and biotransformation studies with (phenoxyphenyl–^{14}C)pyriproxyfen. J Agric Food Chem 43: 235–240

Matter MM, Marei SS, Moawad SM, El–Gengaihi S (1993) The reaction of *Aphis gossypii* and its predator, *Coccinella undecimpunctata* to some plant extracts. Bull Fac Agric, Cairo Univ 44: 417–432

McCoy CW, Bullock RC, Dybas RA (1982) Abamectin B_1: a novel miticide active against citrus mites. Proc Fla State Hortic Soc 95: 51–56

McGaughey WH, Whalon ME (1992) Managing insect resistance to *Bacillus thuringiensis* toxins. Science 258: 1451–1455

Mendel Z, Blumberg D, Ishaaya I (1994) Effects of some insect growth regulators on natural enemies of scale insects (Hom.: Coccidae). Entomophaga 39: 199–209

Miadokova E, Miklovicova M, Duhova V, Garajovga L, Bohmova B, Podstavkova S, Vlcek D (1991) Effects of the insecticide pyrethroid II in the Ames test, and on *Hordeum vulgare* and *Vicia faba*. Biol Plant 33: 156–162

Miller JA, Garris GI, George JE, Oehler DD (1989) Control of lone star ticks (Acari: Ixodidae) on Spanish goats and white-tailed deer with orally administered ivermectin. J Econ Entomol 82: 1650–1656

Mitlin N, Wiygul G, Haynes JW (1977) Inhibition of DNA synthesis in boll weevil, *Anthonomus grandis*, sterilized by Dimilin. Pestic Biochem Physiol 7: 559–563

Mitsui T, Nobusawa C, Fukami J (1984) Mode of inhibition of chitin synthesis by diflubenzuron in the cabbage armyworm, *Mamestra brassicae* L. J Pestic Sci 9: 19–26

Mitsui T, Tada M, Nobusawa C, Yamaguchi I (1985) Inhibition of UDP-*N*-acetylglucosamine transport by diflubenzuron across biomembranes of the midgut epithelial cells in the cabbage armyworm, *Mamestra brassicae* L. J Pestic Sci 9: 55–60

Mittal PK, Adak T, Sharma VP (1994) Comparative toxicity of certain mosquitocidal compounds of larvivorous fish, *Poecilia reticulata*. Indian J Malariol 31: 43–47

Miyamoto J, Hirano M, Takimoto Y, Hatakoshi M (1993) Insect growth regulators for pest control, with emphasis on juvenile hormone analogs. In: Duke SO, Menn JJ, Plimmer JR (eds) Pest control with enhanced environmental safety. ACS Symp Ser 524: 144–168

Mora MA, Auman HJ, Ludwig JP, Giesy JP, Verbrugge DA, Ludwig ME (1993) Polychlorinated biphenyls and chlorinated insecticides in plasma of Caspian terns: relationships with age, productivity, and colony site tenacity in the Great Lakes. Arch Environm Contam Toxicol 24: 320–331

Mordue AJ, Blackwell A (1993) Azadirachtin: an update. J Insect Physiol 39: 903–924

Moreno J, Hawlitzky N, Jimenez R (1993a) Effect of the juvenile hormone analogue fenoxycarb applied via the host on the parasitoid *Phanerotoma ocularis* Kohl (Hym., Braconidae). J Insect Physiol 39: 183–186

Moreno J, Hawlitzky N, Jimenez R (1993b) Morphological abnormalities induced by fenoxycarb on the pupa of *Phanerotoma ocularis* Kohl (Hym., Braconidae). J Appl Entomol 115: 170–175

Morse JG, Bellows TS, Gaston LK, Iwata Y (1987) Residual toxicity of acaricides to three beneficial species on California citrus. J Econ Entomol 80: 953–960

Muir DCG, Hobden BR, Servos MR (1994) Bioconcentration of pyrethroid insecticides and DDT by rainbow trout: uptake, depuration, and effect of dissolved organic carbon. Aquat Toxicol (Amst) 29: 223–240

Mullins JW (1993) Imidacloprid. A new nitroguanidine insecticide. In: Duke SO, Menn JJ, Plimmer JR (eds) Pest control with enhanced environmental safety. ACS Symp Ser 524: 183–198

Murphy SD (1986) Toxic effects of pesticides. In: Casarett and Doull's toxicology. MacMillan, New York, pp 519–581

Nagai K (1990) Effect of insecticides on *Orius* sp., the natural enemy of *Thrips palmi* Karny. JPN J Appl Entomol Zool 34: 321–324

Naqvi SM, Newton DJ (1990) Bioaccumulation of endosulfan (Thiodan insecticide) in the tissues of Louisiana (USA) crayfish, *Procambarus clarkii*. J Environ Sci Health Part B Pestic Food Contam Agric Wastes 25: 511–526

Narayana ML, Babu TR (1992) Evaluation of five insect growth regulators on the egg parasitoid *Trichogramma chilonis* (Ishii) (Hym., Trichogrammatidae) and the hatchability of *Corcyra cephalonica* Staint (Lep., Galleriidae). J Appl Entomol 113: 56–60

Naumann K, Isman MB (1996) Toxicity of neem (*Azadirachta indica* A. Juss) seed extracts to larval honey bees and estimation of danger from field application. Am Bee J (in press)

Naumann K, Currie RW, Isman MB (1994) Evaluation of the repellent effects of a neem insecticide on foraging honey bee and other pollinators. Can Entomol 126: 225–230

Nealis V, Van Frankenhuyzen K (1990) Interactions between *Bacillus thuringiensis* Berliner and *Apanteles fumiferanae* Vier. (Hymenoptera: Braconidae), a parasitoid of the spruce budworm, *Choristoneura fumiferana* (Clem.) (Lepidoptera: Tortricidae). Can Entomol 122: 585–594

Nehéz M, Tóth C, Dési I (1994) The effect of dimethoate, dichlorvos, and parathion-methyl on bone marrow cell chromosomes of rats in subchronic experiments *in vivo*. Ecotoxicol Environ Safety 29: 365–371

Nurzhanova AA, Biyashev GZ, Patakhova AM (1994) On the mutagenic effect and pesticide properties of organophosphorus compounds (with barley and barley aphid as example). Sel'skokhozyaistvennaya Biol 10: 107–109

Ogunba EO, Gemade EII (1992) Preliminary observations on the distribution of ivermectin in Nigeria for control of river blindness. Ann Trop Med Parasitol 86: 649–655

Oksanen A, Nikander S (1989) Ivermectin as a bird anthelmintic: trials with naturally infected domestic fowl. J Vet Med Ser B 36: 495–499

Olszak RW, Pawlik B, Zajac RZ (1994) The influence of some insect growth regulators on mortality and fecundity of the aphidophagous coccinellids *Adalia bipunctata* L. and *Coccinella septempunctata* L. (Col., Coccinellidae). J Appl Entomol 117: 58–63

Orinak A (1993) Determination of unchanged residues of a pyrethroid insecticide, Pyr–Vu–To2, in sheep internal organ tissues. Pestic Sci 37: 1–7

Osman MZ, Bradley J (1993) Effects of neem seed extracts on *Pholeastor glomeratus* L. (Hym., Braconidae), a parasitoid of *Pieris brassicae* L. (Lep., Pieridae). J Appl Entomol 115: 259–265

Osuala FOU, Okwuosa VN (1993) Toxicity of *Azadirachta indica* to freshwater snails and fish, with reference to the physicochemical factor effect on potency. Appl Parasitol 34: 63–68

Paasivirta J, Rantio T (1991) Chloroterpenes and other organochlorines in Baltic, Finnish and Arctic wildlife. Chemosphere 22: 47–56

Pandey N, Gundevia F, Ray PK (1990) Evaluation of the mutagenic potential of endosulfan using the *Salmonella* mammalian microsome assay. Mutat Res 242: 121–126

Pap L, Bajomi D, Székely I (1996) The pyrethroids, an overview. Int Pest Control 38: 15–19

Parella MP, Christie GD, Robb L (1983) Compatibility of insect growth regulators and *Chrysocharis parski* (Hymenoptera: Eulphidae) for control of *Liriomyza trifolii* (Diptera: Agromyzidae). J Eco Entomol 76: 949–951

Patnaik KK, Tripathy NK (1992) Farm-grade chlorpyrifos (Durmet) is genotoxic in somatic and germ-line cells of *Drosophila*. Mutat Res 279: 15–20

Peleg BA (1983a) Effect of three insect growth regulators on larval development, fecundity and egg viability of the coccinelid, *Chilocorus bipustulatus* (Col., Coccinellidae). Entomophaga 28: 117–121

Peleg BA (1983b) Effect of a new insect growth regulator, Ro 13–5223, on hymenopterous parasites of scale insects. Entomophaga 28: 367–372

Peleg BA (1988) Effects of a new phenoxy juvenile hormone analog on California red scale (Homoptera: Diaspididae), Florida wax scale (Homoptera: Coccidae) and the ectoparasite *Aphytis holoxanthus* DeBach (Hymenoptera: Aphelinidae). J Econ Entomol 81: 88–92

Pimentel D, Lehman H (eds) (1993) The pesticide question: environment, economics and ethics. Chapman and Hall, London

Pflüger W, Schmuck R (1991) Ecotoxicological profile of imidacloprid. Pflanzenschutz-Nachr Bayer 44: 145–158

Pikulska B (1988) Teratogenic effect of Tritox on hen embryos. Zool Pol 35: 179–206

Podstavkova S, Miadokova E, Vlcek D, Jablonicka A (1991) Study on mutagenic activity of fosmet and solvent mixture applied to bacterial strains of *Salmonella typhimurium*. Acta Fac Rerum Nat Univ Comenianae Genet 10: 5–12

Polgár L, Darvas B (1991) Effects of a nonsteroidal ecdysteroid agonist, RH 5849 on a host/ parasite system, *Myzus persicae/Aphidius matricariae*. In: Polgár L, Chambers RJ, Dixon AFG, Hodek I (eds) Behaviour and impact of Aphidophaga. SPB Academic The Hague, pp 323–327

Polgár AL, Darvas B (1995) Adaptive strategies in arthropod species III. Hypobiosis: dormancy (quiescence and diapause) and cryptobiosis. Növényvédelem 31: 369–380

Pompa G, Fadini L, Di Lauro F, Caloni F (1994) Transfer of lindane and pentachlorobenzene from mother to newborn rabbits. Pharmacol Toxicol 74: 28–34

Pool-Zobel BL, Guigas C, Klein R, Neudecker C, Renner HW, Schmezer P (1993) Assessment of genotoxic effects by lindane. Food Chem Toxicol 31: 271–283

Post LC, Mulder R (1974) Insecticidal properties and mode of action of 1-/2,6-dihalogenbenzoyl/ -3-phenyl-ureas. In: Kohn GK (ed) Mechanism of pesticide action. ACS Symp Ser 2: 136–143

Putter I, MacConnell JG, Preiser FA, Haidri AA, Ristich SS, Dybas RA (1981) Avermectins: novel insecticides, acaricides and nematicides from a soil microorganism. Experientia 37: 963–964

Rani BEA, Karanth NGK, Krishnakumari MK (1992) Accumulation and embryotoxicity of the insecticide heptachlor in the albino rat (*Rattus norvegicus*). J Environ Biol 13: 95–100

Rembold H, Sharma GK, Czoppelt CH, Schmutterer H (1980) Evidence of growth distruption in insects without feeding inhibition by neem seed fractions. Z Pflanzenkr Pflanzenschutz 87: 290–297

Reynolds SE, Blakey JK (1989) Cyromazine causes decreased cuticle extensibility in larvae of the tobacco hornworm, *Manduca sexta*. Pestic Biochem Physiol 35: 251–258

Rovesti L, Deseő KV (1991) Effect of neem seed kernel extract on Steinernematid and Heterorhabditid nematodes. Nematologica 37: 493–496

Ruder FJ, Kayser H (1993) The carbodiimide product of diafenthiuron inhibits mitochondria *in vivo*. Pestic Biochem Physiol 46: 96–106

Salawu MB, Adedeji SK, Hassan WH (1994) Performance of broilers and rabbits given diets containing full fat neem (*Azadirachta indica*) seed meal. Anim Prod 58: 285–289

Saleem SA, Matter MM (1991) Relative effects of neem seed oil and Deenate on the cotton leafworm, *Spodoptera littoralis* Boids. and the most prevalent predators in cotton fields in Menoufyia Governorate. Bull Fac Agric Cairo Univ 42: 941–952

Samsøe-Petersen L, Moreth L (1994) Initial and extended laboratory tests for the rove beetle *Aleochara bilineata* Gyll. In: Vogt H (ed) Pesticides and beneficial organisms I. OBC WPRS Bull, vol 17, Montfavet, France, pp 89–97

Sanderson JP, Barnes MM (1983) Control of spider mites on almonds. Insecticide and Acaricide Tests 8: 91–92

Sanguanpong U (1992) Zur Wirkung ölhaltiger Neem- und Marrangosamenprodukte auf die Gemeine Spinnmilbe *Tetranychus urticae* Koch sowie Nebenwirkung auf ihren natürlichen Gegenspieler *Phytoseilus persimilis* Athias-Henriot. Institut für Phytopatholgie und Angewandte Zoologie, Universität Giessen, Germany

Sauphanor B, Stäubli A (1994) Evaluation au champ des effets secondaires des pesticides sur *Forficula auricularia* et *Anthocoris nemoralis*: validation des resultats de laboratoire. In: Vogt H (ed) Pesticides and beneficial organisms. IOBC WPRS Bull, vol 17, Montfavet, France, pp 83–88

Sauphanor B, Chabrol L, Faivre D'Archer F, Sureau F, Lenfant C (1993) Side effects of diflubenzuron on a pear psylla predator: *Forficula auricularia*. Entomophaga 38: 163–174

Sauphanor B, Lenfant C, Sureau F (1995) Effect d'un extrait de neem (*Azadirachta indica* A. Juss) sur le développement de *Forficula auricularia* L. (Dermaptera). J Appl Entomol 119: 215–219

Schafer EW, Jacobson M (1983) Repellency and toxicity of 55 insect repellents to red-winged blackbird (*Agelaius phoeniceus*). J Environ Sci Health 18: 493–502

Schauer M (1985) Die Wirkung von Nieminhaltsstoffen auf Blattläuse und die Rübenblattwanze. Doctoral Thesis. Universität Giessen, Germany

Scheuer R, Bourgeois F (eds) (1994) The IPM fitness of fenoxicarb. Group IPM Services, Ciba-Geigy, Basle

Schmutterer H (1992) Influence of azadirachtin, of an azadirachtin-free fraction of an alcoholic neem seed kernel extract and of formulated extracts on pupation, adult emergence and adults of the braconid *Apanteles glomeratus* (L.) (Hym., Braconidae). J Appl Entomol 113: 79–87

Schmutterer H, Holst H (1987) Untersuchungen über die Wirkung des angereicherten, formulierten Niemsamextracts AZT-VR-K auf die Honigbiene *Apis mellifera* L. J Appl Entomol 103: 208–213

Schroeder ME, Flattum RF (1984) The mode of action and neurotoxic properties of the nitromethylene heterocycle insecticides. Pestic Biochem Physiol 22: 148–160

Schröder P (1992) NeemAzal-F in aquatic environment. In: Kleeberg H (ed) Proc 1st Symp Practice oriented results on use and production of Neem ingredients. Druck and Graphic, Giessen, pp 109–121

Schuytema GS, Nebeker AV, Griffis WL, Wilson KN (1991) Teratogenesis, toxicity, and bioconcentration in frogs exposed to dieldrin. Arch Environ Contam Toxicol 21: 332–350

Sechser B (ed) (1994) Polo, Pegazus (diafenthiuron)—IPM fitness and selectivity. Group IPM Services, Ciba-Geigy, Basle

Sechser B, Reber B (1996) Pymetrozine: a case of perfect selectivity. Proc 20th Int Congr Entomol, 25–31, Aug. 1996, Firence (in press)

Sechser B, Bourgeois F, Reber B, Wesiak H (1994) The integrated control of whiteflies and aphids on tomatoes in glasshouses with pymetrozine. Meded Fac Landbouwwet Rijkuniv Gent 59: 579–583

Shahardar RA, Mir AS, Pandit BA , Bandy MAA (1993) Successful treatment of goat warbles with ivermectin. Indian Vet Journal 70: 666–667

Sharma A, Singh RM (1990) Mutagenic effect of carbofuran on *Pisum* and *Vicia*. Geobios (Jodhpur) 17: 273–276

Sharma RL, Bhat TK, Hemaprasanth K (1990) Anthelmintic activity of ivermectin against experimental *Ascaridia galli* infection in chickens. Vet Parasitol 37: 307–314

Siegel JP, Shadduck JA (1990) Clearance of *Bacillus sphaericus* and *Bacillus thuringiensis* ssp. *israelensis* from mammals. J Econ Entomol 83: 347–355

Singh YP, Bagha HS, Vijjan VK (1985) Toxicity of water extract of neem (*Azadirachta indica* A. Juss.) berries in poultry birds. Neem Newsl 2: 1–17

Slamenova D, Dusinska M, Gabelova A, Bohusova T, Ruppova K (1992) Decemtione (Imidan)-induced single-strand breaks to human DNA, mutations at the hgprt locus of V79 cells, and morphological transformations of embryo cells. Environ Mol Mutagen 20: 73–78

Smagghe G, Degheele D (1994) Action of a novel nonsteroidal ecdysteroid mimic, tebufenozide (RH-5992), on insects of different orders. Pestic Sci 42: 85–92

Smagghe G, Degheele D (1995a) Biological activity and receptor-binding of ecdysteroids and the ecdysteroid agonist RH-5849 and RH-5992 in imaginal wing disks of *Spodoptera exigua* (Lepidoptera: Noctuidae). Eur J Entomol 92: 333–340

Smagghe G, Degheele D (1995b) Selectivity of nonsteroidal ecdysteroid agonists RH 5849 and RH 5992 to nymphs and adults of predatory soldier bugs, *Podisus nigrispinus* and *P. maculiventris* (Hemiptera: Pentatomidae). J Econ Entomol 88: 40–45

Smagghe G, Böhm G.-A, Richter K, Degheele D (1995) Effect of nonsteroidal ecdysteroid agonists on ecdysteroid titer in *Spodoptera exigua* and *Leptinotarsa decemlineata*. J Insect Physiol 41: 971–974

Smagghe G, Viñuela E, Budia F, Degheele D (1996a) *In vivo* and *in vitro* effects of the nonsteroidal ecdysteroid agonist tebufenozide on cuticle formation in *Spodoptera exigua*: an ultrastructural analysis. Arch Insect Biochem Physiol (in press)

Smagghe G, Eelen H, Verschelde E, Richter K, Degheele D (1996b) Differential effects of nonsteroidal ecdysteroid agonists in Coleoptera and Lepidoptera: analysis of evagination and receptor binding in imaginal discs. Insect Biochemistry and Molecular Biology (in press)

Solomon MG, Fitzgerald JD (1990) Fenoxycarb, a selective insecticide for inclusion in integrated pest management systems for pear in the UK. Journal of Hortic Sci 65: 535–540

Soltani N, Besson MT, Delachambre J (1984) Effects of diflubenzuron on the pupal–adult development of *Tenebrio molitor* L. (Coleoptera, Tenebrionidae): growth and development, cuticle secretion, epidermal cell density, and DNA synthesis. Pestic Biochem Physiol 21: 256–264

Spollen KM, Isman MB (1996) Acute and sublethal effects of a neem insecticide on the commercial biological control agents *Phytoseiulus persimilis* and *Amblyseius cucumeris* (Acari: Phytoseiidae), and *Aphidolethes aphidimyza* (Rondani) (Diptera: Cecidomyiidae). J Econ Entomol (in press)

Stark JD (1993) Comparison of the impact of a neem seed-kernel extract formulation, Margosan-O and chlorpyrifos on non-target invertebrates inhabiting turf grass. Pestic Sci 36: 293–299

Stark JD, Vargas RI, Messing RH, Purcell M (1992a) Effects of cyromazine and diazinon on three economically important Hawaiian tephritid fruit flies (Diptera: Tephritidae) and their endoparasitoids (Hymenoptera: Braconidae). J Econ Entomol 85: 1687–1694

Stark JD, Wong TTY, Vargas RI, Thalman RK (1992b) Survival, longevity, and reproduction of tephritid fruit fly parasitoids (Hymenoptera: Braconidae) reared from fruit flies exposed to azadirachtin. J Econ Entomol 85: 1125–1129

Sterk G, Creemers P, Merckx K (1994) Testing the side effects of pesticides on predatory mite *Typhlodromus pyri* (Acari, Phytoseiidae) in field trials. In: Vogt H (ed) Pesticides and beneficial organisms. IOBC WPRS Bull, vol 17, Montfavet, France, pp 27–40

Stolz M (1994a) Efficacy of different concentration of seven pesticides on *Phytoseiulus persimilis* A.-H. (Acari: Phytoseiidae) and on *Tetranychus urticae* K. (Acari: Tetranychidae) in laboratory and semifield test. In: Vogt H (ed) Pesticides and beneficial organisms. IOBC WPRS Bull, vol 17. Montfavet, France, pp 49–54

Stolz M (1994b) Side-effects of pyridine azomethine (CGA 215′944), a new insecticide against aphids and whiteflies, on the parasitic wasp Encarsia formosa Gahan (Hymenoptera: Aphelinidae). Z Pflanzenkr Pflanzenschutz 101: 649–653

Strong L (1992) Avermectins: a review of their impact on insects of cattle dung. Bull Entomol Res 82: 265–274

Sulaiman S, Jeffery J, Sohadi AR, Yunus H, Busparani V, Majid R (1990) Evaluation of Bactimos wettable powder, granules and briquets against mosquito larvae in Malaysia. Acta Trop 47: 189–196

Sumer S, Diril N, Izbirak A (1990) A study on the mutagenicity of some insecticides in the *Salmonella* microsome test system. Mikrobiyol Bul 24: 103–110

Sumer S, Izbirak A, Diril N (1991) Mutagenicity determination of some chlorinated insecticides with Ames test. Doga Turk Muhendislik ve Cevre Bilimleri Dergisi 15: 114–121

Surendranath P, Rao KVR (1991) Kelthane residues in tissues of the tropical penaeid prawn, *Metapenaeus monoceros* (Fabricius), under sublethal chronic exposure: a monitoring study. J App Toxicol 11: 219–222

Szent-Györgyi A (1957) Bioenergetics. Academic Press, New York

Szent-Györgyi A (1963) Chemical physiology of contraction in body and heart muscle. Academic Press, New York

Tabashnik BE (1994) Evolution of resistance to *Bacillus thuringiensis*. Annu Rev Entomol 39: 47–73

Talekar NS, Yang JC (1991) Characteristics of parasitism of diamondback moth by two larval parasites. Entomophaga 36: 95–104

Tanabe S, Falandysz J, Higaki T, Kannan K, Tatsukawa R (1993) Polychlorinated biphenyl and organochlorine insecticide residues in human adipose tissue in Poland. Environ Pollut 79: 45–49

Tasheva M, Hristeva V (1993) Comparative study on the effects of five benzoylphenyl-urea insecticides on haematological parameters in rats. J Appl Toxicol 13: 67–68

Tewari GC, Moorthy KPN (1985) Plant extracts as antifeedants against *Henosepilancha vigintioctopunctata* (F.) and their effects on its parasite. Indian J Agric Sci 55: 120–124

Thybaud E (1990) Ecotoxicology of lindane and deltamethrin in aquatic environments. Rev Sci Eau 3: 195–210

Tietze NS, Hester PG, Hallmon CF, Olson MA, Shaffer KR (1991) Acute toxicity of mosquitocidal compounds to young mosquitofish, *Gambusia affinis*. J Am Mosq Control Assoc 7: 290–293

Tietze NS, Hester PG, Shaffer KR, Prescott SJ, Schreiber E.T. (1994) Integrated management of waste tyre mosquitoes utilizing *Mesocyclops longisetus* (Copepoda: Cyclopidae), *Bacillus thuringiensis* var. *israelensis, Bacillus sphaericus*, and methoprene. J Am Mosq Control Assoc 10: 363–373

Tomlin C (ed) (1995) The pesticide manual. 10th edn. BCPC Crop Protection Publication. Unwin, Old Woking, UK

Topaktas M, Rencuzogullar E (1993) Chromosomal aberrations in cultured human lymphocytes treated with Marshal and its effective ingredient carbosulfan. Mutat Res 319: 103–111

Treacy M, Miller T, Black B, Gard I, Hunt D, Hollingworth RM (1994) Uncoupling activity and pesticidal properties of pyrroles. Biochem Soc Trans 22: 244–247

Tripathy NK, Patnaik KK (1991) Mutagenicity of sumithion tested in *Drosophila* somatic and germ cells. Mutat Res 260: 225–231

Trumble JT (1985) Integrated pest management of *Liriomyza trifolii*: influence of avermectin, cyromazine, and methomyl on leaf-miner ecology in celery. Agric Ecosyst Environ 12: 181–188

Uchida M, Izawa Y, Sugimota T (1987) Inhibition of prostaglandin biosynthesis and oviposition an insect growth regulator, buprofezin, in *Nilaparvata lugens* Stål. Pestic Biochem Physiol 27: 71–75

Van Dalen JJ, Meltzer J, Mulder R, Wellinga K (1972) A selective insecticide with a novel *Mode of Action*. Naturwissenschaften 59: 312–313

Van der Staay M (1991) Side-effects of pesticides on predatory mites. Meded Fac Landbouwwet Rijkuniv Gent 56: 355–358

Van de Veire M (1994) Side-effects of pesticides on the parasitic wasp *Encarsia formosa*: comparison of results from laboratory testing methods with practical experiences in glasshouse vegetables. In: Vogt H (ed) Pesticides and beneficial organisms. IOBC WPRS Bull, vol 17, Montfavet, France, pp 41–47

Van de Veire M (1995) Integrated pest management in glasshouse tomatoes, sweet peppers and cucumbers in Belgium. Doctoral Dissertation. Faculteit Landbouwkundige en Toegepaste Biologische Wetenscappen. Universiteit Gent, Gent

Van de Veire M, Degheele D (1993) Side effects of diafenthiuron on the greenhouse whitefly parasitoid *Encarsia formosa* and predatory bug *Orius niger* and its possible use in IPM in greenhouse vegetables. Meded Fac Landbouwwet Rijkuniv Gent 53: 509–514

Vasuki V (1992) Sublethal effects of hexaflumuron, an insect growth regulator, on some non-target larvivorous fishes. Indian J Exp Biol 30: 1163–1165

Velazquez A, Xamena N, Creus A, Marcos R (1990) Mutagenic evaluation of the organophosphorus insecticides methyl parathion and triazophos in *Drosophila melanogaster*. J Toxicol Environ Health 31: 313–326

Verkerk RHJ, Wright DJ (1993) Biological activity of neem seed kernel extracts and synthetic azadirachtin against larvae of *Plutella xylostella* L. Pestic Sci 37: 83–91

Vigano L, Galassi S, Gatto M (1992) Factors affecting the bioconcentration of hexachlorocyclohexanes in early life stages of *Oncorhynchus mykiss*. Environ Toxicol Chem 11: 535–540

Vlckova V (1991) Mutagenic activity of phosmet and a mixture of solvents applied to yeast *Saccharomyces cerevisiae*. Acta Fac Rerum Nat Univ Comenianae Genet 10: 13–18

Vlckova V, Miadokova E, Podstavkova S, Vlcek D (1992) Mutagenic activity of phosmet, the active component of the organophosphorus insecticide Decemtionate EK 20 in *Salmonella* and *Saccharomyces* assays. Mutat Res 302: 153–156

Vogt H (1992) Studies of the side-effects of insecticides and acaricides on *Chrysoperla carnea* Steph. (Neuroptera: Chrysopidae). Meded Fac Landbouwwet Rijkuniv Gent 57: 559–567

Vogt H (1994) Effects of pesticides on *Chrysoperla carnea* Steph. (Neuroptera, Chrysopidae) in the field and comparison with laboratory and semi-field results. In: Vogt H (ed) Pesticides and beneficial organisms. IOBC WPRS Bull, vol 17, Montfavet, France, pp 71–82

Vogt H, Händel U, Viñuela E (1996) Field investigations on the efficacy of Neem Azal-T/

S against *Dysaphis plantaginea* (Passerini) (Homoptera: Aphidae) and its effects on larvae of *Chrysoperla carnea* Stephens (Neuroptera: Chrysopidae). In: Proc 5th Works Practice oriented results on use and production of neem ingredients and pheromones. Wetzlar (in press)

Voss G, Neumann R (1992) Invertebrate neuroscience and its potential contribution to insect control. In: Duce IR (ed) Neurotox '91molecular basis of drug and pesticide action. Elsevier Applied Science, London, pp vii-xx

Wan MT, Watts RG, Isman MB, Strub R (1996) Evaluation of the acute toxicity to juvenile Pacific Northwest salmon of azadirachtin, neem extract, and neem-based products. Bull Environ Contam Toxicol 56: 432–439

Welling W, De Vries JW (1992) Bioconcentration kinetics of the organophosphorus insecticide chlorpyrifos in guppies (*Poecilia reticulata*). Ecotoxicol Environ Safety 23: 64–75

Wellinga K, Mulder R, Van Dalen JJ (1973) Synthesis and laboratory evaluation of 1-(2,6-disubstituted benzoyl)-3-phenyl-ureas, a new class of insecticides. II. Influence of the acyl moiety on insecticidal activity. J Agric Food Chem 21: 993–998

West AJ, Morcıue AJ (1992) The influence of azadirachtin on the feeding behaviour of cereal aphids and slugs. Entomol Exp Appl 62: 75–79

Whitworth JAG, Hay CRM, McNicholas AM, Morgan D, Maude GH, Taylor DW (1992) Coagulation abnormalities and ivermectin. Ann Trop Med Parasitol 86: 301–305

Williams CM (1967) Third generation pesticides. Sci Am 217: 13–17

Wills LE, Mullens BA, Mandeville JD (1990) Effects of pesticides on filth fly predators (Coleoptera: Histeridae, Staphylinidae; Acari: Macrochelidae, Uropodidae) in caged layer poultry manure. J Econ Entomol 83: 451–457

Wing KD (1988) RH 5849, a nonsteroidal ecdysone agonist: effects on a *Drosophila* cell line. Science (Wash DC) 241: 467–469

Wipfli MS, Merritt RW (1994) Disturbance to a stream food web by a bacterial larvicide specific to black flies: feeding responses of predatory macroinvertebrates. Freshwater Biol 32: 91–103

Wipfli MS, Merritt RW, Taylor WW (1994) Low toxicity of the black fly larvicide *Bacillus thuringiensis* var. *israelensis* to early stages of brook trout (*Salvelinus fontinalis*), brown trout (*Salmo trutta*), and steelhead trout (*Oncorhynchus mykiss*) following direct and indirect exposure. Can J Fish Aquat Sci 51: 1451–1458

Wislocki PG, Grosso LS, Dybas RA (1989) Environmental aspects of abamectin use in crop protection. In: Campbell WC (ed) Ivermectin and abamectin. Springer Berlin Heidelberg New York, pp 182–200

Yokoyama Y, Ohmori I, Suzuki H, Kohda K, Kawazoe Y (1993) Hyperthermic potentiation of bleomycin cytotoxicity in the presence of *Bacillus thuringiensis* ssp. *israelenis* delta-endotoxin. Anticancer Res 12: 1079–1081

Yoshida S, Moriguchi Y, Konishi Y, Taguchi S, Yakushiji T (1992) Analysis of organophosphorus termiticides in human milk. JPN J Toxicol Environ Health 38: 52–56

Zayed SMAD, Amer SM, Nawito MF, Farghaly M, Amer HA, Fahmy MA, Mahdy F (1993) Toxicological potential of malathion residues in stored soybean seeds. J Environ Sci Health Part B Pestic Food Contam Agric Wastes 28: 711–729

Zhang L, Shonu T, Yamanaka S, Tanabe H (1994) Effects of insecticides on the entomopathogenic nematode *Steinernema carpocapsae* Weiser. Appl Entomol Zool 29: 539–547

CHAPTER 12

Management of Resistance to Novel Insecticides

I. Denholm[1], A.R. Horowitz[2], M. Cahill[1] and I. Ishaaya[2]
[1]Department of Biological and Ecological Chemistry, IACR-Rothamsted, Harpenden, Hertfordshire AL5 2JQ, UK
[2]Department of Entomology, Agricultural Research Organization, The Volcani Center, Bet Dagan 50250, Israel

1 Introduction

Taking new insecticides from the laboratory to the marketplace is a time-consuming, costly and decidedly risky process. Long-standing challenges of generating the large body of efficacy, toxicity, persistence and metabolism data required by regulatory authorities are now increasingly supplemented by a need to anticipate and counter the threat of pests rapidly acquiring resistance to new molecules. Faced with the unequivocal fact that no insecticide, however novel, is immune to resistance, far greater emphasis is being placed on evaluating resistance risks prior to the approval of new toxicants, and on formulating recommendations for combating resistance following their introduction. Whether viewed from the standpoint of continued profitability or sustained pest susceptibility, resistance management has become a requirement that agrochemical companies and public sector researchers igonore at their peril.

From previous chapters in this book, it is clear that the potential for discovering and developing novel insecticides has by no means diminished. Indeed, the diversity of insecticides available for controlling key agricultural and public health pests is probably greater now than at any time in the past. Opportunities for minimizing the selection of specific resistance genes through pre-planned alternations and/or restricting the use of specific chemicals should have increased accordingly. Countering this optimism, many of the major targets of new insecticides, especially mites, whiteflies, aphids and lepidopteran pests, have proved extremely adaptable to virtually all insecticides inroduced in the past. Contravention of management guidelines, leading to even localized over-reliance on particular products, can and has hastened the loss of entire classes of pesticide chemistry.

This chapter focuses on challenges faced when confronting the threat of resistance to new insecticides. Our intention is not to deal exsaustively with the theoretical and practical basis of resistance management; this has been addressed in several review articles (e.g. Denholm and Rowland 1992; Roush 1989; Tabashnik 1990) and applies equally to novel agents and to conventional insecticides. Instead, we shall investigate some constrainst and practical considerations with identifying and implementing management tactics for compounds approaching or newly introduced into the insecticide market, drawing on successes and failures encountered with resistance management so far. In keeping with the authors' own research interests, most examples cited refer to pests of agricultural and/or horticultural importance. However, most (if not all) of the principles explored apply to public health and veterinary pests as well.

2 Resistance Risk Assessment

The importance of attempting to evaluate the risk of resistance to new insecticides, in order to pre-empt its appearance in the field, is well established in the resistance literature. There has, however, been considerable debate over both the procedures to be employed for this purpose, and the interpretation and relevance of ensuing data. In Europe, the need to resolve this debate is heightened by the impending introduction of a revised European Union (EU) directive for pesticide approval, which seems certain to include a requirement for data on the status or potential of resistance to new active ingredients. The appropriate guidelines are currently being prepared by the European and Mediterranean Plant Protection Organization (EPPO) and are expected to be finalized and implemented during 1997 (C. Furk, pers. comm. 1996).

Although determined partly by the physico-chemical properties of an insecticide and its mode of action, the risk of resistance depends far more on interactions between the bionomics of target pests and usage of the chemical under consideration. This is exemplified well by the contrasting fates of the insect growth regulator buprofezin applied against the cotton whitefly, *Bemisia tabaci* (Gennadius), in two distinct cropping systems. Introduced into intensive glasshouse horticulture in the Netherlands, buprofezin quickly established itself as the compound of choice against whitefly populations already strongly resistant to conventional insecticide groups. Its over-use against genetically isolated populations in greenhouses led rapidly to reduced efficacy, subsequently attributed to buprofezin resistance (Cahill et al. 1996a). Introduced against more dispersed and mobile field populations on Israeli cotton within the framework of an existing insecticide resistance strategy (see Sect. 4.3), buprofezin has been used only in strict moderation and has retained its effectiveness over several seasons (Horowitz and Ishaaya 1992, 1994).

Factors determining the selection (and hence risk) of resistance to insecticides can, for convenience, be classified into genetic or ecological ones relating to the intrinsic properties of pests and resistance mechanisms, and operational ones relating to the chemical itself and how it is applied (Georghiou 1983). Some of these factors most relevant to new insecticides are considered below.

2.1 Genetic Factors

Resistance can originate through structural alterations of genes encoding target-site proteins or detoxifying enzymes, or through processes (e.g. amplification) affecting gene expression (Soderlund and Bloomquist 1990). In either case, it is generally considered that resistance mutations occur independently of insecticide exposure, and are therefore as likely to occur before an insecticide is introduced as they are during its use in the field.

One major advantage with many of the novel molecules described in this volume is that, in the absence of prior selection, resistance genes may still be present at frequencies approaching mutation rates. Estimates of these vary from 10^{-3} to 10^{-16} depending on the nature of the mutational event (McKenzie 1996). This should assist resistance management considerably, since virtually all proposed management tactics are theoretically far more effective when resistance genes are still rare than when they have reached readily detectable or economically damaging frequencies (Comins 1977; Georghiou and Taylor 1977; Tabashnik 1990). However, one major threat facing new insecticides is that resistance genes are present at frequencies significantly higher than expected at the time of their introduction. The most likely cause of this is cross-resistance from genes selected by compounds already in widespread use against the same target organism(s).

Cross-resistance, which usually reflects either a shared target site or a common detoxification pathway, can usually be rationalized in hindsight, but remains very difficult to predict in advance without supporting experimentation. Even between closely related molecules, levels of resistance conferred by the same mechanism can vary substantially. As an example, target-site resistance to pyrethroids in houseflies (*Musca domestica* L) conferred by the *kdr* (knockdown resistance) allele affects almost all compounds in this class to a similar extent (ca. tenfold resistance). However, resistance due to the more potent *super-kdr* allele is highly dependent on the alcohol moiety of pyrethroid molecules, ranging from ca. tenfold to virtual immunity (Farnham and Khambay 1995a,b). Significant variation in resistance levels to different benzoylphenylureas has also been reported in the leafworm, *Spodoptera littoralis* (Boisduval) (Ishaaya 1992), and in the diamondback moth, *Plutella xylostella* (L) (Furlong and Wright 1994).

Cross-resistance between chemical groups is even harder to anticipate. An Israeli strain of *S. littoralis* with strong resistance to organophosphates and pyrethroids was found to exhibit weak cross-resistance to teflubenzuron, despite its structural and functional dissimilarity to any insecticide used previously (Ishaaya and Klein 1990). Similarly, testing of nicotine-resistant strains of the aphids *Myzus persicae* (Sulzer) and *M. nicotianae* Blackman disclosed consistent cross-tolerance to the novel nitroguanidine, imidacloprid (Devine et al. 1996). This was expressed as lowered mortality of adults, but more markedly as reduced inhibition of feeding by sublethal concentrations of imidacloprid, and is most likely attributable to nicotine and imidacloprid both acting as acetylcholine agonists at the nicotinic acetylcholine receptor in post-synaptic nerve membranes (Bai et al. 1991).

Such considerations highlight a need for rigorous empirical testing to establish the risk of cross-resistance to new insecticides, exploiting strains with isolated and well-characterized resistance mechanisms and/or recently collected field strains representative of resistance in contemporary populations. Lack of cross-resistance, as documented for laboratory and field strains of *M. domestica* tested with diflubenzuron and cyromazine (Keiding et al. 1991), field strains of *Bemisia tabaci* tested with diafenthiuron and pymetrozine (Denholm et al. 1995), and a pyriproxyfen-resistant strain of *B. tabaci* tested with buprofezin and diafenthiuron (Ishaaya and Horowitz 1995), is clearly an encouraging result that enables novel compounds to be deployed more confidently and managed more effectively. Confirmation of cross-resistance by no means precludes the release of a new molecule, but forewarns of a potential problem and the need for careful deployment in relation to cross-resisted chemicals.

Problems encountered with the development of fipronil and triazamate exemplify this well. By blocking GABA-regulated chloride channels, fipronil mimics the mode of action of cyclodiene insecticides, and is therefore potentially vulnerable to a widespread gene conferring target-site resistance to cyclodienes in insects (ffrench-Constant et al. 1993; Thompson et al. 1993). Indeed, tests against susceptible and cyclodiene-resistant strains of *M. domestica* disclosed ca. 90-fold resistance to fipronil in the latter (Colliot et al. 1992). However, in other major pests this association has proved less or not apparent (N. Hamon, pers. comm. 1996), implying that fipronil should retain its efficacy against some populations intensively exposed to cyclodienes in the past. Triazamate is a novel acetylcholinesterase (AChE) inhibitor active against several economically important aphids including *M. persicae*. In laboratory and field trials, it retained good efficacy against *M. persicae* populations possessing a widespread resistance mechanism based on overproduction of a carboxylesterase, and conferring broad-spectrum protection against organophosphorous, carbamate and pyrethroid insecticides (Dewar et al. 1988; Murray et al. 1988). During the last 2 years, however, a new resistance mechanism based on insecticide-insensitive AChE has been identified in *M. persicae*. This confers particularly strong resistance to pirimicarb, and a high level of cross-resistance to triazamate (Moores et al. 1994). The prospect of exploiting triazamate to control this pest in the future will depend critically on the occurrence and spread of the new mechanism. This in turn is likely to reflect the scale of pirimicarb usage in different countries.

Another approach available for investigating genetic risks is to subject pest populations to selection with new insecticides in the laboratory, and monitor at regular intervals for changes in response to the selecting agent. Aside from disclosing genetic variation within the selected population, this has the potential attraction of providing access to novel resistance genes at a sufficiently early stage for work on their dominance, expression and fitness to contribute to managing them under field conditions. Experiments of this kind have yielded varied results. On the one hand, repeated selection of a field strain of *B. tabaci* with imidacloprid led within 1 year to a ca. 50-fold reduction in susceptibility, the first demonstration of resistance to this chemical in any pest species (Prabhaker et al. 1995). In contrast, selection of *M. domestica* strains with diflubenzuron and cyromazine at concentrations causing 56–94% mortality per generation caused no significant increase in tolerance to either chemical (Keiding et al. 1992).

A major drawback with selection experiments is that both positive and negative results are vulnerable to conflicting interpretations, particularly in view of the vested interests involved. A positive response can be construed as an artifact of exposure conditions unlikely to be encountered in practice. This argument has some scientific support, since it is possible for repeated laboratory exposure to concentrations within the tolerance range of an experimental population to generate polygenic mechanisms less likely to evolve under the harsher selection regimes imposed in the field (Whitten and McKenzie 1982). A negative response merely implies that the selected population lacked the appropriate resistance gene(s), which is hardly surprising given their probable rarity prior to field selection. If conducting such experiments, every effort must be made to maximize genetic variation by initiating populations with large numbers of insects sampled from as wide a geographical area as possible. Long-standing laboratory strains whose gene pool has been depleted through founding effects and/or rearing bottlenecks are extremely unlikely to yield any relevant information at all.

A simpler and less ambiguous approach to studying the genetic potential for resistance, already adopted successfully to isolate cyclodiene-resistant mutants of *Drosophila melanogaster* Meigen (ffrench-Constant et al. 1990) and pyriproxyfen-resistant mutants of *B. tabaci* (Ishaaya and Horowitz, 1995), is to screen large numbers of field-caught insects at concentrations or doses exceeding the tolerance range of existing populations. Any survivors are likely to possess at least one major gene for resistance, and could form the nucleus of an experimental resistant colony.

2.2 Ecological Factors

The importance of incorporating ecological factors into resistance risk assessment is self-evident from a large number of studies implicating the dynamics, bionomics and dispersal capabilities of pest organisms as primary determinants of resistance development (reviews by Roush and McKenzie 1987; Denholm and Rowland 1992; McKenzie 1996). Collectively, these studies leave no doubt as to the ecological conditions under which resistance is expected to develop fastest. In the agricultural sector, the 'worst case scenario' would be a species that is relatively sedentary, monophagous, highly fecund, and a serious pest of a crop of considerable commercial importance. One insect meeting these criteria to a large extent is the damson-hop aphid, *Phorodon humuli* (Shrank), which is now the most damaging pest of hops in the Paleoarctic region (Hrdy et al. 1986). During the summer, *P. humuli* is virtually restricted to wild and cultivated hops, on the latter of which insecticides remain essential for effective aphid control. The result of repeatedly exposing a large proportion of *P. humuli* individuals present in any region to a limited range of chemicals has been to select for strong resistance to almost all available aphicides (Lorriman and Llewellyn 1983; Lewis and Madge 1984; Hrdy et al. 1986). Wild hops acting as untreated refugia have clearly contributed little to retarding the buildup of resistance. Indeed, the primary role of the holocyclic life-cycle, whereby aphids retreat to primary hosts (*Prunus* spp.) during the winter and redistribute the following spring, appears to have been to enhance the spread of resistance to both wild and cultivated hop varieties (Hrdy et al. 1986).

The ecology of *P. humuli*, coupled with its history of resistance development, identifies this species as presenting a severe resistance risk against which any new insecticide (e.g. imidacloprid) should be used with utmost care. The major aphid pests of cereals in Europe, including *Sitobion avenae* (F) and *Rhopalosiphum padi* (L), appear to present a contrasting situation in which migration between treated and untreated hosts has effectively prevented the evolution of resistance. Despite a long history of control with insecticides in the UK, these species have retained susceptibility in bioassays (Furk et al. 1983), and no field control failures have been reported.

Ecological differences with a major bearing on resistance development occur even between closely related and coexisting species. Of the two bollworms attacking Australian cotton, only *Helicoverpa armigera* (Hübner) has acquired strong resistance to insecticides. *Helicoverpa punctigera* Wallengren, despite being an equally important pest and occurring alongside *H. armigera* on several intensively treated crops, has remained fully susceptible to all chemical groups (Forrester et al. 1993). The most plausible explanation is that *H. punctigera* occurs in greater abundance on a far larger range of unsprayed hosts than *H. armigera* (Zalucki et al. 1986; Fitt 1989). This has enabled any resistance genes present to be swamped by susceptible insects inhabiting refugia, thereby precluding a significant increase in their frequency on treated crops (Forrester et al. 1993).

Relatively subtle differences in pest ecology can sometimes have profound implications for resistance development by the same species. In the cotton/ vegetable cropping systems of the arid southwestern USA, *B. tabaci* has recently become a devastating pest and a major target of insecticide sprays. However, the consequences in terms of resistance have varied substantially on a regional basis. Resistance problems have proved much more severe and persistent on cotton in southcentral Arizona than in the extreme southwest of this State and the adjacent Imperial Valley of California (Castle et al. 1996; Dennehy et al. 1996). It is hypothesized that this is attributable to the higher proportion of unsprayed hosts, especially alfalfa, acting as a buffer to resistance in the latter areas.

Although very selective, these few examples serve to illustrate some challenges of relating resistance risks to pest bionomics, and of tailoring resistance management recommendations accordingly. The implications of the aphid example are the most clear-cut, since there is no doubt that any new insecticide must be managed much more stringently to sustain its efficacy against hop aphids than against cereal aphids. Since usage recommendations are generally specified on a crop-by-crop basis, this should be feasible in practice provided a sufficient supply of alternative and effective aphicides is available. Differences in resistance risks associated with the two *Helicoverpa* species imply less need to restrict the use of a new chemical against *H. punctigera* than against *H. armigera*. In practice this is almost impossible due to their coincidence on the same crops and the extreme difficulty of distinguishing the species other than as adults. Thus, in this case the only viable option is to direct management tactics formulated for the most 'at risk' species (i.e. *H. armigera*) against the bollworm complex as a whole. For *B. tabaci* in Arizona, there may be scientific support for adjusting management guidelines to reflect regional variation in resistance risks, but this would be a very difficult concept to convey convincingly to grower communities. Again, the most workable approach to managing new whitefly insecticides, as adopted recently in the USA (see Sect. 4.3), is to formulate recommendations appropriate to the most vulnerable part of the species' range and to implement these on a Statewide basis.

Enclosed environments such as greenhouses and glasshouses, which preclude migration and escape from insecticide exposure under climatic regimes favouring rapid and continuous population growth, provide the most ideal ecological conditions of all for selecting resistance genes (Parrella 1987; Sanderson and Roush 1995). Very low or zero damage tolerance thresholds for high value ornamental or vegetable produce accentuate the problem by promoting over-frequent spraying and hence intensifying selection for resistance. Over the years these environments have, not surprisingly, proved potent sources of novel resistance mechanisms for a diverse range of control agents. As an example, resistance to buprofezin (Horowitz and Ishaaya 1994; Cahill et al. 1996a), pyriproxyfen (Horowitz and Ishaaya 1994) and imidacloprid (Cahill et al. 1996b) in *B. tabaci* were all first documented in protected horticulture. An added complication is that extensive national and international trade in glasshouse produce risks transferring highly resistant insects over long distances and into field populations (Frey 1993; Denholm et al. 1996). Over-use of insecticides against *M. domestica* and other livestock pests in intensive animal units, accompanied by the dispersal of resistant individuals from these sites in the spring and summer (Denholm et al. 1985), represents an analogous problem in the veterinary pest control sector.

Protected cropping systems still heavily dependent on chemical control probably constitute the most serious and intractable threat to the continued effectiveness of new insecticides. When reviewing options for resistance management in glasshouses, Sanderson and Roush (1995) emphasized the extreme importance of adopting cultural and biological control practices to minimize growers' reliance on insecticides. This message has been aired on many occasions, but with such limited success that one questions whether highly valued molecules should be registered for use on protected crops other than within the framework of an established and clearly defined integrated control strategy. Denying glasshouse growers unrestricted access to novel classes of insecticide chemistry may seem a radical step, but could have the beneficial consequence of forcing the implementation of relatively simple and well proven non-chemical tactics that are far less exploitable by those attempting to combat resistance development under open field conditions.

2.3 Operational Factors

Although closely linked to aspects of pest genetics and ecology, operational factors are best distinguished as ones which, in principle at least, are fully at an operator's discretion and hence directly exploitable for combating resistance. Much of the work conducted on this subject is still largely theoretical but nonetheless very relevant to new insecticides since it focuses on instances when resistance genes are still rare and more likely to occur in heterozygous than homozygous conditions (reviews by Roush 1989; and Tabashnik 1990; Denholm and Rowlund 1992; McKenzie 1996). Factors that potentially exert a major influence in this respect include the rate, method and frequency of applications, their biological persistence, and whether insecticides are used singly or as mixtures of active ingredients.

Equating operational factors with resistance risks for novel molecules is fraught with difficulty, since at the time of their introduction it is impossible to test empirically many of the assumptions on which population genetics models of resistance development are based. Anticipating the selection pressure imposed by a particular application dose is a case in point. If resistance alleles are present, the only entirely non-selecting doses will be ones sufficiently high to overpower all individuals, regardless of their genetic composition, or ones sufficiently low to kill no insects at all (Roush and Daly 1990). The latter is obviously a trivial option. Prospects of achieving the former depend critically on the potency and dominance of (as yet) unrecognized and uncharacterized resistance mechanisms. A pragmatic approach to this dilemma is to set application doses as far above the tolerance range of susceptible insects as economic and environmental constraints permit, in the hope that at least heterozygotes will be effectively controlled (McKenzie 1996). Even this approach can backfire badly if resistance turns out to be more common than suspected (resulting in the presence of resistant homozygotes) or resistance alleles exhibit a high degree of dominance (Tabashnik and Croft 1982; McKenzie 1996; Roush 1989). Unless a high proportion of insects escape exposure altogether, the consequence could then be to select very rapidly for strong resistance, despite creating a temporary impression of outstanding control efficacy. Such uncertainties imply that recommended application doses for new insecticides should continue to be based, as at present, on field efficacy, cost and environmental toxicity rather than on considerations of resistance management *per se*.

In practice, concerns over optimizing dose rates to avoid resistance are secondary to ones regarding the application process itself. Delivery systems and/or habitats promoting uneven or inadequate coverage will generally be more prone to selecting for resistance since pests are more likely to encounter exposure conditions under which selection is most intense. This was elegantly demonstrated through experiments assessing the relative survival of endosulfan-susceptible and -resistant phenotypes of the coffee berry borer (*Hypothenemus hampei* (Ferrari) in coffee plantations treated with this chemical in New Caledonia, southwest Pacific (Parkin et al. 1992). The practice of spraying plantations from roadsides with vehicle-mounted mistblowers generated gradients in the concentration of endosulfan that resulted in differential selection pressures in different parts of a field. Employing alternative technology to ensure higher and more uniform coverage may not have prevented resistance entirely in this case, but could have contributed to delaying its development. Similarly, underdosing with phosphine in inadequately sealed grain stores has been implicated as a primary cause of resistance to this chemical in a range of stored product pests (Taylor and Halliday 1986).

The timing of insecticide applications relative to the life cycle of a pest can also be a critical determinant of selection pressure. The best documented example of this relates to the selection of pyrethroid resistance in *H. armigera* in Australia. On cotton foliage freshly treated with the manufacturer's recommended dose, pyrethroids were shown to kill most larvae up to 3–4 days old irrespective of whether they were resistant or not by laboratory criteria (Daly et al. 1988; Daly 1993). The greatest discrimination of phenotypes occurred when larvae achieved a threshold size. Targeting of insecticides against neonates, as is generally advocated for bollworm control, not only increases the likelihood of contacting larvae at the most exposed stage in their development, but also offers the greatest prospect of retarding resistance by overpowering its expression. Other reports of age-, life stage- or size-dependent expression of resistance to organophosphates in the leafworm, *S. littoralis* (Dittrich et al. 1980), to pyrethroids in the Colorado beetle, *Leptinotarsa decemlineata* (Say) (Follet et al. 1993) and the tobacco budworm, *Heliothis virescens* (F) (Roush and Luttrell 1989), and to pyriproxyfen in *B. tabaci* (Horowitz and Ishaaya 1994) enhance the importance of considering the implications of pest population structure when assessing resistance risks and deploying new insecticides.

From the standpoint of resistance management, pesticide persistence is commonly viewed as a necessary evil. Some degree of persistence is often essential to ensure an acceptable standard of control, especially when contending with disease vectors or continued invasion of crop pests from alternative host plants. On the other hand, it is almost intuitive that persistent applications accentuate resistance risks by exposing a larger number of individuals for longer periods to the selecting agent. This conclusion is supported by theoretical (eg. Mani and Wood 1984) and experimental (e.g. Denholm e.t al. 1983) studies showing residual treatments to select more effectively than ones that give immediate control but then dissipate rapidly.

The effects of persistence are, in reality, more complex than this simplistic argument suggests and deserve further scrutiny. The first of two major resistance risks persistence imposes is that selection may continue to operate long after population densities have been reduced to an acceptable level and control is no longer required (Roush 1989). Since it is generally more feasible to prolong the intrinsic persistence of a chemical (e.g. by developing photostable formulations) than to reduce it, the most effective way to address this concern is to distinguish pest problems that demand extended control from those that do not, and to tailor insecticide use recommendations accordingly. The second risk relates less to persistence *per se* than to the fact that residual applications decay or become diluted to concentrations that can select for resistance more strongly than the original application dose. This again has empirical support; studies on mosquitoes (Rawlings et al. 1981), bollworms (Daly et al. 1988) and blowflies (McKenzie and Whitten 1984), among others, confirmed that aged deposits discriminate more readily between genotypes or phenotypes than ones freshly applied. The critical factor determining this risk is the treatment 'decay curve' (McKenzie, 1996), defined in terms of its biological effects rather than the amount of chemical detectable by residue analysis. Unlike most effects of operational parameters, this can at least be anticipated in advance of resistance appearing by measuring the survival of susceptible insects exposed to ageing deposits at varying time intervals (Roush 1989). Applications resulting in a protracted and very gradual decline in the kill of test insects arguably pose the greatest resistance risk.

One new insecticide for which these considerations are particularly relevant is imidacloprid. Its efficacy (and availability) against crop pests as both a long-lasting systemic treatment and as a shorter-lived foliar spray (Elbert et al. 1990, 1996) has generated much speculation as to which mode of application is more likely to promote resistance. Although there are still insufficient data to resolve this debate categorically, it is nonetheless instructive to consider factors on which the answer depends. The versatility of imidacloprid can be viewed as an advantage or as a drawback, according to one's perspective. It is advantageous in that imidacloprid treatments can be matched more precisely than is usually possible to the needs of different cropping systems. In cases where homopteran or coleopteran pests are a predictable and persistent early season problem, use of imidacloprid as a seed treatment or soil application appears justified and should relieve pressure on chemicals required later in the season. Its comparatively low impact on natural enemies when applied in this manner can also contribute to delaying any subsequent sprays and may alleviate the risk of resistance to imidacloprid itself. However, the inherent risk of selecting for resistance with such prolonged exposure remains considerable, and must ultimately be judged in the light of pest ecology—in particular the likely proportions of individuals on treated and untreated hosts. For pests that are more erratic, or are only damaging for short time intervals, prophylactic systemic treatments are best avoided in favour of foliar sprays applied when insect numbers exceed defined density thresholds. Treatment for treatment, foliar sprays probably present the lower resistance risk, although repeated spraying of imidacloprid against the same pest is likely to negate this advantage entirely (Cahill et al. 1996b).

The drawback with imidacloprid (and any equally versatile chemical) is that without strict regulation it can be perceived by growers as a 'cure-all', offering continuous control through a succession of systemic and foliar applications. When reviewing management options for imidacloprid, Elbert et al. (1996) correctly placed greatest emphasis on avoiding this scenario. Whatever the relative risks arising from the two modes of application, the risk from combining these approaches is both unacceptable and unnecessary.

One of the prerequisites for mixing insecticides to combat resistance is that genes for resistance to each ingredient must still be rare, so that individuals possessing both traits are unlikely to occur at all (Comins 1986; Curtis 1985; Tabashnik 1989). Novel (or new) chemicals are clearly more amenable to this tactic than ones to which resistance may already have been selected. However, there are several other operational, economic and political factors affecting resistance risks with mixtures and/or constraining their use in practice. The underlying principle of 'redundant killing' (Comins 1986; Roush 1989) not only requires genes to be rare, but also depends critically on the ingredients conferring mutual protection through the effective life of an application. Failure to ensure that they exhibit similar biological persistence risks exposing one compound to more intense selection than the other, thereby accelerating the selection of doubly resistant phenotypes. Two potentially conflicting challenges of choosing ideal mixture partners are therefore to (1) ensure maximum similarity in efficacy and persistence against the target pest(s), and (2) ensure maximum dissimilarity in chemical structure and function to minimize the likelihood of cross-resistance. More practical constraints on mixtures include their added cost, (potentially) greater environmental impact, difficulty of registration, and contravention of current marketing philosophies, especially since the best-suited partners are more likely than not to be owned by different manufacturers. We also recognize the additional challenge of recommending particular combinations for combating resistance, while continuing to warn of the dangers of exploiting mixtures generally for this purpose. For these reasons, mixtures of toxicants appear unlikely to assume a significant role for preventive resistance management in the near future, despite their potential to reduce resistance risks under specialized circumstances.

3 Challenges with Resistance Monitoring

Accurate and regular monitoring for changes in susceptibility is essential for anticipating resistance problems and for assessing the effectiveness of resistance management tactics (Denholm 1990; ffrench-Constant and Roush 1990). It is highly desirable to evaluate, define and standardize test methods for novel insecticides prior to their release in the field. This requirement is now receiving increasing attention from agrochemical companies who, alone or in collaboration with public sector scientists, are playing a leading role in developing and advocating methods appropriate to particular products, and in training staff in their implementation (e.g. Leonard and Perrin 1994; Denholm et al. 1995; Elbert et al. 1996; Wege, 1996). Two aspects the monitoring of immediate relevance to new insecticides are: (1) the choice of an appropriate bioassay method and (2) ways of improving the sensitivity and capacity of such tests.

3.1 Design of Bioassays

To be fully effective, monitoring tests must aspire to several criteria relating to their precision and/or practical utility. They should ideally be as rapid and simple as possible to encourage their consistent use by staff who may lack access to specialized apparatus or sophisticated holding facilities. They should yield repeatable results with an unambiguous endpoint, preferably based on death since sublethal responses such as 'moribund' or 'affected' are susceptible to subjective interpretation (Welty et al. 1988). Whenever possible they should also be targeted against life-stage(s) in which resistance would be of greatest concern in the field, since there is no *a priori* guarantee that economically significant resistance will be equally manifest at all stages of development (e.g. Horowitz and Ishaaya 1994). Since the expression of resistance can also vary substantially between exposure methods (e.g. Dennehy et al. 1983), it is also important to anticipate which bioassay is likely to discriminate most effectively between susceptible and resistant phenotypes. In practice this is difficult to predict before resistance has been documented and its expression studied in detail. However, attributes of a test method such as the steepness of dose-response relationships can be useful indicators of discriminatory power (Dennehy et al. 1993; ffrench-Constant and Roush 1990).

Most importantly of all, test methods need to be designed to contend with the toxicological, physico-chemical and behavioural characteristics of the chemical under consideration. Novel insecticides can prove particularly challenging in this respect since they often, by definition, exhibit unusual features or elicit unconventional symptoms rendering traditional bioassays inappropriate. Problems with evaluating some novel compounds against *Bemisia tabaci* illustrate this point well. Since most chemicals used to control *B. tabaci* in the past have exhibited rapid contact activity against adults, a residual bioassay exposing adults for a maximum of 72 h to leaf discs dipped in formulated insecticide has proved of broad applicability (Cahill et al. 1995; Dittrich et al. 1985). The advent of the insect growth regulators buprofezin and pyriproxyfen, neither of which inflict direct mortality on whitefly adults (Ishaaya et al. 1988; Ishaaya and Horowitz 1992), has necessitated the exposure and monitoring of eggs or nymphs to disclose resistance to these agents (Cahill et al. 1996a; Horowitz and Ishaaya 1992, 1994). Tests involving nymphs in particular take much longer (ca. 15 days) than ones against adults since the cumulative effects on nymphal development are not reliably apparent until the pre-adult stage.

Even compounds with pronounced activity against adults, e.g. imidacloprid, diafenthiuron and pymetrozine, can prove problematic. The unconventional effects of imidacloprid on whitefly behaviour in leaf-dip bioassays have led to erratic and inconsistent results, most effectively overcome by exposing adults to foliage treated systemically with imidacloprid either through plant roots or the petiole of an excised leaf (Cahill et al. 1996b; Elbert et al. 1996; Prabhaker et al. 1995). Equivalent problems encountered with aphids were circumvented by immersing adults in solutions of imidacloprid and transferring them to untreated foliage, thereby yielding repeatable estimates of mortality after a 48-h holding period (Devine et al. 1996).

The toxicity of diafenthiuron against insects depends on desulphuration in the presence of light to a carbodiimide derivative (CGA 140408) that is considered to inhibit ATP-ase activity in mitochondria (Ruder et al. 1991). The poor repeatability encountered when testing this chemical against *B. tabaci* most likely reflected the variable and inefficient photoconversion achieved under laboratory lighting regimes (Denholm et al. 1995). This difficulty was overcome effectively by conducting bioassays following exposure of treated plants to sun (Ishaaya et al. 1993), or with the carbodiimide itself instead of the parent compound, a modification that has proved equally applicable to aphids and spider mites (unpubl. data). Since photoconversion of diafenthiuron is considered to occur on leaves before pickup by insects, monitoring with CGA 140408 is unlikely to compromise detection of resistance to the marketed product.

Pymetrozine is thought to act primarily by suppressing stylet penetration and hence feeding by whiteflies and other homopteran pests, leading to protracted death through starvation (Kayser et al. 1994). Testing of pymetrozine against *B. tabaci* and the cotton aphid, *Aphis gossypii* Glover, showed a holding period of at least 96 h (and preferably 120 h) to be essential for obtaining reliable dose-response data for whitefly adults or aphid nymphs (Denholm et al. 1995). Results after shorter periods were very erratic and could readily yield spurious indications of resistance to this chemical.

3.2 Monitoring Procedures

In addition to developing bioassays, attention has been paid to the statistical design of monitoring programmes (Halliday and Burnham 1990; Martinson et al. 1991; Roush and Miller 1986; Sawicki et al. 1989), and to obtaining representative samples from field populations (e.g. Dennehy and Granett 1984). These aspects are again especially relevant to new insecticides, since the statistical and logistical challenges of detecting incipient resistance are far greater than when resistance is well established and causing obvious shifts in response to insecticides.

Full dose-response assays are invaluable for assessing the precision and repeatability of a bioassay method, for quantifying and comparing baseline responses of laboratory and field strains, and for investigating the occurrence of cross-resistance. Once resistance monitoring is underway, however, this approach constitutes a wasteful and inefficient use of resources. A low frequency of resistant phenotypes may have an imperceptible effect on parameters of fitted probit lines, and will be detected much more readily by testing as many insects as possible at one or more diagnostic doses or concentrations near the upper tolerance limit of susceptible individuals (ffrench-Constant and Roush 1990; Roush and Miller 1986). The choice of diagnostic concentrations for novel insecticides demands a degree of pragmatism, since it is impossible to anticipate the potency of any resistance that may subsequently appear. One compromise at this stage is to utilize two concentrations that collectively overcome possible pitfalls with interpretation. A concentration lying between the LC_{90} and LC_{99} of baseline populations is well placed to anticipate low-level resistance but, if used alone, is vulnerable to temporal or spatial variation in response unconnected to resistance (Martinson et al. 1991). A second concentration, five- to ten-fold higher than the first, guards against this possibility and should still be sufficiently low to disclose 'true' resistance in the majority of cases.

No aspect of a monitoring programme should be regarded as immutable, since there will undoubtedly be opportunities to fine-tune procedures in the light of experience or events. Confirmation of resistance, followed by even preliminary studies of its potency and expression, offers substantial scope to reassess or refine monitoring criteria. Investigation of the first report of pyriproxyfen resistance in *B. tabaci* quickly showed resistance to be much more strongly expressed as enhanced egg hatch (550-fold at LC_{50}) than as increased pupal survival and adult formation (ten-fold) (Horowitz and Ishaaya 1994). Bioassays against eggs therefore offer a more sensitive (and rapid) means of monitoring the occurrence and spread of this mechanism. Preliminary research into the expression of resistance may also enable the development and promotion of simpler and more 'user-friendly' bioassays optimized specifically for resistance detection (e.g. Dennehy 1987). One increasingly popular approach to resistance monitoring that will remain much less applicable to novel insecticides than to conventional products is the use of *in vitro* diagnostics for particular resistance mechanisms (Brown and Brogdon 1987; ffrench-Constant and Roush 1990). These not only require detailed, long-term research into the biochemical and/ or molecular biological nature of resistance, but also they are hampered by ignorance of the exact mode of action of most novel insecticides.

4
Designing and Implementing Management Strategies

As stressed by Roush (1989), outlining a 'best-bet' resistance management strategy for a new insecticide need not be a complex process. Only relatively few of the resistance management tactics that have been proposed are sufficiently risk-free and intuitive to stand a reasonable chance of success in the majority of circumstances. Foremost among these are: (1) restricting the number of applications over time and/or space; (2) creating or exploiting refugia; (3) avoiding unnecessary persistence; (4) alternating between chemically unrelated molecules; and (5) ensuring that each treatment is applied as strategically as possible against the most vulnerable stage(s) in a pest's life cycle. In all cases, the process of weaving these tactics into a coherent strategy will benefit from a sound knowledge of pest population biology, which can be studied in advance of resistance appearing and is applicable to conventional and novel insecticides alike.

Based on experience to date, the greatest challenge with managing resistance is not the identification of appropriate resistance countermeasures, but the task of ensuring compliance with recommendations from growers and pest control operatives. Factors constraining the implementation of resistance management are discussed elsewhere (e.g. Forrester 1990; Sawicki and Denholm 1987; Wege 1996) and will not be reiterated in this chapter. It should be appreciated, however, that well-recognized difficulties with communication and education are not peculiar to resistance management; they limit all pest management reforms to the same extent. In the case of resistance, the primary constraint is probably not the challenge of extending advice on its management, but a widespread lack of appreciation of the problem in the first place. The number of growers, advisors and consultants who remain poorly informed of resistance and its consequences, even in highly developed countries, highlights a continued failure by researchers and industrialists to convey the gravity of the threat through all accessible channels (e.g. television, radio and popular publications) and in layperson's terms. This situation needs to be rectified as a matter of urgency. It is certainly no coincidence that the greatest successes with resistance management have occurred within an infrastructure promoting (and resourcing) extensive and direct communication with growers and pest management consultants (e.g. Forrester 1990; Horowitz et al. 1994).

4.1
Contribution of the Agrochemical industry

One significant recent development has been the increasing commitment of agrochemical producers, individually or through their Insecticide Resistance Action Committee (IRAC), to confronting resistance problems. The concern of individual companies is apparent from several contributions to this book by product manufacturers referring explicitly to the importance of proactive monitoring and resistance management. IRAC's research and extension activities have been reported regularly as this organization has evolved technically and expanded geographically (Jackson 1986; Leonard and Perrin 1994; Voss 1988; Wege 1996). Unlike single companies, IRAC can potentially influence the deployment of chemicals produced by several manufacturers to address and minimize cross-resistance risks. The best example to date relates to the use of acaricides against spider mites on fruit crops in Europe. Recommendations for alternating conventional acaricides based on the likelihood of cross-resistance (Lemon 1988; Sterk and Highwood 1992) have recently been extended to encompass a new class of four mitochondrial electron transport inhibitor (METI) acaricides (tebufenpyrad, fenazaquin, fenpyroximate and pyridaben), almost certainly acting at the same target site, and introduced simultaneously into the European market by different manufacturers or distributors. Dialogue between these companies has led to standardized protocols for resistance monitoring and a consensus to recommend only one application of only one METI compound per crop per season (Leonard and Perrin 1994; Wege 1996). This agreement, reached in the face of fierce competition for the same market 'niche', represents an outstanding level of inter-company collaboration to safeguard new molecules from the threat of resistance.

The task of managing resistance to the chloronicotinyl insecticides has some interesting parallels with that for METI acaricides, but also differs in important respects. In this case, one member of the class (imidacloprid) has already been introduced and gained widespread use against homopteran and coleopteran pests in many parts of the world. Others (e.g. acetamiprid; see Ishaaya and Horowitz, this Vol.) seem destined to be released for a similar range of pests in the near future. Although it is premature to assume a common site of action, the structural resemblance of chemicals announced to date implies a strong probability of cross-resistance if conferred by target-site modification or a shared detoxification pathway. Recognizing this threat, the manufacturers of imidacloprid have advocated that their preliminary guidelines for managing this chemical be extended to all chloronicotinyls, in order to conserve the effectiveness of the class as a whole (Elbert et al. 1996). In practice, this ideal may be more vulnerable to commercial conflict than was the case with the acaricides, since newcomer molecules will initially be competing for part of a market already dominated by imidacloprid. Whether or not this perception is correct, management of the chloronicotinyls would appear to present the collaborative spirit of IRAC with its sternest test so far.

4.2
Importance of Chemical Diversity

The most important principle of any management strategy, that of avoiding prolonged selection of the same resistance mechanism, demands access to a range of distinct chemicals that can be alternated with a periodicity of at least one pest generation. Provided that other effective insecticides are still available, introducing a new molecule should enhance the prospects of resistance management by increasing the diversity of chemicals available for this purpose. The corollary to this statement is that if existing compounds have already been rendered ineffective due to resistance, a single new insecticide is under dire threat of being over-used and hence succumbing rapidly to resistance in its own right.

The release of a new insecticide into an environment already beset with resistance problems unquestionably presents the greatest management challenge of all. The primary factor promoting excessive reliance on (and therefore resistance to) buprofezin against *B. tabaci* and *Trialeurodes vaporariorum* (Westwood) in the Netherlands (Cahill et al. 1996a; De Cock et al. 1995) was a lack of effective alternatives; the salvation that buprofezin offered to glasshouse growers outweighed the distributor's recommendations to restrict its usage. Similarly, the introduction of benzoylphenylureas to control multi-resistant *P. xylostella* in some areas of Southeast Asia led, within 18 months to extensive control failures caused by resistance and the virtual loss of this novel and invaluable class of insecticide chemistry (Cheng et al. 1991). These and numerous other examples highlight a fundamental dilemma confronting researchers, insecticide producers and, increasingly, registration authorities, i.e. those situations in which new chemicals are most urgently needed (and potentially most profitable) are ones in which resistance is most likely to be selected rapidly. The only effective and long-term solution in such cases is to coordinate the introduction of at least two distinct molecules and take every measure to ensure their complementary use from the outset. Given the number of new insecticides currently becoming available, this is not an impossible ideal. It is exemplified, as outlined below, by recent progress with whitefly control in the USA, which in turn was inspired by the approach pioneered to manage new insecticides against *B. tabaci* on cotton in Israel.

4.3 Preventive Resistance Management in Israel and the USA

The current cotton resistance management in Israel, first implemented in 1987, has been described in detail elsewhere (Horowitz et al. 1993, 1994, 1995). Although it incorporates a range of chemical and non-chemical countermeasures, one primary objective has been to conserve the effectiveness of novel insecticides, especially pyriproxyfen and buprofezin against *B. tabaci*, and the benzoylphenylureas against the Egyptian leafworm, *S. littoralis*. Under recommendations coordinated by the Israeli Cotton Board and promoted through extensive education programmes and insecticide pricing incentives, these chemicals are restricted to a single application per season within an alternation strategy optimized to contend with the entire pest complex and exploit natural enemies to the greatest extent possible. Regular monitoring initiated at the time of their introduction has disclosed no significant directional increase in tolerance of *S. littoralis* to benzoylphenylureas or of *B. tabaci* to buprofezin, despite minor within-season shifts in response to the latter (Horowitz and Ishaaya 1994). In most areas the response of *B. tabaci* to pyriproxyfen has also remained stable, despite the proven potential of whiteflies to develop strong resistance to this chemical in glasshouses and the laboratory (Horowitz and Ishaaya 1994). In 1995 and 1996, however, there were localized outbreaks of pyriproxyfen resistance on cotton that are presently being investigated further.

One major achievement of the Israeli strategy has been a dramatic decrease in the number of insecticide applications against the whole range of cotton pests, but especially against *B. tabaci*. Sprays against whiteflies now average less than 2 per growing season compared with over 14 per season in 1986 (Horowitz et al. 1994). Through integrated pest management and the strategic use of novel insecticides, Israel has escaped the pesticide 'treadmill' that threatened the future of its cotton industry in the early 1980s, and in so doing has reduced *B. tabaci* to a much less important and fully manageable component of the pest complex. Most importantly of all, the strategy has created the ideal environment for introducing additional new insecticides to contend with changing circumstances, e.g. if further restrictions on the use of pyriproxyfen prove necessary.

Drawing extensively on the Israeli experience, researchers, extension staff and cotton industry officials in the southwestern USA achieved in 1996 an unprecedented success with the introduction and regulation of new insecticides for use on cotton in that country. In so doing they took a substantial step towards combating a whitefly problem that appeared intractable as recently as the previous season. In 1994, the most effective control treatments against *B. tabaci* were certain combinations of pyrethroids and organophosphates, one of which (fenpropathrin plus acephate) formed the mainstay of whitefly control in 1994 (Dennehy et al. 1995), but which also became strongly resisted during 1995 (Dennehy et al. 1996). Faced with potential disaster, and recognizing the danger of releasing and relying on a single new insecticide, the cotton industry in collaboration with the respective manufacturers of buprofezin and pyriproxyfen applied to the Environmental Protection Agency (EPA) for a Section 18 Emergency Exemption to approve both chemicals simultaneously (Dennehy et al. 1996). This application was justified to the EPA on the grounds that limited use of these two products in alternation would assist with sustaining the efficacy of both and reduce selection for resistance to other insecticides used on cotton (P.C. Ellsworth, cited in Anonymous 1996). The Exemption was granted, and buprofezin and pyriproxyfen were released for use in Arizona in 1996 with the stipulation that each could only be used once per season. This represents the first major emergency approval by EPA for a combined package of two products, for the same use, in the same crop, during the same season.

5 Conclusions

Although supported by contemporary examples whenever possible, the overall theme of this chapter is far from original. In closing, it is illuminating to quote words written in 1975 by the late R.M. Sawicki when reviewing the implications of his research on housefly resistance for the future of a then novel group of insecticides—the synthetic pyrethroids. Sawicki (1975) wrote:

Lessons learned from the development of sequential resistance [to organochlorines and organophosphates] in the housefly should not be lost; although resistance cannot be avoided with current pest control methods and there is no way of foretelling the cross-resistance of newly introduced compounds, resistance can be delayed and costly and unnecessary mistakes prevented when experience already gained is put to good use.

Perhaps the most useful additional lesson learned since 1975, is that no individual or organization, however well informed or motivated, is likely to succeed with managing resistance singlehandedly. Of the many challenges confronting insect pest management in the years ahead, that of securing sufficient cooperation and mutual commitment to safeguard novel chemicals from the fate suffered by their predecessors could well prove the most crucial of all.

6 Acknowledgements

We thank T. Dennehy, C. Furk, N. Hamon and P. Wege for advice, and A. McCaffery and P. Weintraub for reviewing and commenting on the manuscript.

References

Anonymous (1996) Arizona's insecticide resistance model. California-Arizona Cotton Sept/ Oct 1996: 12–13

Bai D, Lummis SCR, Leicht W, Breer H, Sattelle DB (1991) Actions of imidacloprid and a related nitromethylene on cholinergic receptors of an identified insect motor neurone. Pestic Sci 33: 197–204

Brown TM, Brogdon WG (1987) Improved detection of insecticide resistance through conventional and molecular techniques. Annu Rev Entomol 32: 145–162

Cahill M, Byrne FJ, Gorman K, Denholm I, Devonshire AL (1995) Pyrethroid and organophosphate resistance in the tobacco whitefly *Bemisia tabaci* (Homoptera: Aleyrodidae). Bull Entomol Res 85: 181–187

Cahill M, Jarvis W, Gorman K, Denholm I (1996a) Resolution of baseline responses and documentation of resistance to buprofezin in *Bemisia tabaci* (Homoptera: Aleyrodidae). Bull Entomol Res 86: 117–122

Cahill M, Denholm I, Gorman K, Day S, Elbert A, Nauen R (1996b) Baseline determination and detection of resistance to imidacloprid in *Bemisia tabaci* (Homoptera: Aleyrodidae). Bull Entomol Res 86: 343–393

Castle S, Henneberry T, Toscano N, Prabhaker N, Birdsall S, Weddle R (1996) Silverleaf whiteflies show no increase in insecticide resistance. Calif Agric 50: 18–23

Cheng EY, Kao CH, Chiu CS (1991) Insecticide resistance study in *Plutella xylostella*: the IGR-resistance and the possible management strategy. J Agric Res China 39: 208–220

Colliot F, Kukorowski KA, Hawkins DW, Roberts DA (1992) Fipronil: a new soil and foliar broad spectrum insecticide. Proc Brighton Crop Prot Conf, British Crop Protection Council, Farnham, pp 29–34

Comins HN (1977) The development of insecticide resistance in the presence of migration. J Theor Biol 64: 177–197

Comins HN (1986) Tactics for resistance management using multiple pesticides. Agric Ecosyst Environ 16: 129–148

Curtis CF (1985) Theoretical models of the use of insecticide mixtures for the management of resistance. Bull Entomol Res 75: 259–265

Daly JC (1993) Ecology and genetics of insecticide resistance in *Helicoverpa armigera*: interactions between selection and gene flow. Genetica 90: 217–226

Daly JC, Fisk JH, Forrester NW (1988) Selective mortality in field trials between strains of *Heliothis armigera* (Lepidoptera: Noctuidae) resistant and susceptible to pyrethroids: functional dominance of resistance and age class. J Econ Entomol 81: 1000–1007

De Cock A, Ishaaya I, Van de Viere, M, Degheele D (1995) Response of buprofezin-susceptible and -resistant strains of *Trialeurodes vaporariorum* (Homoptera: Aleyrodidae) to pyriproxyfen and diafenthiuron. J. Econ Entomol 88: 763–767

Denholm I (1990) Monitoring and interpreting changes in insecticide resistance. Funct Ecol 4: 601–608

Denholm I, Rowland MW (1992) Tactics for managing pesticide resistance in arthropods: theory and practice. Annu Rev Entomol 37: 91–112

Denholm I, Farnham AW, O' Dell K, Sawicki RM (1983) Factors affecting resistance to insecticides in house-flies, *Musca domestica* L. (Diptera: Muscidae). I. Long-term control with bioresmethrin of flies with strong pyrethroid-resistance potential. Bull Entomol Res 73: 481–489

Denholm I, Sawicki RM, Farnham AW (1985) Factors affecting resistance to insecticides in house-flies, *Musca domestica* L. (Diptera: Muscidae). IV. The population biology of flies on animal farms in south-eastern England and its implications for the management of resistance. Bull Entomol Res 75: 143–158

Denholm I, Rollett AJ, Cahill M, Ernst GH (1995) Response of cotton aphids and whiteflies to diafenthiuron and pymetrozine in laboratory assays. Proc 1995 Beltwide Cotton Prod Conf. National Cotton Council of America, Memphis, pp 991–994

Denholm I, Cahill M, Byrne FJ, Devonshire AL (1996) Progress with documenting and combating insecticide resistance in *Bemisia.* In: Gerling D, Mayer RT (eds) Bemisia 1995: taxonomy, biology, damage, control and management. Intercept, Andover, pp 577–603

Dennehy TJ (1987) Decision-making for managing pest resistance to pesticides. In: Ford MG, Holloman DW, Khambay BPS, Sawicki RM (eds) Combating resistance to xenobiotics: biological and chemical approaches. Ellis Horwood, Chichester, pp 118–126

Dennehy TJ, Granett J (1984) Monitoring dicofol-resistant spider mites (Acari: Tetranychidae) in California cotton. J Econ Entomol 77: 1386–1392

Dennehy TJ, Granett J, Leigh TF (1983) Relevance of slide–dip and residual bioassay comparisons to detection of resistance in spider mites. J Econ Entomol 76: 1225–1230

Dennehy TJ, Farnham AW, Denholm I (1993) The microimmersion bioassay: a novel method for the topical application of pesticides to spider mites. Pestic Sci 39: 47–54

Dennehy TJ, Simmons A, Russell J, Akey D (1995) Establishment of a whitefly resistance documentation and management program in Arizona. 1995 Arizona Cotton Report, University of Arizona Cooperative Extension, Tucson, pp 287–296

Dennehy TJ, Williams L, Russell JS, Li X, Wigert M (1996) Monitoring and management of whitefly resistance to insecticides in Arizona. Proc 1996 Beltwide Cotton Prod Conf. National Cotton Council of America, Memphis, pp 135–140

Devine GJ, Harling ZK, Scarr AW, Devonshire AL (1996) Lethal and sublethal effects of imidacloprid on nicotine-tolerant *Myzus nicotianae* and *Myzus persicae.* Pestic Sci 48: 57–62

Dewar A, Read LA, Thornhill WA (1988) The efficacy of novel and existing aphicides against resistant *Myzus persicae* on sugar beet in the laboratory. Proc 1988 Brighton Crop Prot Conf. British Crop Protection Council, Farnham, pp 477–482

Dittrich V, Luetkemeier N, Voss G (1980) OP-resistance in *Spodoptera littoralis*: inheritance, larval and imaginal expression, and consequences for control. J Econ Entomol 73: 356–362

Dittrich V, Hassan SO, Ernst GH (1985) Sudanese cotton and the whitefly: a case study of the emergence of a new primary pest. Crop Prot 4: 161–176

Elbert A, Iwaya K, Overbeck H, Tsuboi S (1990) Imidacloprid, a novel systemic nitromethylene analogue insecticide for crop protection. Proc 1990 Brighton Crop Prot Conf. British Crop Protection Council, Farnham, pp 21–28

Elbert A, Nauen R, Cahill M, Devonshire AL, Scarr AW, Sone S, Steffens R (1996) Resistance management with chloronicotinyl insecticides using imidacloprid as an example. Pflanzenschutz-Nachr Bayer 49: 5–53

Farnham AW, Khambay BPS (1995a) The pyrethrins and related compounds. Part XXXIX—Structure–activity relationships of pyrethroidal esters with cyclic side chains in the alcohol component against resistant strains of housefly (*Musca domestica*). Pestic Sci 44: 269–275

Farnham AW, Khambay BPS (1995b) The pyrethrins and related compounds. Part XL— structure-activity relationships of pyrethroidal esters with acyclic side chains in the alcohol component against resistant strains of housefly (*Musca domestica*). Pestic Sci 44: 277–281

ffrench-Constant RH, Roush RT (1990) Resistance detection and documentation: the relative roles of pesticidal and biochemical assays. In: Roush RT, Tabashnik BE (eds) Pesticide resistance in arthropods. Chapman and Hall, New York, pp 1–38

ffrench-Constant RH, Roush RT, Mortlock D, Dively GP (1990) Isolation of dieldrin resistance from field populations of *Drosophila melanogaster* (Diptera: Drosophilidae). J Econ Entomol 83: 1733–1737

ffrench-Constant RH, Rocheleau TA, Steichen JC, Chalmers, AE (1993) A point mutation in a *Drosophila* GABA receptor confers insecticide resistance. Nature 363: 449–451

Fitt GP (1989) The ecology of *Heliothis* species in relation to agroecosystems. Annu Rev Entomol 34: 17–52

Follett PA, Gould F, Kennedy GG (1993) Comparative fitness of three strains of Colorado potato beetle (Coleoptera: Chrysomelidae) in the field: spatial and temporal variation in insecticide selection. J Econ Entomol 86: 1324–1333

Forrester NW (1990) Designing, implementing and servicing an insecticide resistance management strategy. Pestic Sci 28: 167–179

Forrester NW, Cahill M, Bird L, Layland JK (1993) Management of pyrethroid and endosulfan resistance in *Helicoverpa armigera* (Lepidoptera: Noctuidae) in Australia. Bull Entomol Res (Special Supplement No.1) pp 132

Frey JE (1993) The analysis of arthropod pest movement through trade in ornamental plants. In: Ebbels D (ed) Plant health and the European single market. Br Crop Prot Counc Monogr no 54, British Crop Protection Council, Farnham, pp 157–165

Furk C, Cotten J, Gould HJ (1983) Monitoring for insecticide resistance in aphid pests of field crops in England and Wales. Proc 10th Int Congr Plant Protection, Brighton, Nov 1983, vol 2, pp. 20–25

Furlong MJ, Wright DJ (1994) Examination of stability of resistance and cross-resistance patterns to acylurea insect growth regulators in field populations of the diamondback moth, *Plutella xylostella*, from Malaysia. Pestic Sci 42: 315–326

Georghiou GP (1983) Management of resistance in arthropods. In: Georghiou GP, Saito T (eds) Pest resistance to pesticides. Plenum, New York, pp 769–792

Georghiou GP, Taylor CE (1977) Operational influences in the evolution of insecticide resistance. J Econ Entomol 70: 653–658

Halliday WR, Burnham KP (1990) Choosing the optimal diagnostic dose for monitoring insecticide resistance. J Econ Entomol 83: 1151–1159

Horowitz AR, Ishaaya I (1992) Susceptibility of the sweetpotato whitefly (Homoptera: Aleyrodidae) to buprofezin during the cotton season. J Econ Entomol 85: 318–324

Horowitz AR, Ishaaya I (1994) Managing resistance to insect growth regulators in the sweetpotato whitefly (Homoptera: Aleyrodidae). J Econ Entomol 87: 866–871

Horowitz AR, Seligman IM, Forer G, Bar D, Ishaaya I (1993) Preventive insecticide resistance strategy in *Helicoverpa* (*Heliothis*) *armigera* (Lepidoptera: Noctuidae) in Israeli cotton. J Econ Entomol 86: 205–212

Horowitz AR, Forer G, Ishaaya I (1994) Managing resistance in *Bemisia tabaci* in Israel with emphasis on cotton. Pestic Sci 42: 113–122

Horowitz AR, Forer G, Ishaaya I (1995) Insecticide resistance management as a part of an IPM strategy in Israeli cotton fields. In: Constable GA, Forrester NW (eds) Challenging the future Proc World Cotton Research Conference 1 CSIRO, Melbourne, Australia, pp 537–544

Hrdy I, Kremheller HT, Kuldová J, Luders W, Sula J (1986) Insecticide resistance of the hop aphid *Phorodon humuli* in the Bohemian, Bavarian and Baden Wurttemberg hop-growing areas. Acta Entomol Bohemoslov 83: 1–9

Ishaaya I (1992) Insect resistance to benzoylphenylureas and other IGRs: mechanisms and countermeasures. In: Mullin CA, Scott JG (eds) Molecular mechanisms of insecticide resistance: diversity among insects. American Chemical Society Sympo Ser no 505, Washington, pp 235–250

Ishaaya I, Horowitz AR (1992) Novel phenoxy hormone analog (pyriproxyfen) suppresses embryogenesis and adult emergence of sweetpotato whitefly (Homoptera: Aleyrodidae). J Econ Entomol 85: 2113–2117

Ishaaya I, Horowitz AR (1995) Pyriproxyfen, a novel insect growth regulator for controlling whiteflies: mechanism and resistance management. Pestic Sci 43: 227–232

Ishaaya I, Klein M (1990) Response of susceptible laboratory and resistant field strains of *Spodoptera littoralis* (Lepidoptera: Noctuidae) to teflubenzuron. J Econ Entomol 83: 59–62

Ishaaya I, Mendelson Z, Melamed-Madjar V (1988) Effect of buprofezin on embryogenesis and progeny formation of sweetpotato whitefly (Homoptera: Aleyrodidae). J Econ Entomol 81: 781–784

Ishaaya I, Mendelson Z, Horowitz AR (1993) Toxicity and growth-suppression exerted by diafenthiuron in the sweetpotato whitefly, *Bemisia tabaci*. Phytoparasitica 21: 199–204

Jackson GJ (1986) Insecticide resistance—what is industry doing about it? Proc 1986 Brit Crop Protection Conf, British Crop Protection Council, Thorton Heath, pp 943–949

Kayser H, Kaufmann L, Schurmann F, Harrewijn P (1994) Pymetrozine (CGA 215944): A novel compound for aphid and whitefly control. An overview of its mode of action. Proc Brighton Crop Protect Conf, British Crop Protection Council, Farnham, pp 737–742

Keiding J, Jespersen JB, El-Khodary AS (1991) Resistance risk assessment of two insect development inhibitors, diflubenzuron and cyromazine, for control of the housefly *Musca domestica*. Part I: larvicidal tests with insecticide-resistant laboratory and Danish field populations. Pestic Sci 32: 187–206

Keiding J, El-Khodary AS, Jespersen JB (1992) Resistance risk assessment of two insect development inhibitors, diflubenzuron and cyromazine, for control of the housefly *Musca domestica* L. Part II: effect of selection pressure in laboratory and field populations. Pestic Sci 35: 27–37

Lemon RW (1988) Resistance monitoring methods and strategies for resistance management in insect and mite pests of fruit crops. Proc 1988 Brighton Crop Prot Conf, British Crop Protection Council, Farnham, UK, pp 1089–1096

Leonard PK, Perrin RM. (1994) Resistance management—making it happen. Proc 1994 Brighton Crop Protect Conf, British Crop Protection Council, Farnham, pp 969–974

Lewis GA, Madge DS (1984) Esterase activity and associated insecticide resistance in the damson-hop aphid, *Phorodon humuli* (Schrank) (Hemiptera: Aphididae). Bull Entomol Res 74: 227–238

Lorriman F, Llewellyn M (1983) The growth and reproduction of hop aphid (*Phorodon humuli*) biotypes resistant and susceptible to insecticides. Acta Entomol Bohemoslov 80: 87–95

Mani GS, Wood RJ (1984) Persistence and frequency of application of an insecticide in relation to the rate of evolution of resistance. Pestic Sci 15: 325–336

Martinson TE, Nyrop JP, Dennehy TJ, Reissig WH (1991) Temporal variability in repeated bioassays of field populations of European red mite (Acari: Tetranychidae): implications for resistance monitoring. J Econ Entomol 84: 1119–1127

McKenzie JA (1996) Ecological and evolutionary aspects of insecticide resistance. Academic Press, San Diego.

McKenzie JA, Whitten MJ (1984) Estimation of the relative viabilities of insecticide resistance genotypes of the Australian sheep blowfly *Lucilia cuprina*. Aust J Biol Sci 37: 45–52

Moores GD, Devine GJ, Devonshire AL (1994) Insecticide-insensitive acetylcholinesterase can enhance esterase-based resistance in *Myzus persicae* and Myzus nicotianae. Pestic Biochem Physiol 49: 114–120

Murray A, Siddi G, Jacobson RM, Vietto M, Thirugnanam M (1988) RH-7988: a new selective systemic aphicide. Proc 1988 Brighton Crop Protection Conf, British Crop Protection Council, Farnham, pp 73–80

Parkin CS, Brun LO, Suckling DM (1992) Spray deposition in relation to endosulfan resistance in coffee berry borer (*Hypothenemus hampei*) (Coleoptera: Scolytidae) in New Caledonia. Crop Prot 11: 213–220

Parrella MP (1987) Biology of *Liriomyza*. Annu Rev Entomol 32: 201–224

Prabhaker N, Toscano NC, Castle S, Henneberry T (1995) Assessment of a hydroponic bioassay for evaluation of imidacloprid against whiteflies. Proc 1995 Beltwide Cotton Prod Conf National Cotton Council of America, Memphis, pp 72–73

Rawlings P, Davidson G, Sakai RK, Rathor HR, Aslamkhan KM, Curtis CF (1981) Field measurement of the effective dominance of an insecticide resistance in anopheline mosquitoes. Bull WHO 59: 631–640

Roush RT (1989) Designing resistance management programs: how can you choose? Pestic Sci 26: 423–441

Roush RT, Daly JC (1990) The role of population genetics in resistance research and management. In: Roush RT, Tabashnik BE (eds) Pesticide resistance in arthropods. Chapman and Hall, New York, pp 97–152

Roush RT, Lutterell RG (1989) Expression of resistance to pyrethroid insecticides in adults and larvae of tobacco budworm (Lepidoptera: Noctuidae): implications for resistance monitoring. J Econ Entomol 82: 1305–1310

Roush RT, McKenzie JA (1987) Ecological genetics of insecticide and acaricide resistance. Annu Rev Entomol 32: 361–380

Roush RT, Miller GL (1986) Considerations for the design of insecticide resistance monitoring programs. J Econ Entomol 79: 293–298

Ruder FJ, Guyer W, Benson JA, Kayser H (1991) The thiourea insecticide/acaricide diafenthiuron has a novel mode of action: inhibition of mitochondrial respiration by its carbodiimide product. Pestic Biochem Physiol 41: 207–219

Sanderson JP, Roush RT (1995) Management of insecticide resistance in the greenhouse. In: Bishop A, Hansbeck M, Lindquist R (eds) Proc 11th Conf Insect Dis Manag Ornam, Fort Myers, Florida, pp 18–20

Sawicki RM (1975) Effects of sequential resistance on pesticide management. Proc 8th Brit Insectic Fungic. Conf British Crop Protection Council, Thornton Heath, pp 799–811

Sawicki RM, Denholm I (1987) Management of resistance to pesticides in cotton pests. Trop Pest Manag 33: 262–272

Sawicki RM, Denholm I, Forrester NW, Kershaw CD (1989) Present insecticide management strategies on cotton. In: Green. MB, de B Lyon DJ (eds) Pest management on cotton. Ellis Horwood, Chichester, pp 31–43

Soderlund DM, Bloomquist JR (1990) Molecular mechanisms of insecticide resistance. In: Roush RT, Tabashnik BE (eds) Pesticide resistance in arthropods. Chapman and Hall, New York, pp 58–96

Sterk G, Highwood DP (1992) Implementation of IRAC anti-resistance guidelines with IPM programmes for Belgian apple and pear orchards. Proc 1992 Brighton Crop Protection Conf British Crop Protection Council, Farnham, pp 517–526

Tabashnik BE (1989) Managing resistance with multiple pesticide tactics: theory, evidence and recommendations. J Econ Entomol 82: 1263–1269

Tabashnik BE (1990) Modelling and evaluation of resistance management tactics. In: Roush RT, Tabashnik BE (eds) Pesticide resistance in arthropods. Chapman and Hall, New York, pp 153–182

Tabashnik BE, Croft BA (1982) Managing pesticide resistance in crop–arthropod complexes: interactions between biological and operational factors. Environ Entomol 11: 1137–1144

Taylor RWD, Halliday D (1986) The geographical spread of resistance to phosphine by coleopterous pests of stored products. Proc 1986 Brit Crop Prot Conf British Crop Protection Council, Thornton Heath, pp 607–612

Thompson M, Steichen JC, ffrench-Constant RH (1993) Conservation of cyclodiene insecticide resistance–associated mutations in insects. Insect Mol Biol 2: 149–154

Voss G (1988) Insecticide/acaricide resistance: industry's efforts and plans to cope. Pestic Sci 23: 149–156

Wege PJ (1996) The role of industry in successful resistance management. In: Bourdot GW, Suckling DM (eds) Pesticide resistance: prevention and management. New Zealand Plant Protection Society, Lincoln, pp 101–107

Welty C, Reissig WH, Dennehy TJ, Weires RW (1988) Comparison of residual bioassay methods and criteria for assessing mortality of cyhexatin-resistant European red mite (Acari: Tetranychidae). J Econ Entomol 81: 442–448

Whitten MJ, McKenzie JA (1982) The genetic basis for pesticide resistance. In: Lee KE (ed) Proc 3rd Aust Conf Grassland Invertebr Ecol South Australian Government Printers, Adelaide, pp 1–16

Zalucki MP, Daglish G, Firempong S, Twine PH (1986) The biology and ecology of *Heliothis armigera* (Hübner) and *H. punctigera* Wallengren (Lepidoptera: Noctuidae) in Australia: what do we know? Aust J Zool 34: 799–814

Subject Index